Intelligent Multimedia Signal Processing for Smart Ecosystems

Shabir A. Parah • Nasir N. Hurrah • Ekram Khan
Editors

Intelligent Multimedia Signal Processing for Smart Ecosystems

Springer

Editors
Shabir A. Parah
Department of Electronics and Instrumentation
University of Kashmir
Srinagar, Jammu and Kashmir, India

Nasir N. Hurrah
Aligarh Muslim University
Aligarh, India

Ekram Khan
Department of Electronics Engineering
Aligarh Muslim University
AMU Aligarh, India

ISBN 978-3-031-34875-4 ISBN 978-3-031-34873-0 (eBook)
https://doi.org/10.1007/978-3-031-34873-0

This Springer imprint is published by the registered company Springer Nature Switzerland AG
The registered company address is: Gewerbestrasse 11, 6330 Cham, Switzerland

Contents

Part I
Smart Ecosystems: Opportunities and Challenges

Smart Ecosystems for Sustainable Development: Opportunities, Challenges, and Solutions

Nasir N. Hurrah, Ekram Khan, and Shabir A. Parah

1 Introduction

The world is changing and with the rapid advancement in technology, the boundary between the physical and digital worlds is beginning to disappear. These advancements are envisioned to bring together hardware and software with the aim to collect and process data across various platforms, facilitating high flexibility, efficient, and faster processes to yield cost effective high-quality goods [1]. As a result, productivity will rise, encouraging industrial development and boosting economic benefits. The current digital platforms employ state-of-the-art technologies like Artificial Intelligence (AI), Machine Learning (ML), Cloud Computing, 5G internet, Blockchain Technology, Internet of Things (IoTs), Big Data, and Cyber Physical Systems (CPS) [2, 3]. The state-of-art-the systems used in these applications share huge data in real-time between different devices and modules to carry out several automatic or monitored tasks in various application areas like consumer electronics, healthcare, games, telecommunication, industry, automotive, aeronautics and military applications [4]. All the present technologies have made life easier but it is still a long way to go in order to develop a strong, robust, reliable and collaborative autonomous system [5].

Although technology is moving leaps and bounds in every domain, the recent scare of COVID-19 gave a strong signal for the development of collaborative and autonomous systems [6]. With the rapid expansion of smart and connected devices, we must reconsider the function of our products and how they interact with other

N. N. Hurrah · E. Khan
Department of Electronics Engineering, Aligarh Muslim University, Aligarh, India

S. A. Parah (✉)
Department of Electronics and Instrumentation Technology, University of Kashmir, Srinagar, India

S. A. Parah et al. (eds.), *Intelligent Multimedia Signal Processing for Smart Ecosystems*, https://doi.org/10.1007/978-3-031-34873-0_1

objects. It's not good enough to think about how your existing product works within a given smart system: instead, it is better to assign an active role to a system, train the modules to interact within and may be across ecosystems, and train to adopt to the changes as ecosystems mature. In this scenario, the concept of smart ecosystems is helpful, as it can add value to every element of daily life and have the freedom to innovate in fresh ways by building new digital revenue-generating ecosystems [7]. This idea has been prominent to push for the applications like smart healthcare, E-commerce, Industry 4.0, smart cities, smart grids, smart homes, smart contracts, smart transportation, smart agriculture, smart airports, to name a few [8–10]. In recent years, the term "smart ecosystem" has gained increasing popularity among researchers, policymakers, and practitioners. Smart ecosystems are complex systems that are characterized by their ability to adapt and self-organize, as well as their capacity to generate and utilize information in real-time [11]. Contrary to conventional manual or semi-autonomous systems, software and hardware must be designed concurrently and synergistically for intelligent system design, so that the system's functional and nonfunctional requirements are automatically met. This new design approach is collaborative and inventive, requiring specific hardware, software, and system engineering skills [12].

Also, there is great scope for global collaboration between various smart systems to present a unified centralized smart ecosystem. A collaborative approach for wise management of our environment, cities, buildings, factories, offices, and personal lives is necessary to ensure the sustainability of humanity's high standard of living [13, 14]. However, every area of life has its own community, and they are somewhat isolated from one another; cross-communication is possible but not common. Traditionally, the majority of devices belong to a specific domain, and each domain frequently restricts devices from sharing data outside of its sphere. In a smart home, for instance, an air-handling unit is part of the HVAC (heating, ventilation, and air conditioning) system, and one or more sensors are used to capture the actual state (such as the temperature) of the room or area in which it is installed [15]. The control system determines what action needs to be taken on its own (for example, increasing the speed of a fan), but it may not always share that information with parties outside of its own realm. This has worked for a long time, but the need to optimize and apply real-time cloud-based/automatic decision making is driving change. Also, there is need for interconnectedness and creation of dependence between the systems to achieve the best performance through a centralized ecosystem as depicted in Fig. 1.

This chapter provides an overview of the concept of smart ecosystems, including their definition, characteristics, and benefits. This chapter summarizes all of the specifics and fundamental ideas associated with smart ecosystems. Our focus is on the application areas of this technology and the latest methods that need to be adapted for its full-scale implementation. Following that, we define what a smart and intelligent ecosystem is and discuss the various models and methods that can be used to model a smart ecosystem, particularly using technologies such as IoTs, CPS, AI, ML, Cloud/Edge computing, neural networks, and security models. Finally, we present main challenges and some future perspectives.

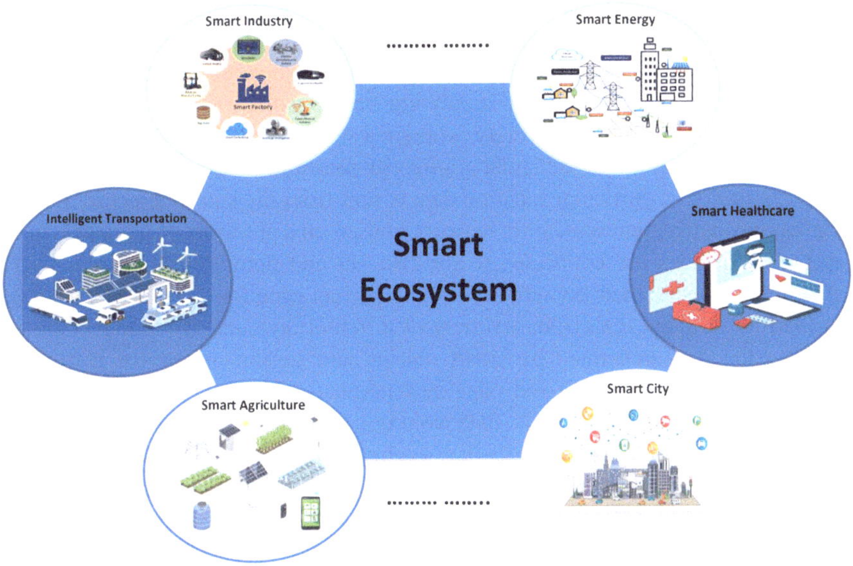

Fig. 1 Smart ecosystems

2 Smart Ecosystems

Smart ecosystems are defined as a collection of intelligent entities that are interconnected and interdependent, and that interact with each other to achieve common goals. These entities can be both living (such as humans, animals, and plants) and non-living (such as sensors, devices, and infrastructure). A smart ecosystem resembles a natural biological ecosystem in that it is a distributed, adaptive, open sociotechnical system that is self-organizing, scalable, and sustainable [16]. In addition, it is a dynamic entity influenced by external and internal factors, susceptible to periodic disturbances, and frequently recovering from a past event. The assumption of a smart ecosystem is that of a system that is actively interacting, and simply put, smart ecosystems use technology to deliver high quality of services and facilitate users for better efficiency and effectiveness [17]. Hence, the key characteristic of smart ecosystems is their ability to sense and respond to changes in their environment, as well as their ability to learn and adapt over time. It is established to make people's daily life more comfortable and transform the ways people interact with everyday tasks through the ecosystem. The goal is to learn more about the environment so that you can comprehend, manage, and act on this knowledge [18]. For example, in an industry environment a Smart Ecosystem is envisioned to provide more precise and real-time visibility in the flow of materials, parts, and products in a manufacturing line and warehouse management. Production systems are also spending money on such systems to restructure production workflows, enhance tool and material tracking, and reduce production costs.

A number of technologies are involved at various levels that permit the development of a smart ecosystem. There is a discernible trend that appears when one compares the supply chains studied in the smart home, smart city, and Industrial IoT research projects. In each case, the supply line begins with objects that have sensors built into them. Consider a smart city, wherein a sensor embedded in a traffic signal gathers information about vehicular traffic. To determine the amount of traffic at any given time in a particular location data is sent from the smart object to a smart program through a communication network housed on a platform. In a typical smart ecosystem, devices are connected to the Internet and permitted to communicate with entities outside their domain [19, 20]. A perfect state for an ecosystem can be achieved by utilizing data from a variety of different domains. The most common route for the data generated by smart objects and sensors is service platforms. These could be locally deployed edge systems that perform preliminary analysis and data extraction. As a result, data volumes are reduced—which is crucial if every device is sending data—while data quality is enhanced. The vital information would be transmitted to the cloud platform, which is located in a data center to the intermediate distance. On the cloud platform, additional analytics enabled by AI may be utilized to generate insights and value from the collected data.

A smart ecosystem, say in a tall building can be truly scalable, smart, where devices such as fire alarms, surveillance cameras, chillers, thermostats, lighting systems, air conditioners and appliances are all connected and controlled through a central hub or smartphone or a cloud-based management system. This allows a user to control and automate various systems within the ecosystem, such as adjusting the temperature or turning off lights remotely. Consider the case where one of the functions is "optimal indoor climate for everyone with minimal energy usage." The HVAC system may be informed about people who are about to enter a particular level and information regarding the number of people who have entered a specific area using data from an access control system or an elevator controller [15]. As an example of a component in a smart ecosystem [21], provides the HVAC system for controlling the energy flow to an electric motor that powers pumps, fans, compressors, and other equipment. Consequently, the cloud-based system is facilitating everyone's day by helping to reserve and set up meeting rooms, alerting about guests' arrival, and optimizing the indoor climate based on internal and external factors like humidity levels, the outside temperature, and the number of people on a floor and where they are located. Similar operations are carried out in other applications, such as smart factories, where the use of robots for automation is growing, and where there is a push to collect data from all probable sources inside a factory. Collecting data can be used to improve a factory in a variety of ways, such as switching from preventive to predictive maintenance to increase a conveyor's uptime. Smart buildings and smart factories are just few among many smart ecosystems: they are interlinked, and one may belong to another. For example, a smart city ecosystem consists of sub-ecosystems, such as smart transportation, smart water utilities, smart buildings, smart factories, smart grids, smart elevators, and smart pumps. In a similar way, an intelligent variable speed drive (VSD), which is a small ecosystem in and of itself, could also be part of a smart pump or smart elevator ecosystem.

A collaborative ecosystem can be created by forming alliances between various businesses, organizations, and individual experts to jointly identify problems, find solutions, and develop new standards. For instance, a recent report by Deloitte on the IoT ecosystem stated that in order to have a significant impact on the market, IoT providers must collaborate [22]. Successful participants in various smart systems are collaborating with one another more frequently in the smart ecosystem.

3 Characteristics of Smart Ecosystems

There are several key characteristics that define smart ecosystems. These characteristics highlight the necessary traits of smart ecosystems to address every aspect of the application for which they are envisioned. Some of the important characteristics are discussed in this section.

3.1 Interconnectedness

Smart ecosystems are composed of interconnected entities that can communicate with each other, share information, and collaborate to achieve common goals. Smart ecosystems are highly connected, with devices, sensors, and applications linked together via the internet or other communication channels. This allows for the seamless flow of data and information between different components, enabling real-time monitoring and control of the system [23]. This needs a high-speed communication network and the bunch of rules for every process.

In smart homes, for example, a range of devices such as smart thermostats, lighting systems, and security cameras are connected to a central hub, which can be controlled using a smartphone or other device. This enables users to monitor and control their home environment remotely, adjust the temperature, lighting, and security settings as required. Similarly, in smart cities, sensors and devices are placed throughout the city to monitor traffic, air quality, and other environmental factors. This data is then analyzed in real-time using high-speed connectivity to provide insights and inform decisions around city planning and management.

3.2 Interoperability

Smart ecosystems are designed to be interoperable, allowing different components and technologies to work together seamlessly. This is achieved through the use of standardized protocols and interfaces, ensuring that data and information can be shared across different systems.

In smart homes, for example, devices and sensors from different manufacturers can be integrated using protocols such as Wi-Fi, Bluetooth, and Zigbee. This enables users to control their home environment using a single app or platform, regardless of the devices they are using. In smart transportation, interoperability is crucial to enable seamless communication between different modes of transportation, such as buses, trains, and taxis. This can help to optimize travel times and reduce congestion in cities.

3.3 Adaptability

Smart ecosystems are capable of adapting to changes in their environment in real-time, and can reconfigure themselves to maintain optimal performance. Smart ecosystems are highly adaptable, with the ability to respond to changes in the environment and adjust their operations accordingly. This is achieved through the use of real-time data analytics and ML, which enables the system to make informed decisions and take actions based on the current situation.

In smart agriculture, for example, sensors and drones can be used to monitor crop health, soil moisture, and other environmental factors. This data can be analyzed in real-time to optimize irrigation, fertilization, and other aspects of crop management. Similarly, in smart energy, real-time data analytics can be used to manage the grid and balance energy supply and demand. This can help to reduce energy waste and improve the efficiency of the energy system.

3.4 Automation

Smart ecosystems are designed to be intelligent, with the ability to sense, process, analyze, learn, and act on information in real-time. Advanced algorithms and ML techniques are used to process the massive amounts of data generated by the system, providing valuable insights and predictive capabilities [11].

In smart healthcare, for example, sensors and wearables can be used to monitor patients' vital signs, track medication adherence, and detect changes in their health status. This data can be analyzed in real-time to provide insights and inform healthcare providers' decision-making. In smart manufacturing, ML algorithms can be used to analyze data from sensors and machines to identify patterns and detect potential issues before they occur. This can help to optimize manufacturing processes, reduce downtime, and increase productivity.

3.5 Security and Privacy

Security and privacy are critical aspects of every digital ecosystem, given the vast amount of data and information generated in every application [24]. One of the biggest design considerations in smart ecosystems is existence of various mechanisms to ensure the security and privacy of user data and information, as well as the security of the system as a whole. The security of the data acquisition modules, cloud platforms, interconnecting sub-systems, and other systems is also necessary in the design of a reliable smart ecosystem [25, 26].

In smart cities, for example, security and privacy are critical given the large amount of data generated by the system. The devices like smart locks and cameras can be controlled remotely, and can be programmed to automatically lock and unlock the doors at certain times of the day, or when certain conditions are met, such as when the owner arrives or leaves home. Advanced security measures such as secure data transfer, encryption, and authentication protocols must be implemented to ensure the safety and privacy of user data [27]. Every device and sub-system must be secured against cyber threats and unauthorized access. Strong authentication and encryption measures must be implemented to ensure that user data and information is protected [28].

3.6 Self-Organization

Self-organization is a key characteristic of smart ecosystems, and it refers to the ability of a system to organize and regulate itself without the need for external control or intervention. Smart ecosystems are able to organize themselves, and can optimize their behaviour to achieve their objectives. Smart ecosystems are typically decentralized, meaning that there is no single entity or central authority controlling the system. Instead, the system is made up of multiple nodes or agents that interact with each other in a distributed manner. This allows the system to adapt and respond to changing conditions in real-time. Smart ecosystems are also robust, meaning that they can maintain their functionality and performance even when individual nodes or agents fail or are removed from the system [19]. This is because the system as a whole is resilient and able to recover from disruptions. Through the emergence of new properties and behaviors, as well as robustness and learning, smart ecosystems can continue to evolve and improve their functionality over time.

Self-organization in smart ecosystems leads to the emergence of new properties or behaviors that are not present in the individual nodes or agents. These emergent properties can be beneficial for the ecosystem, as they enable it to perform complex tasks and solve problems more effectively than individual nodes or agents would be able to do alone.

4 Architecture of Smart Ecosystems

Smart ecosystems are composed of a range of interconnected components in digital and physical worlds that work together to achieve their objectives. The components of a smart ecosystem include hardware devices, software applications, cloud infrastructure, communication protocols, and decision-making algorithms [29, 30]. Additionally, advances in AI and ML are allowing devices to become more intelligent and adaptable, making it possible for them to make decisions on their own and learn from their surroundings. The systems/mechanisms usually perform very few and basic tasks in an application, and hence there is a collaboration to construct a robust and reliable system with the ability to automate all tasks in an application with accurate decision making and control. This will form a smart ecosystem with all these technologies at one place connected and inter-dependent to perform all the necessary tasks with need of no human intervention in any application area.

Figure 2 shows a simple design hierarchy of a smart ecosystem. To build a smart ecosystem one needs to design a smart object aided with sensors with capability of communication and/or control. So at the design level a smart ecosystem refers to a network of sensory modules, machines, tools, and equipment's that can communicate with one another to automatically monitor or control various tasks. Then there need to have connectivity between the sensors and the edge platform, the provision and maintenance of the edge platform, and the connectivity from the edge platform to the cloud platform and the data center. And then you need someone to operate the data center, and someone to host the cloud platform. AI is an essential component of smart ecosystems. AI can be used to identify patterns and trends in data, and automate complex tasks. Low latency connections are then required for services like IoT in order to provide real-time responses. Lastly, there is requirement of connectivity to major internet exchanges and international carriers on a global scale.

4.1 *Sensors and Actuators*

A smart system gathers information from the environment about status/changes of interest using a variety of sensors and devices. Sensors are an essential component of smart ecosystems, as they enable the collection of data about the environment and the entities within it. Sensors can measure a range of parameters, including temperature, humidity, pressure, sound, and motion, and can be deployed in a variety of environments, from buildings and transportation systems to natural ecosystems. Depending on the data of interest, sensors can take the shape of passive sensors, active sensors, optical, thermal, mechanical, electrical, etc. sensors [31]. These components form the foundation of a smart ecosystem by providing real-time

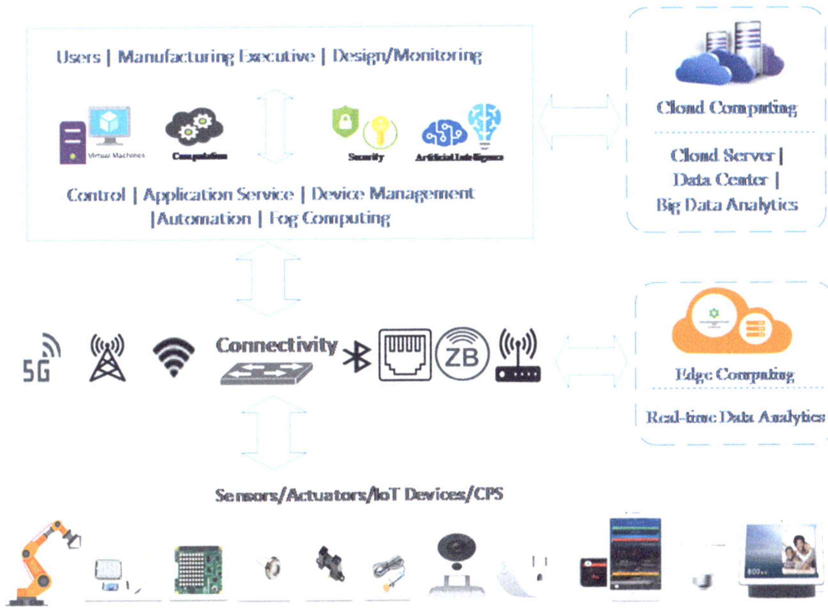

Fig. 2 Smart ecosystem design hierarchy

data and enabling automated responses. A smart ecosystem transforms data it collects from various sensors throughout the environment into actionable outputs and tangible knowledge via different services. Camera is an excellent example of sensor that is an essential component of many smart ecosystems. They provide real-time video feeds that can be used for surveillance, monitoring, and security purposes. They can be used to detect movement, recognize faces, and identify objects. Cameras can also be used in combination with other sensors and alarms, to create a comprehensive security system.

Actuators are devices that can act on the environment or other entities within the smart ecosystem. Data is collected by a gadget and processed by other subsystems using different techniques like filtering, denoising, amplification, etc. [32]. The gadget transmits the data to the cloud network using a gateway. Due to the cloud storage of this data, several services and functions are now available. Actuators are used to control the behavior of the smart ecosystem and can be programmed to respond to changes in the environment or the behavior of other entities within the ecosystem. Actuators perform actions based on the data collected by sensors, such as turning on/off lights or adjusting the temperature of a room. Examples of actuators include motors, valves, and pumps.

4.2 Internet of Things Devices

With the design of very small sensory modules and high-speed wireless connectivity involving 5G and other communication networks, along with cloud computing, one of the latest additions in the technological domain has been that of IoT [33]. The devices in an IoT environment are connected to the internet and can communicate with other devices and systems within the smart ecosystem [34]. Examples of IoT devices include smart thermostats, smart locks, smart appliances, and wearable devices. These devices enable remote monitoring and control, as well as data exchange between different systems. IoT devices can be used to optimize production processes and improve supply chain efficiency. For example, IoT-enabled traffic lights can adjust their timing based on real-time traffic data, helping to reduce congestion. IoT sensors can be used to monitor air quality in cities, optimize energy usage in buildings, and aid in real-time control of other processes. IoT-enabled waste management systems can optimize collection routes based on the level of fullness of each bin. In the healthcare industry, IoT devices can be used to monitor vital signs and track medication adherence [35]. In the home, for example, IoT devices can be used to control lighting and temperature, monitor energy usage, and even order groceries automatically when supplies run low. These devices are designed to collect data and communicate with each other through the IoT technology.

4.3 Communication Networks

Communication networks are essential for enabling the exchange of information between the various entities within the smart ecosystem. Communication networks can be wired or wireless and can use a range of protocols, including Wi-Fi, Bluetooth, and Zigbee. The choice of communication network will depend on the requirements of the smart ecosystem and the characteristics of the environment in which it is deployed. The open digital ecosystem ensures smooth collaboration between service providers, its customers, and its partners, enabling joint cloud-to-cloud solutions thanks to standardized interfaces and encrypted communication via the Internet or VPN connections.

With faster speeds, lower latency, and greater bandwidth, 5G enables a variety of new use cases, including ultra-high definition video streaming, augmented and virtual reality experiences, and real-time gaming. Additionally, 5G supports massive machine-to-machine communication, enabling the deployment of a wide range of IoT devices and services [36]. With its improved reliability and connectivity, 5G also has the potential to enhance critical infrastructure, such as healthcare, transportation, and public safety. Wi-Fi is the most common communication protocol locally used in smart ecosystems. It provides a high-speed, reliable connection that can be used to connect a wide range of devices. It is also easy to set up and use, making it ideal for consumer applications. Bluetooth is another commonly used communication protocol in smart ecosystems. It is ideal for low-power devices, such as wearables, and can be used to connect devices in close proximity. It is also a popular protocol for use in audio and entertainment devices such as headphones and speakers. Zigbee is a low-power, wireless communication protocol that is specifically designed for use in IoT devices. It is ideal for devices that need to operate for long periods without the need for frequent battery changes. Zigbee is also a popular choice for devices that require a mesh network, where devices can communicate with each other and extend the range of the network.

4.4 Data and Predictive Analytics

Data analytics is the process of extracting insights and knowledge from the data collected by sensors within the smart ecosystem. Data analytics can be performed using a range of techniques, including statistical analysis, ML, and AI. Data analytics enables the smart ecosystem to make informed decisions based on real-time data, and can help to optimize its performance in various applications like manufacturing, weather forecasting, social networking, supply chain management etc. [37–39].

Predictive analytics involves using data, ML, and AI to predict future outcomes or events [40, 41]. These technologies are particularly useful in creating predictive models, which can be used to anticipate problems and prevent them from happening. It involves the use of algorithms and statistical models to analyze and understand data, and make predictions or decisions without human intervention. This can be used in smart ecosystems to optimize operations and reduce waste, such as predicting when equipment will fail or forecasting energy demand in a building. It can also be used to diagnose diseases, detect fraud, improve supply chain efficiency, and optimize manufacturing processes.

4.5 Cloud/Edge Computing

With cloud computing, it is possible to use the data specifically to manage processes, increase availability, or improve products. Data exchange and collaboration are made possible on a global scale and in accordance with the same standards thanks to the cloud infrastructure. A smart ecosystem system works on a large amount of data that needs to be processed, stored, cached, transmitted, and computed. Some applications, particularly industrial ones, require quick reactions (real-time), as in the case of Industrial IoT based systems [42]. The network is put under a significant deal of strain when processing a lot of data. This is the major driver behind the introduction of edge computing, cloud computing, and fog computing, all of which are crucial to the growth of the smart ecosystem [43, 44]. Cloud is the network of servers and storage devices that allow the devices and applications in the ecosystem to communicate with each other. The data collected by sensors and IoT devices can be processed and stored in the cloud, which allows for more efficient and scalable data analysis. Cloud computing also enables access to data and services from anywhere and at any time, which is essential for creating a seamless and integrated smart ecosystem. The cloud infrastructure is also responsible for providing the computing power necessary to run the various applications in the ecosystem. It allows for the deployment of software updates and security patches, ensuring that the smart ecosystem remains up-to-date and secure.

Edge computing involves processing data locally, closer to the source, rather than sending it to a central server for processing. This can reduce latency and increase efficiency in smart ecosystems, making it possible to process data in real-time and respond quickly to changing conditions.

4.6 Decision-Making Algorithms

Decision-making algorithms are used to process data and make decisions about how the smart ecosystem should behave. Decision-making algorithms can be rule-based, heuristic, or based on ML techniques [45, 46]. AI is considered as one of biggest technology achievement and it is a rapidly evolving with the potential to revolutionize many industries and improve our daily lives. AI is the simulation of human intelligence processes by machines, especially computer systems. These processes include learning (the ability to improve from experience), reasoning (the ability to understand complex ideas and make decisions), and self-correction. ML is a type of AI that enables systems to learn and improve from experience without being explicitly programmed. AI and ML algorithms can be used to analyze the vast amounts of data collected by sensors and IoT devices. These algorithms can identify patterns and make predictions, which can then be used to optimize various processes and systems within the smart ecosystem. These algorithms can be used to optimize the working of the smart ecosystem in real-time, and to enable it to adapt to changes in the environment or the behavior of other entities within the ecosystem.

5 Benefits of Smart Ecosystems

Smart ecosystems are designed to be human-centric, with a focus on improving the quality of life for individuals and communities. This is achieved through the use of technologies such as smart homes, smart cities, and connected healthcare, which enable people to live and work more efficiently and effectively. Smart ecosystems offer a number of benefits to individuals, organizations, and society as a whole. These benefits are described in detail in this section.

5.1 Increased Efficiency

Smart ecosystems rely on automation and real-time data to optimize operations, which can result in significant efficiency gains [47]. For example, a smart building could use sensors to detect occupancy levels and adjust lighting and temperature settings accordingly, which can reduce energy consumption and lower operating costs. Similarly, a smart factory could use connected sensors and machines to monitor production processes and identify opportunities for optimization, leading to higher output and better product quality. Smart cities can also benefit from increased efficiency by optimizing traffic flow, reducing congestion, and improving public transportation systems.

5.2 Improved Sustainability

Smart ecosystems are designed to be sustainable, with a focus on energy efficiency, resource conservation, and waste reduction, leading to a more sustainable future. This is achieved through the use of advanced technologies such as smart grids, renewable energy sources, and intelligent transportation systems. For example, a smart grid could use real-time data to optimize energy production and distribution, reducing the reliance on fossil fuels and improving overall energy efficiency. Smart buildings and homes can also reduce their environmental footprint by using sensors to monitor energy use in lighting, heating, and cooling, water consumption, and waste generation, and adjusting usage patterns accordingly. Renewable energy sources such as solar panels and wind turbines can be integrated into the building's energy system to further reduce its environmental impact Additionally, smart agriculture can help reduce water consumption and optimize crop growth, leading to a more sustainable and efficient food system. In smart cities, intelligent transportation systems can help to reduce traffic congestion and emissions by optimizing traffic flow and promoting the use of public transportation. Smart waste management systems can help to reduce waste and improve recycling rates, while smart energy systems can help to reduce energy consumption and promote the use of renewable energy sources.

5.3 Enhanced Safety and Security

Smart ecosystems can leverage advanced technologies such as AI and ML to improve safety and security. For example, a smart city could use connected sensors to monitor traffic patterns and identify potential hazards, such as accidents or road closures, and provide real-time alerts to drivers and emergency services. Similarly, a smart home could use connected cameras and sensors to detect intruders or other potential security threats and alert homeowners and law enforcement. Smart healthcare systems can also improve patient safety and reduce medical errors by using connected devices to monitor vital signs and provide early warning alerts [48]. For security of digital information in particular, the techniques like Blockchain technology, Federated Learning, homomorphic encryption, cryptography, watermarking etc. are currently employed [49, 50].

5.4 Better Quality of Life

Smart ecosystems can improve quality of life by providing convenient and personalized services that meet the needs and preferences of individuals. For example, a smart home could use voice-activated assistants and connected devices to automate daily tasks such as turning on the lights, adjusting the thermostat, or ordering groceries. Smart homes are also designed to provide added convenience, comfort and security. Smart lighting systems can be programmed to automatically turn on and off, or to adjust to different levels of brightness depending on the time of day, or the presence of people in the room. Smart home devices can also be integrated with other home systems, such as home entertainment systems, to create a more seamless experience for the homeowner. Similarly, a smart city could provide real-time information about public transportation schedules, events, and weather, allowing residents to plan their day more efficiently. Smart healthcare systems can also improve quality of life by providing remote monitoring and care, enabling patients to receive high-quality care in the comfort of their own homes.

5.5 Increased Innovation

Smart ecosystems can facilitate innovation by providing a platform for collaboration and experimentation. For example, a smart city could use open data platforms to share data and insights with startups, researchers, and entrepreneurs, creating opportunities for new technologies and services to emerge. Similarly, a smart factory could use advanced analytics and automation to optimize processes and create new products and services. Smart ecosystems can also provide a platform for experimentation, allowing organizations to test new technologies and approaches in a safe and controlled environment.

Overall, smart ecosystems have the potential to transform the way we live, work, and interact with our environment, leading to more sustainable, efficient, and connected communities. The benefits of smart ecosystems are numerous and varied, and their potential impact on our lives is immense. As such, governments, businesses, and individuals should continue to invest in and promote the development of smart ecosystems to realize their full potential.

6 Applications of Smart Ecosystems

Smart ecosystems have a wide range of applications across different industries and sectors. The scope of smart ecosystem is in number of applications ranging from personal care to mass production, small module to a giant factory, simple monitoring to complete control etc. In this section we will briefly describe some of the possible areas where a smart ecosystem will do wonders. The natural environment, smart city (or bigger area), smart buildings, smart offices, smart manufacturing, smart healthcare and smart life are all interrelated parts of a smart ecosystem. Various case studies are being performed in these application areas to illustrate the potential of smart ecosystems to deliver significant benefits across different industries and sectors [51].

6.1 *Smart City*

Cities are relying more and more on their intelligence in an effort to attain sustainability. Smart cities are defined as those that can use cutting-edge technologies and resources in an integrated manner to respond to daily changes in order to achieve sustainable social and environmental economic growth [52, 53]. Smart cities are one of the most promising applications of smart ecosystems, and there are several successful examples around the world. In Singapore, for example, the government has developed a range of initiatives to promote the use of smart technologies in areas such as transportation, healthcare, and energy. Similar developments are found in many other countries. These initiatives have resulted in significant improvements in efficiency, sustainability, and quality of life.

A smart city ecosystem consists of sub-ecosystems, such as smart transportation, smart water utilities, smart buildings, smart elevators, smart pumps, and intelligent VSDs. A single intelligent VSD, which is a small ecosystem in itself, could be part of a smart pump ecosystem or a smart elevator ecosystem. Smart traffic management systems can be used to optimize traffic flow, reduce congestion, and improve air quality by reducing emissions from vehicles [54]. Smart lighting systems can be used to reduce energy consumption and costs, while also improving safety and security. Smart waste management systems can be used to optimize the collection and disposal of waste, reducing the environmental impact of the city. Another area where cities are interested in improving their appeal to retailers and attracting investment is inner city retailing and the integration of online and offline stores. E-government is another component of the smart city ecosystem with goals and the ability to interact with citizens more easily. These provisions and endless possibilities in the smart city ecosystem forms the basis for multiple Smart ecosystems coexisting under one umbrella. These systems use a variety of technologies, such as sensors and cameras, to gather data on things like traffic flow, air quality, and energy usage. This data is then used to improve the efficiency and liveability of the city, by optimizing traffic flow and reducing pollution, for example.

6.2 Smart Healthcare

Smart ecosystems have the potential to revolutionize the healthcare industry by enabling remote patient monitoring, personalized medicine, and the development of new treatments. Smart healthcare is an area where there is significant potential for smart ecosystems to deliver significant benefits [55, 56]. Smart ecosystems can be used to collect data about patients in real-time, and to analyze this data to identify patterns and make informed decisions about treatment. They can also be used to support the development of new drugs and therapies.

In most of the developed countries, national health services currently include a range of initiatives to promote the use of smart technologies in areas such as remote patient monitoring, predictive analytics, and personalized medicine. These initiatives have the potential to improve patient outcomes and reduce the cost of healthcare.

6.3 Smart Energy

Energy utilization is the biggest concern for the functioning of various technologies as the consumption is increasing day by day even though resources are limited. The emergence of new business and financing models, new evidence on climate change, favourable regulations, declining renewable energy costs, advancements in technology and materials, and the increased appeal of investments in clean energy, are expected to attract about $11 trillion in total investments by 2050 [57]. Smart ecosystems are envisioned to optimize the generation and use of energy in various forms from every private to public place. These systems collect data about energy consumption and generation, and to use this data to optimize energy use and reduce waste [58, 59]. For example, smart homes often use sensors to gather data on things like occupancy and energy usage, which can be used to improve energy efficiency and provide personalized experiences for the homeowner. Smart ecosystems can also be used to support the development of renewable energy sources, such as solar and wind power. Smart grids are considered to be the best alternative to every energy generation/distribution system to avoid the energy crisis and environmental degradation [60].

Smart ecosystems ensure energy-savings and low carbon emissions through mobility services, intelligent routing systems, and other similar technologies, as well as by creating novel energy provisioning ideas. For instance, if you are aware that there are more people in the city center at particular times of the day or due to a significant event, you may need to increase the power supply for some services, such as night-time street lighting, but not on other days. In addition to energy conservation, there is high provision of collaboration with other systems to improve the environment in the globe as a whole [61]. One of the frameworks for developing a smart energy network ecosystem for integrated energy services in metropolitan areas is presented by the authors in [62]. Additionally, they offer preliminary findings for the innovation ecosystem's test platform, Smart Otaniemi, which links experts, businesses, technologies, and pilots.

6.4 Smart Agriculture

Smart ecosystems can be used to optimize agricultural production and reduce waste [63]. Precision agriculture is another area where smart ecosystems are already delivering significant benefits. They can be used to collect data about soil moisture, nutrient levels, and weather conditions, and to use this data to optimize crop growth and reduce the use of pesticides and fertilizers. Smart ecosystems can also be used to monitor animal health and welfare, and to support sustainable farming practices [64]. In the United States for example, the Agtech startup Farmers Edge has developed a system that uses sensors and satellite imagery to provide farmers with real-time data about soil moisture, temperature, and other factors. This data can be used to optimize crop growth and reduce the use of pesticides and fertilizers [65].

6.5 Smart Industry

Smart Industry is a transformative approach that integrates digital technology and automation into manufacturing processes to create a smarter and more efficient system. It involves the integration of sensors, automation, and software systems to create a more efficient and productive manufacturing environment. A smart industry involves use of advanced technologies such as the IOT, AI, and big data analytics to improve the efficiency and productivity of industrial processes. Industry 4.0, also referred to as the Fourth Industrial Revolution and used interchangeably with smart industry, is a comprehensive and holistic approach that involves smart manufacturing and encompasses the entire value chain. The goal is to streamline and optimize the entire manufacturing process, from design and planning to production and distribution, while reducing costs and increasing productivity. Industry 4.0 also places a strong emphasis on sustainability and resource efficiency. By utilizing renewable energy sources, implementing circular economy principles, and optimizing resource usage in real-time, manufacturers can reduce waste and lower carbon emissions.

Industry 4.0 is driving the development of smart factories, which are highly automated and connected to manufacturing systems that utilize the IOT, also known as Industrial-IoTs. The IoT allows smart factories to gather and analyze real-time data from sensors and connected devices throughout the production process. This data can be used to optimize and improve operations, reduce costs, and enhance the overall customer experience [66, 67]. Additionally, IoT systems designed for this application often use predictive maintenance techniques, which use data and analytics to predict when equipment is likely to fail, allowing for proactive maintenance and reducing downtime [68].

6.6 Intelligent Transportation

The transition to intelligent transport systems has been accelerated by technological breakthroughs, mainly with regard to the AI, IoT, big data, wireless sensor networks, and Global Positioning System (GPS) [69]. Smart ecosystems can be used to optimize transportation systems, including public transit, ride-sharing, and logistics. They can be used to collect data about traffic patterns, road conditions, and vehicle performance, and to use this data to optimize the performance of the transportation system [70, 71]. Smart ecosystems can also be used to improve the safety and efficiency of transportation systems, and to reduce environmental impact [72].

Intelligent transportation can be linked to other ecosystems to develop an even bigger and comprehensive ecosystem. For example, a connected car industry can be transformed into a bigger smart ecosystem by incorporation of smart manufacturing and big data analytics. This entails offering a plethora of data collection services, including those for traffic, weather, emergency services, navigation, entertainment, and communication. This is just the beginning of how these services connect with the aforementioned smart services. Add to that sensors for predictive maintenance; platforms for car sharing; and intelligence for autonomous driving. Smart cities often use connected transportation systems, such as bike-sharing programs and autonomous vehicles, to provide citizens with more efficient and sustainable transportation options [73].

6.7 Intelligent Language Processing

Recent advancements in Digital Signal Processing and AI tools have led to significant progress in speech, text, and language processing [74]. Smart ecosystems designed for intelligent language processing (ILP) aids in a wide range of applications, including virtual assistants, chatbots, speech recognition, machine translation, sentiment analysis and security [75]. These systems use a combination of statistical models, ML algorithms, and linguistic rules to analyze text and speech data. They can be used to automate customer service interactions, analyze social media sentiment, extract insights from large volumes of text data, and much more. Some common techniques used in language processing systems include part-of-speech tagging, named entity recognition, speech synthesis, Text-to-Speech (TTS) conversion, and machine translation. These techniques allow the system to understand the grammatical structure of a sentence, identify key entities or concepts, determine the emotional tone of a piece of text, and translate text from one language to another. These advancements are due to the use of deep learning techniques, such as RNNs and CNNs, generative adversarial networks (GANs) deep clustering, deep attractor networks, and DNN models like WaveNet, and Tacotron.

7 Challenges in Smart Ecosystem Design

No organization, regardless of location or size, can claim ownership of the full value creation process, as was amply demonstrated by the eco Association research into the smart city, smart home, and Industrial IoT industries. Instead, businesses must carve out a position for themselves in the connected smart ecosystem, build a new network of alliances, and perhaps even learn to work with rivals. Smart ecosystems offer a range of benefits, but they also present a number of challenges that need to be addressed. For smooth functionality of smart ecosystems, there is strong need for the development of new technologies and the need for greater collaboration between different stakeholders [76, 77]. Also, the technology is unreliable in many applications and its development and deployment needs further boost by other technologies for complete reliance In this section we will highlight few issues that need to be taken care of.

7.1 Integration

Smart ecosystems are made up of different technologies, devices, and systems, and it is important that these components work together to create a seamless user experience. However, integrating different systems can be challenging, especially when they use different protocols, data formats, or communication interfaces [78]. For example, some smart devices may use Bluetooth for communication, while others may use Wi-Fi or Zigbee. To integrate these devices into a smart ecosystem, developers may need to create custom interfaces or middleware that can translate between different protocols and data formats.

7.2 Data Privacy and Security

Smart ecosystems collect and process vast amounts of data from different sources, including sensors, IoT devices, and social media. This data must be protected from unauthorized access, manipulation, and misuse to ensure user privacy and prevent cyber-attacks. As more devices are connected to the internet, the risk of cyber-attacks increase. It's important that manufacturers design devices with security in mind and that consumers take steps to secure their devices [79]. It's important that companies handling this data implement strict privacy policies and that consumers are aware of what data is being collected and how it will be used. To achieve this, smart ecosystems must have robust security and privacy measures, such as encryption, authentication, and access controls. There has been a good work going on various Cybersecurity mechanisms like Data encryption, cryptography, Data hiding, Differential privacy, Federated learning and Blockchain technology.

7.3 Interoperability

Smart ecosystems must be interoperable, which means they should work seamlessly with different technologies, devices, and systems, regardless of their origin or vendor. However, achieving interoperability can be challenging due to the lack of standardization and the diversity of technologies. For example, there are many different standards for IoT devices, such as Zigbee, Z-Wave, and Bluetooth, which can make it difficult to create an ecosystem that works with all devices. To address this challenge, industry groups and standards organizations are working to create more interoperable standards, such as the Open Connectivity Foundation (OCF).

7.4 Scalability

Smart ecosystems must be scalable to accommodate new technologies, devices, and users without compromising their performance or functionality. However, scaling up smart ecosystems can be costly and complex, especially when they rely on legacy systems or infrastructure. For example, adding more sensors to a smart building may require upgrading the building's electrical system or installing additional network infrastructure. To address this challenge, developers may need to create modular and flexible systems that can be easily expanded or modified.

7.5 Energy Efficiency

The energy crisis is one of the biggest challenges before every technology. Smart ecosystems are itself designed to be energy-efficient in order to minimize their environmental impact and reduce their operational costs. However, achieving energy efficiency can be challenging, especially when the ecosystem relies on power-hungry devices or systems [80]. For example, an IoT device that is always connected to the internet may consume more energy than a device that only connects when necessary. To address this challenge, developers may need to design devices and systems that are optimized for energy efficiency, such as by using low-power wireless technologies like Bluetooth Low Energy (BLE) or Zigbee. Also, design of efficient algorithms to monitor and control various operations can help.

7.6 User Adoption and Acceptance

Smart ecosystems must be user-friendly and easy to use to encourage user adoption and acceptance. However, achieving user acceptance can be challenging, especially when the ecosystem requires significant behaviour changes or involves complex interactions with different devices or systems [76]. For example, a smart home system that requires users to interact with multiple apps or devices may be overwhelming and difficult to use. To address this challenge, developers may need to create intuitive and user-friendly interfaces, as well as provide training and support to users. Scientists are devising various mechanisms to ensure the acceptance of newly designed systems through a continuous feedback and upgradation from every corner of that particular environment.

7.7 Governance and Regulation

Smart ecosystems must be governed and regulated to ensure compliance with legal, ethical, and social norms. However, achieving effective governance and regulation can be challenging, especially when the ecosystem involves different stakeholders with conflicting interests or agendas. For example, a smart city project may involve multiple government agencies, private companies, and community groups, each with their own priorities and goals. To address this challenge, smart ecosystems may need to have clear governance structures and regulations that balance the interests of all stakeholders [81]. Additionally, ensuring that they comply with relevant laws and regulations, such as data privacy laws, environmental regulations, and safety standards. This may involve creating new regulations or adapting existing ones to the unique characteristics of smart ecosystems, such as the use of IoT devices or big data analytics.

In addition to these challenges, there are several other factors that can impact the development and deployment of smart ecosystems. For example, the availability of funding and resources, the level of public awareness and support, and the level of collaboration and cooperation among stakeholders can all impact the success of a smart ecosystem project. Another challenge is the sheer amount of data generated by various devices in the system. This data needs to be collected, stored, and analyzed, which requires significant investment in infrastructure and expertise. With seamless connectivity of numerous devices in an application the data sharing is currently enormous and a good amount of research is going in the areas like

Big data, cloud computing and edge computing. To address these challenges and create successful smart ecosystems, it is important to take a holistic approach that considers the technical, social, and economic factors involved. This may involve working with a diverse set of stakeholders, including technology companies, government agencies, community groups, and academic researchers, to identify and address the most pressing challenges and opportunities. It may also involve using innovative approaches, such as open data initiatives, crowdsourcing, and citizen science, to engage users and promote transparency and accountability. In order to reduce development costs and boost productivity, designers must be efficiently and interactively assisted throughout the design process. This is necessary due to the time-to-market pressure, unique customer requirements, rapid technological change, increasing application complexity, and diversity of design styles, methodologies, and associated tools [82, 83].

8 Conclusion

This chapter gives a brief overview of the latest technological advances towards realization of smart ecosystems that offers an environment with all system functional elements or components designed to be integrative and collaborative. Smart ecosystems represent a significant opportunity to deliver real-world benefits across a range of industries and sectors. A smart ecosystem offers an integrated environment to facilitate user friendly and attractive life with the aid of technology. The primary goal of a smart ecosystem is to provide a seamless, integrated experience for users by connecting multiple devices and services. While there are still significant challenges to be addressed, including data privacy and security, interoperability, and governance, the potential benefits of smart ecosystems make them a compelling area for research and development. As smart technologies continue to evolve and new applications emerge, it is likely that smart ecosystems will become an increasingly important part of our daily lives, delivering greater efficiency, sustainability, and quality of life. Overall, a smart ecosystem, be it big or small, has the potential to bring many benefits to society, but it is important that we address the challenges associated with it to ensure that it is used in a responsible and secure manner. It is important to consider the security, privacy and the infrastructure challenges that comes with it, in order to fully realize the potential of this technology.

References

1. Hurrah NN, Loan NA, Parah SA, Sheikh JA, Muhammad K, de Macedo ARL, de Albuquerque VHC (2020) INDFORG: industrial forgery detection using automatic rotation angle detection and correction. IEEE Trans Industr Inform 17(5):3630–3639
2. Kaw JA, Loan NA, Parah SA, Muhammad K, Sheikh JA, Bhat GM (2019) A reversible and secure patient information hiding system for IoT driven e-health. Int J Inf Manag 45:262–275

3. Hurrah NN, Parah SA, Sheikh JA (2019) A secure medical image watermarking technique for e-healthcare applications. In: Handbook of multimedia information security: techniques and applications. Springer, Cham, pp 119–141
4. Qureshi I, Parah SA, Lone NA, Hurrah N, Qureshi GJ (2021) Realization of a robust watermarking system in spatial domain. In: Recent innovations in computing: proceedings of ICRIC 2020. Springer, Singapore, pp 369–379
5. Locatelli P, Gastaldi L, Zacchetti D, Pacelli V, Cirilli F (2021) From network approach to ecosystem approach: a new framework for change management. In: European conference on innovation and entrepreneurship. Academic Conferences International Limited, Reading, p 503-XIX
6. Nimmagadda S, Namugenyi C, Mani N, Reiners T (2020) Managing embedded digital ecosystems in pandemic big data contexts. Springer Nature Switzerland AG
7. Martins VF, Tomczyk Ł, Amato C, Eliseo MA, Oyelere SS, Akyar ÖY et al (2020) A smart ecosystem for learning and inclusion: an architectural overview. In: Computational science and its applications–ICCSA 2020: 20th international conference, Cagliari, Italy, July 1–4, 2020, proceedings, part I 20. Springer, Cham, pp 601–616
8. Riđić O, Jukić T, Riđić G, Ganić M, Bušatlić S, Karamehić J (2022) The smart city, smart contract, smart health care, internet of things (IoT), opportunities, and challenges. In: Blockchain technologies for sustainability. Springer, Singapore, pp 135–149
9. Keskin B (2021) A building information modeling (BIM)-centric digital ecosystem for smart airport life cycle management. Doctoral dissertation, Syracuse University
10. Lee D, Kim K (2022) National investment framework for revitalizing the R&D collaborative ecosystem of sustainable smart agriculture. Sustainability 14(11):6452
11. Baranovski I, Stankovski S, Ostojić G, Nikolić V, Simonović M, Stanojević M (2022) Toward a smart ecosystem with automated services, Series: mechanical engineering. Facta Universitatis
12. Yus R, Bouloukakis G, Mehrotra S, Venkatasubramanian N (2022) The semiotic ecosystem: a semantic bridge between IoT devices and smart spaces. ACM Trans Internet Technol (TOIT) 22(3):1–33
13. Baranovski I, Stankovski S, Ostojić G, Horvat S (2020) Support for self-service automated parking system. In: 2020 19th international symposium Infoteh - Jahorina (Infoteh), pp 1–5
14. Eltawil A, Mostafa NA, Matsushita Y (2021) Toward a smart and sustainable campus: future vision, opportunities, and challenges. In: Resilient and responsible smart cities, vol 1. Springer, Cham, pp 233–242
15. He T, Jazizadeh F, Arpan L (2022) AI-powered virtual assistants nudging occupants for energy saving: proactive smart speakers for HVAC control. Build Res Inf 50(4):394–409
16. Briscoe G, Sadedin S, De Wilde P (2011) Digital ecosystems: ecosystem-oriented architectures. Nat Comput 10:1143–1194
17. Jensen JJ (2020) Applying a "smart ecosystem" mindset to rethink your products. Computer 53(12):98–101
18. Protasenko OF, Mygal GV, Manakova HO (2021) Safety of man-machine interaction in the digital environment
19. Stankovski S, Ostojic G et al (2020) Edge computing for fault detection in smart systems ICIST 2020 proceedings, pp 22–26
20. Kohtamäki M, Rabetino R, Parida V, Sjödin D, Henneberg S (2022) Managing digital servitization toward smart solutions: framing the connections between technologies, business models, and ecosystems. Ind Mark Manag 105:253–267
21. Afzal I, Parah SA, Hurrah NN, Song OY (2020) Secure patient data transmission on resource constrained platform. Multimed Tools Appl. https://doi.org/10.1007/s11042-020-09139-3
22. Dickson R, Lisachuk H, Cotteleer M (2018) Growing internet of things platforms: an ecosystem of value cocreation. https://www2.deloitte.com/us/en/insights/focus/internet-of-things/iot-ecosystem-platforms-value-creation.html
23. Basatneh R, Najafi B, Armstrong DG (2018) Health sensors, smart home devices, and the internet of medical things: an opportunity for dramatic improvement in care for the lower extremity complications of diabetes. J Diabetes Sci Technol 12(3):577–586

24. Mansoor S, Parah SA (2023) HAIE: a hybrid adaptive image encryption algorithm using Chaos and DNA computing. Multimed Tools Appl. https://doi.org/10.1007/s11042-023-14542-7
25. Burhan M, Rehman RA, Khan B, Kim BS (2018) IoT elements, layered architectures and security issues: a comprehensive survey. Sensors 18:2796
26. Kunal S, Saha A, Amin R (2019) An overview of cloud-fog computing: architectures, applications with security challenges. IEEE Secur Priv 2(4):1–14
27. Parah SA, Sheikh JA, Loan NA, Ahad F, Bhat GM (2018) Utilizing neighborhood coefficient correlation: a new image watermarking technique robust to singular and hybrid attacks. Multidim Syst Sign Process 29:1095–1117
28. Parah SA, Sheikh JA, Bhat GM (2014) A secure and efficient spatial domain data hiding technique based on pixel adjustment. Am J Eng Technol Res 14(2):33
29. Zagorskiy A (2022) Principles for implementing the ecosystem of next-generation intelligent computer systems
30. Bhat GM, Mustafa M, Parah SA, Ahmad J (2010) Field programmable gate array (FPGA) implementation of novel complex PN-code-generator-based data scrambler and descrambler. Maejo Int J Sci Technol 4(1):125–135
31. Liu X, Lam KH, Zhu K, Zheng C, Li X, Du Y, Liu C, Pong P (2019) Overview of spintronic sensors with internet of things for smart living. IEEE Trans Magn 55(11):1–22
32. Malik SA, Parah SA, Aljuaid H, Malik BA (2023) An iterative filtering based ECG denoising using lifting wavelet transform technique. Electronics 12(2):387. https://doi.org/10.3390/electronics12020387
33. Pourqasem J (2018) Cloud-based IoT: integration cloud computing with Internet of things. Int J Res Ind Eng 7:482–494
34. Gill SS, Garraghan P, Buyya R (2019) Router: fog enabled cloud based intelligent resource management approach for smart home IoT devices. J Syst Softw 154:125–138
35. Parah SA, Sheikh JA, Ahad F, Bhat GM (2018) High capacity and secure electronic patient record (EPR) embedding in color images for IoT driven healthcare systems. In: Internet of things and big data analytics toward next-generation intelligence. Springer, Cham, pp 409–437
36. Sheikh JA, Akhter S, Parah SA, Bhat GM (2018) Blind digital speech watermarking using filter bank multicarrier modulation for 5G and IoT driven networks. Int J Speech Technol 21:715–722
37. Wang J, Xu C, Zhang J, Zhong R (2022) Big data analytics for intelligent manufacturing systems: a review. J Manuf Syst 62:738–752
38. Fathi M, Haghi Kashani M, Jameii SM, Mahdipour E (2022) Big data analytics in weather forecasting: a systematic review. Arch Comput Methods Eng 29(2):1247–1275
39. Akter S, Michael K, Uddin MR, McCarthy G, Rahman M (2022) Transforming business using digital innovations: the application of AI, blockchain, cloud and data analytics. Ann Oper Res 308:7–39. https://doi.org/10.1007/s10479-020-03620-w
40. Dev S, Wang H, Nwosu CS, Jain N, Veeravalli B, John D (2022) A predictive analytics approach for stroke prediction using machine learning and neural networks. Healthcare Anal 2:100032
41. Deloitte (2019) Predictive analytics in health care: emerging value and risks. https://www2.deloitte.com/xe/en/insights/topics/analytics/predictive-analytics-health-care-value-risks.html
42. Qiu T, Chi J, Zhou X, Ning Z, Atiquzzaman M, Wu DO (2020) Edge computing in industrial internet of things: architecture, advances and challenges. IEEE Commun Surv Tutor 22(4):2462–2488
43. Shi W, Cao J, Zhang Q, Li Y, Xu L (2016) Edge computing: vision and challenges. IEEE Internet Things J 3(5):637–646
44. Wang K, Tan Y, Shao Z, Ci S, Yang Y (2019) Learning-based task offloading for delay-sensitive applications in dynamic fog networks. IEEE Trans Veh Technol 68(11):11399–11403
45. Parah SA, Rashid M, Vijaykumar V (2022) Artificial intelligence for innovative healthcare informatics. Springer. isbn:978-3-030-96568-6

46. Parah SA, Sheikh JA, Bhat GM (2012) On the realization of a secure, high capacity data embedding technique using joint top-down and down-top embedding approach. Comput Sci Eng 49:10141–10146
47. Rzevski G, Skobelev P, Zhilyaev A (2022) Emergent intelligence in smart ecosystems: conflicts resolution by reaching consensus in resource management. Mathematics 10(11):1923
48. Hurrah NN, Khan E, Khan U (2023) CADEN: cellular automata and DNA based secure framework for privacy preserving in IoT based healthcare. J Ambient Intell Humaniz Comput 14:2631–2643. https://doi.org/10.1007/s12652-022-04510-8
49. Parah SA, Sheikh JA, Dey N, Bhat GM (2017) Realization of a new robust and secure watermarking technique using DC coefficient modification in pixel domain and chaotic encryption. J Glob Inform Manage 25(4):80–102
50. Hurrah NN, Parah SA, Sheikh JA (2020) Embedding in medical images: an efficient scheme for authentication and tamper localization. Multimed Tools Appl 79:21441–21470
51. Angelidou M, Politis C, Panori A, Barkratsas T, Fellnhofer K (2022) Emerging smart city, transport and energy trends in urban settings: results of a pan-European foresight exercise with 120 experts. Technol Forecast Soc Chang 183:121915
52. Khang A, Rani S, Sivaraman AK (eds) (2022) AI-centric smart city ecosystems: technologies, design and implementation. CRC Press, Boca Raton. https://doi.org/10.1201/9781003252542
53. Komninos N, Panori A (2019) The creation of city smartness: architectures of intelligence in smart cities and smart ecosystems. In: Smart cities in the post-algorithmic era. Edward Elgar Publishing, Cheltenham, pp 101–127
54. Komninos N, Kakderi C, Collado A, Papadaki I, Panori A (2021) Digital transformation of city ecosystems: platforms shaping engagement and externalities across vertical markets. J Urban Technol 28(1–2):93–114
55. Hlongwane SN (2021) Development of a management tool for value creation supported by information systems in healthcare: an ecosystem perspective. Doctoral dissertation, Stellenbosch: Stellenbosch University
56. Ahad F, Parah SA, Sheikh JA, Bhat GM (2015) On the realization of robust watermarking system for medical images. In: 2015 Annual IEEE India conference (INDICON), New Delhi, India, pp 1–5. https://doi.org/10.1109/INDICON.2015.7443363
57. Deloitte (2020) Extracting value from decarbonisation: an integrated approach to reducing emissions. https://www2.deloitte.com/za/en/pages/energy-and-resources/articles/extracting-value-from-decarbonisation.html
58. Jiang Y, Zheng W (2021) Coupling mechanism of green building industry innovation ecosystem based on blockchain smart city. J Clean Prod 307:126766
59. Bedi P, Goyal SB, Rajawat AS, Shaw RN, Ghosh A (2022) Application of AI/IoT for smart renewable energy management in smart cities. In: AI and IoT for smart city applications, Vol. 1002. Springer, Singapore, pp 115–138
60. Mourtzis D, Angelopoulos J, Panopoulos N (2022) Smart grids as product-service systems in the framework of energy 5.0-a state-of-the-art review. Green Manuf Open 1(1):5
61. Jiang B, Fei Y (2014) Smart home in smart microgrid: a cost-effective energy ecosystem with intelligent hierarchical agents. IEEE Trans Smart Grid 6(1):3–13
62. Dai C, Tang M, Liu Y, He J, Yang Z, Yang Y (2020) Designing smart energy network ecosystem for integrated energy services in urban areas. In: Proceedings of 2020 IEEE 16th international conference on automation science and engineering (CASE), pp 305–310
63. Smania GS, de Sousa Mendes GH, Godinho Filho M, Osiro L, Cauchick-Miguel PA, Coreynen W (2022) The relationships between digitalization and ecosystem-related capabilities for service innovation in agricultural machinery manufacturers. J Clean Prod 343:130982
64. Chukkapalli SSL, Mittal S, Gupta M, Abdelsalam M, Joshi A, Sandhu R, Joshi K (2020) Ontologies and artificial intelligence systems for the cooperative smart farming ecosystem. IEEE Access 8:164045–164064
65. Akhtar MN, Shaikh AJ, Khan A, Awais H, Bakar EA, Othman AR (2021) Smart sensing with edge computing in precision agriculture for soil assessment and heavy metal monitoring: a review. Agriculture 11(6):475

66. Mourtzis D, Angelopoulos J, Panopoulos N (2022) Digital manufacturing: the evolution of traditional manufacturing toward an automated and interoperable smart manufacturing ecosystem. In: The digital supply chain. Elsevier, Amsterdam, pp 27–45
67. Kamili A, Hurrah NN, Parah SA, Bhat GM, Muhammad K (2020) DWFCAT: dual watermarking framework for industrial image authentication and tamper localization. IEEE Trans Industr Inform 17(7):5108–5117
68. García Á, Bregon A, Martínez-Prieto MA (2022) Towards a connected Digital Twin Learning Ecosystem in manufacturing: enablers and challenges. Comput Ind Eng 171:108463
69. Visan M, Negrea SL, Mone F (2022) Towards intelligent public transport systems in smart cities; collaborative decisions to be made. Procedia Comput Sci 199:1221–1228
70. Chehri A, Sharma T, Debaque B, Duclos N, Fortier P (2022) Transport systems for smarter cities, a practical case applied to traffic management in the city of Montreal. In: Sustainability in energy and buildings 2021. Springer Singapore, pp 255–266
71. Hensher DA, Nelson JD, Mulley C (2022) Electric car sharing as a service (ECSaaS)–acknowledging the role of the car in the public mobility ecosystem and what it might mean for MaaS as eMaaS? Transp Policy 116:212–216
72. Mora L, Wu X, Panori A (2020) Mind the gap: developments in autonomous driving research and the sustainability challenge. J Clean Prod 275:124087
73. Ahmed ST, Basha SM, Ramachandran M, Daneshmand M, Gandomi AH (2023) An edge-AI enabled autonomous connected ambulance route resource recommendation protocol (ACA-R3) for eHealth in smart cities. IEEE Internet Things J 10:11497–11506
74. Iliev AI, Stanchev PL (2020) Smart ecosystems through voice and images. In: CATA, proceedings of 35th international conference on computers and their applications, EPiC series in computing, vol. 69. EasyChair, Bishop's Stortford, pp 256–263
75. Apturkar A, Iliev AI, Anand A, Oli A, Siddenki SR, Meka VR (2020) Sentiment analysis of speech with application to various languages. Digit Present Preserv Cult Sci Herit 10:103–118
76. Panori A, Kakderi C, Komninos N, Fellnhofer K, Reid A, Mora L (2021) Smart systems of innovation for smart places: challenges in deploying digital platforms for co-creation and data-intelligence. Land Use Policy 111:104631
77. El Moussa N, Molinelli D, Pezzè M, Tappler M (2021) Health of smart ecosystems. In: Proceedings of the 29th ACM joint meeting on European software engineering conference and symposium on the foundations of software engineering, pp 1491–1494
78. Cioroaica E, Kuhn T, Bauer T (2018) Prototyping automotive smart ecosystems. In: 2018 48th annual IEEE/IFIP international conference on dependable systems and networks workshops (DSN-W). IEEE, Luxembourg, pp 255–262
79. Bhat GM, Mustafa M, Ahmad S, Ahmad J (2009) VHDL modeling and simulation of data scrambler and descrambler for secure data communication. Indian J Sci Technol 2(10):41–44
80. Ahmed RA, Hemdan EED, El-Shafai W, Ahmed ZA, El-Rabaie ESM, Abd El-Samie FE (2022) Climate-smart agriculture using intelligent techniques, blockchain and Internet of Things: concepts, challenges, and opportunities. Trans Emerg Telecommun Technol 33(11):e4607
81. Badran A (2021) Developing smart cities: regulatory and policy implications for the State of Qatar. Int J Public Adm 46:1–14
82. Philipp J, Thees H, Olbrich N, Pechlaner H (2022) Towards an ecosystem of hospitality: the dynamic future of destinations. Sustainability 14(2):821
83. Borges-Tiago T, Veríssimo J, Tiago F (2022) Smart tourism: a scientometric review (2008-2020). Euro J Tour Res 30:3006–3006

Big Data in Smart Ecosystems: Trends, Challenges and Future Prospectus

Nusrat Rouf, Saima Saleem, Majid Bashir Malik, and Khalid Bashir Dar

1 Introduction

The current global population surpasses eight billion and more than five billion people globally are connected to internet which is equivalent to 63.1% of the world's population [1]. The current global smartphone subscriptions exceed six billion, with an additional several hundred million expected to sign up in the following few years [2]. Through the increased use of electronic gadgets, billions of people are producing enormous amounts of data [3]. Big data refers to data that is disseminated online through various sources. Big data refers not only to large and varied data sets but also to the technologies that collect, handle, store, and analyze them in order to address complex problems [4]. Sam Madden from Massachusetts Institute of Technology (MIT) composed the term Big Data and defines it as, "Big data means too big, too fast, or too hard for existing tools to process" [5]. TechAmerica Foundation has defined it as, "Big data is a term that narrates voluminous data of high velocity, complex and changing data that require advanced techniques and technologies to enable the capture, storage, distribution, management, and analysis of the information" [6]. Laney introduced a definition (3Vs) to give the insight of what "big" data is where 3V's represent volume, velocity, and variety [7]. The definition of 3Vs explains that the data size is voluminous, the data is disseminated

N. Rouf (✉) · M. B. Malik
Department of Computer Sciences, Baba Ghulam Shah Badshah University, Rajouri, India
e-mail: nusratrouf@bgsbu.ac.in

S. Saleem
Department of Computer Science, Jamia Millia Islamia, New Delhi, India

K. B. Dar
Department of Computer Science, Chaudhary Charan Singh University, Meerat, India

S. A. Parah et al. (eds.), *Intelligent Multimedia Signal Processing for Smart Ecosystems*, https://doi.org/10.1007/978-3-031-34873-0_2

with tremendous speed, and the data is present in different formats and is gathered from different sources, respectively.

1.1 Characteristics of Big Data

(a) **Volume:** Volume is described as the amount of data that is produced from diverse sources. IBM conducted a survey in 2013 that revealed the results that from 1144 respondents 572 considered datasets more than one terabyte to be big [8]. One terabyte store as much data that is enough to store 16 million (approx.) Facebook photographs. In 2021, there were over 4.26 billion people using social media globally, and by 2027, this number is expected to reach nearly six billion as shown in Fig. 1 [9]. The data volume will reach a total of 163 zettabytes by the year 2025 as per IDC report published in 2017 [2]. However, what may be assumed as big data today may not be declared as big data in future due to advancement in storage capacity.

(b) **Velocity:** Velocity describes the rate at which data is produced and analyzed. Big data technologies allow firms to create real-time intelligence from voluminous data. Number of active users have increased exponentially. Figure 2 presents number of active users (in billions) per month on different social media platforms [10].

(c) **Variety:** Variety implies several forms of data that is disseminated through various online platforms. Data generated is mainly of three types: "Structured Data", "Semi-Structured Data" and "Unstructured Data". Structured data follows a proper format like data in databases, .csv files and spread sheets [4]. Only 5% of total data disseminated constitutes of structured data. Semi-structured data doesn't have a proper format like email data, log files, word documents and

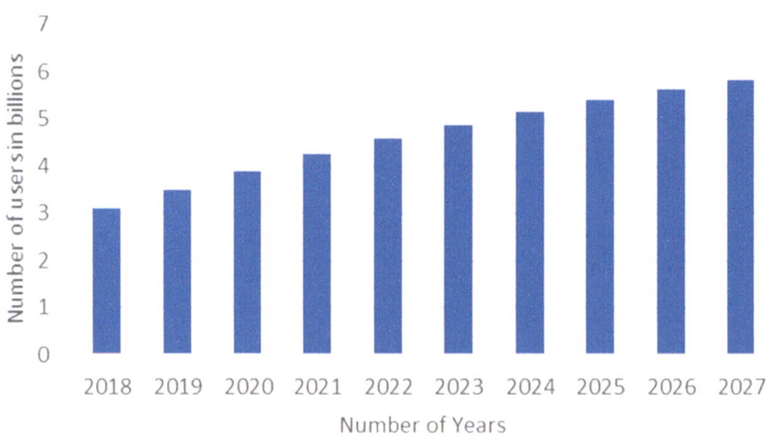

Fig. 1 Number of social media users worldwide from 2018 to 2027 (in billions)

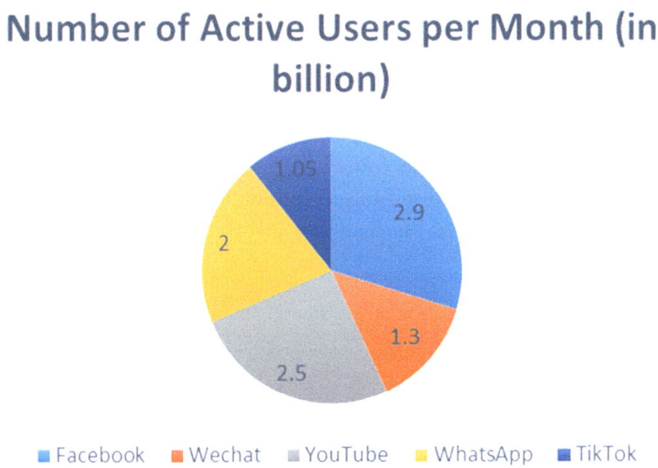

Fig. 2 Statistics of social media user generated content in one minute

website data. Unstructured data lacks a format like image files, audio files and video files.

Studies such as [11] have made it clear that the definition of 3Vs is not sufficient to define the big data. As a result, the attributes of big data are expanded to include veracity, value, and variability.

(d) **Veracity:** IBM coined it as a fourth V which relates to data being unreliable. It answers the question: 'Is mined data meaningful for the problem that is being analyzed'. For example, interpreting sentiments of customer on social media that are ambiguous. Data with high signal-to-noise-ratio (SNR) has more veracity [12].
(e) **Variability:** Variability refers to lack of consistency in data. It is the change in rate of data flow [13]. This characteristic leads to a major challenge in data processing such as the necessity to match, filter and transform data obtained from many sources.
(f) **Value:** Value refers to usefulness of data. Oracle defined value as a defining characteristic of big data. Data received in raw form has a poor value in comparison to its volume. It is a process of discovering concealed values from huge amounts of data [14].

The remainder of this chapter are framed as: Section 2 presents the big data framework that gives insights about different data sources, transformations, the big data platforms and tools. Section 3 discusses big data applications areas such as smart cities, healthcare and smart grids. Section 4 discusses the technologies for big data analytics where the key focus is on the storage and security technologies. Section 5 presents the major challenges associated with big data analytics.

2 Big Data Framework

With a significant change in the way processing is carried out, the conceptual framework of big data analytics is comparable to that of the conventional approach to analyzing small datasets. Various frameworks have been developed for analyzing data. Multiple layered architecture is used to handle big data efficiently. Figure 3 presents the generalized layered big data framework and each layer is discussed in following section.

2.1 Big Data Sources

These are storehouses that hold enormous volumes of data. Data source can be internal (data that is self-generated), external (data that is extracted from external sources), in multiple formats (data that can have variety of formats e.g. web log data), and residing at different locations in various business and other applications. The data from all sources is integrated to perform data analytics.

2.2 Big Data Integration and Transformation

After collection of data from different sources, data is transferred to storage infrastructure for processing and analysis [15]. It can be done using middleware which is a service-oriented approach. Here data continues to be in same state and services such as call data, process data and retrieve data are used to transform data.

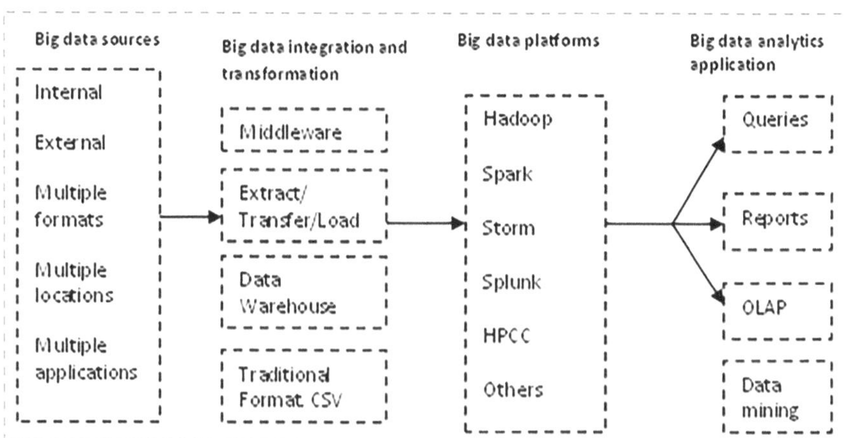

Fig. 3 Big Data Framework

Data warehousing involves a process called Extract, Transform and Load (ETL). Extracting data incorporates connecting multiple data sources for data collection and processing the data generated from those sources. Transformation is implementation of certain rules to transform extracted data to a particular format. Loading is importing transformed data to a storage infrastructure [16].

2.3 *Big Data Platforms and Tools*

The platforms and tools employed for big data analytics are discussed below:

(a) **Apache Hadoop:** Hadoop is java based open-source framework developed by Doug cutting [17]. It is horizontally scalable platform. It is designed to store and process enormous amounts of data. The two generally accepted modules of Hadoop are:

- *Hadoop Distributed File System (HDFS):* HDFS takes care of data management and storage within the Hadoop cluster. As the data arrives in Hadoop cluster for analytics, it breaks the data and redistributes it among various nodes in a cluster [18]. It is fault-tolerant and scalable. It consists of two main types of nodes viz., NameNode (master) and SlaveNodes (slaves). NameNode stores metadata and SlaveNodes store application data.
- *MapReduce:* It is a component of Hadoop given by Dean and Ghemawat at Google. MapReduce is based on divide and conquer method. Here the problem is recursively broken down into smaller ones until all of them are resolved. MapReduce divides data processing task into two parts: mappers and reducers. Mappers process the data and give their intermediate results to reducers. Reducers integrate the results provided by mappers and produce some output which is again fed to HDFS. This concept of MapReduce removes bottleneck of traditional system thus providing greater fault tolerance and availability. However, there exist few limitations of MapReduce such as, it performs poorly with iterative algorithms. Mappers repeatedly read the same data from the disc after each iteration's results are written back to the disc to be sent to the subsequent iteration. This lowers the performance of system. Hadoop is an extension of MapReduce which supports iterative algorithms. It includes cache component which improves efficiency. CGL MapReducer is another advancement that works on producing better MapReduce repetitious tasks [19].

(b) **Spark:** It is an in-memory data processing engine, open-source cluster computing and horizontally scalable framework developed to overcome the limitations of Hadoop. It is used for the real-time processing of huge amounts of data. Initially spark was developed by researchers at university of California; Berkeley's AMP Lab and later donated it to Apache Software Foundation [20]. Spark allows reuse of working data set across multiple parallel operations. Resilient

Distributed Datasets (RDD) is the primary abstraction of Spark. RDD is a parallel, read-only collection of objects that can be spread over several cluster nodes. It integrates well with the databases including ones from the NoSQL family, for example, Cassandra, Hbase and ones from relational database family, e.g., MySQL. Spark developers proposed Berkeley data analytics stack (BDAS) which is a data processing stack [4].

(c) **Storm:** Apache Storm is an open-source, real-time, distributed stream data processing system. It provides a framework to build processing applications that use resources of all nodes within a cluster. It is fault-tolerant, scalable and is highly reliable. The main abstraction of storm is stream which is a sequence of tuples created and processed in parallel in a distributed fashion. Streams of tuple flow through topologies which are digraphs where vertices denote computations and edges denote flow of data. Vertices are divided into Spout (supplier of stream) and Bolt (processing unit). Storm architecture is same as Hadoop. There are two distinct kinds of nodes in it viz., Nimbus (master) and one or a couple of worker nodes (supervisor). Node management in storm cluster is done by ZooKeeper. Zookeeper provides status of supervisors to Nimbus and then Nimbus assigns work to supervisors [21].

(d) **Splunk:** Splunk is a platform for processing massive data generated by machines in real-time [22]. It integrates Big data and cloud technologies. It usually assists users in searching, monitoring, and analysing their machine-generated data through web interface. Splunk deals with any kind of data coming from different sources such as web logs, credit card data, social data, and structured database data. It provides results in a perceptive way such as graphs, alerts and reports in order to get better insights.

(e) **High Performance Computing Clusters (HPCC):** HPCC is a data intensive computing platform where HPC clusters are vertically scalable systems with thousands of processors. Clusters use high-powered hardware that enhance efficiency and speed. Such systems are not scalable but still can process terabytes of data. High level Enterprise Control Language (ECL) is capable of handling complex issues. HPCC makes sure that ECL is run for the shortest possible period and that node processing is carried out in parallel. However, this platform doesn't need any third party software like Oozie, RDBMS, Cassendra, etc. [23].

(f) **Hbase:** Instead of the Google file system, HBase employs the HDFS. Updates are periodically written to disc files after being stored in memory. To prevent frequent seeks, updates are placed at the end of a data file and periodically, the files are compressed. In order to perform recovery in the event of a server crash, updates also go to the end of a write ahead log. A comparison of various Big Data Technologies is shown in Table 1.

Table 1 Comparison of various Big data technologies

Big data technology	Processing	Latency	Throughput	Model	In-memory computing
Apache Hadoop	Batch	High	Low	Open-source	No
Storm	Real-time	Low	High	Open-source	Yes
Splunk	Real-time	Low	High	Commercial	–
Spark	Real-time	Medium	High	Open-source	Yes
HPCC	Batch	High	High	Open-source	–

3 Big Data Analytics Applications

Big data analytics is the process of analysing vast amounts of diverse data in order to find previously unknown patterns, correlations, market trends, customer preferences, and other helpful information that can assist businesses in making better-informed business decisions [24]. The next generation of smart infrastructures will be powered by the collection, management, storage, and analytics of Big Data. The use of big data analytics can help to collect, organize, and analyse data that has been gathered from a variety of sources, including environmental sensors, historical weather and mobility data, and other sources. It can also be used to forecast future environmental conditions and visualize current conditions [25]. This section highlights the application of big data analytics in areas including smart city and its various components like Intelligent Transport Systems, smart grids, and smart healthcare. The application of big data in these areas is explained as follows:

(a) **Smart City:** Big data analytics, a crucial disruptive technology that has lately come to light, enables the "smartness" of smart cities. It is the main technology in the field of "smart cities" for enabling evidence-based decision making, creating and offering new public services, and allowing citizens to contribute on establishing new public services. Big Data analytics has several applications in various facets of smart cities, including planning, traffic control and transportation, crime analysis, energy, and environment. Due to the dynamic nature of smart city life, stream data analytics must receive a lot of attention in order to allow online and real-time services. Moreover, historical data or batch data analytics are crucial for planning (short- and long-term) and decision-making purposes in smart cities [26].

(b) **Intelligent Transport Systems:** Intelligent Transport Systems (ITS) is a group of communications and technology that increase the efficiency and safety of roads, provide precise traffic information in real-time, reduce the amount of congestion brought on by accidents, construction projects, or other events, and, in the end, provide information to help drivers. CCTV, traffic detection, and other monitoring and sensing systems are all part of ITS solutions [27]. Data for ITS can be acquired from a heterogenous sources, like from smart cards, GPS, sensors, social media, and more. Using accurate and effective data analytics, it is possible to better provide ITS with data that appears to be in a disordered state. The volume of data that ITS produces is increasing from a Trillion-byte level to a Petabyte with the advancement of ITS. Because of the sheer magnitude of the data required for analytics, conventional data processing techniques simply cannot keep up. Big data analytics can manage the vast quantities of data with varying degrees of complexity that are generated by ITS. Three issues: data management, analysis, and storage—have been solved by big data analytics. Massive amounts of data may be processed via big data systems like Apache Hadoop and Spark, which are widely utilized in both industry and academics. Efficiency in ITS operations can be increased by employing big data analytics. The traffic management department is able to determine flow of traffic in real time by quickly gathering and analysing vast amounts of historical and current traffic data. The level of ITS safety can be raised with big data analytics. A vast amount of real-time transportation data can be gathered using sophisticated sensor and detection techniques. When emergency rescue is required, the Big Data analytics-based system's real-time response capabilities can considerably boost the emergency rescue capability [28].

The authors in [29] provide a big data analytical based approach for capturing the semantics of traffic accidents and congestion using ontology in order to decrease congestion and road accidents. The study employs machine learning techniques and big data to deliver qualitative information about traffic ontology. The Clustering machine learning technique is utilized to analyze various traffic and congested conditions on several weekdays and at various times of the day to demonstrate the influence of school and business travel.

To control the entire traffic systems, for instance [30], suggest a traffic management system that aims to address traffic issues primarily in India, [31] describe a low-cost smart traffic system (STS) to deliver traffic updates quickly.

To address the issue of traffic management in India, particularly congestion, the authors in [31] proposes a smart traffic control system. The low-cost STMS seeks to improve service with traffic indicators that are rapidly updated in response to traffic conditions. In order to gather various forms of data from the vehicles, it is suggested that sensors be positioned in the middle of the road every five hundred meters on motorways. The gathered big data is analysed for traffic density and used to forecast traffic patterns taking into account the volume of cars using the road and its capacity. The proposed solution includes a smartphone app that provides drivers with real-time traffic information and alternative routes. To avoid the congested areas, drivers may choose to use alternate routes.

(c) **Smart Grid:** With the extensive deployment of sensors and smart meters, the standard electrical transmission and distribution network has now been enhanced to incorporate an information layer for data gathering, storage, and analysis due to the advancements in Information and Communication Technology (ICT). Conventional power meters in distribution systems only generate a minimal quantity of data that can be gathered and examined for billing purposes manually. While today's massive data acquired from two-way smart grids at various temporal resolutions requires advanced data analytics, it may also be used to extract vital information about the condition of the energy network. For example, demand forecasting, consumer behavior analysis, and energy generation optimization can all be done with the use of high-resolution user consumption data. The security of the power system is more dependent on predictive maintenance and problem identification based on data analytics and modern metering infrastructure [32]. Therefore, the rapid development of ICT offers engineers a fresh perspective on how to perceive, manage, and smarten up the conventional electrical system. Huge volumes of data, containing measurements and control instructions, are generated by an information layer that is integrated into the energy network and are stored in the grid for efficient acquisition, transmission, storage, and analysis.

For a microgrid that is connected to the grid, extreme weather or grid blackouts may result in an inadvertent islanding event, endangering the safety operation and creating technical problems. In order to recognize islanding accidents, in [33] a neural network is trained using features taken from the differential transient of the rate of change of frequency signal. As an islanding detection method, a support vector machine (SVM) classifier is developed in [34] with a number of characteristics derived from system variables. A sliding window is used to implement the feature extraction method, and its width is optimized for the maximum rate of detection.

Social media sites like Twitter and Facebook may serve as a real-time social sensor for the smart grid, providing information on when and where power outages occur. In [35], a probabilistic framework is developed for identifying a specific event from the fragmented and noisy tweets. In a trial, the technique performs well in identifying areas of accumulating outages, which may be incorporated into a social data-driven power outage management.

Electric power quality (PQ) is a global issue that concerns the amount, frequency, and waveform of voltage and current in a power system. It is closely tied to both the secure operation of the power grid and consumer happiness. Harmonic distortions and unstable conditions commonly develop in the power grid due to the growing use of nonlinear and power electronics-based loads and generators. Deep learning is effective at classifying PQ events in electrical networks [36].

(d) **Smart Healthcare:** Medical data digitization changes the dimensions of data, expands data size, and elevates the importance of data analytics. Electronic Health Records (EHR) have replaced traditional hard copy medical data as a result of the development of ICT and the advent of digitized systems [37, 38]. These systems led to the exponential rise of data, which hastened the development of big data analytics, particularly in the field of healthcare [39]. Healthcare Analytics Market Size Industry Report: 2021–2028, 2021 Grand View Research, Inc. estimates that the global market for healthcare analytics was valued at USD 23.6 billion in 2020 and will increase at a Compound Annual Growth Rate (CAGR) of 23.8 percent from 2021 to 2028. This enormous growth meets the expanding demand for better healthcare, helped by cutting-edge technology [40]. Potential sources of medical big data in the healthcare industry include EHR, administrative claim reports, hospital registries, prospective cohort studies, biometric data, medical imaging, patient data, the internet, biomarkers, and clinical trials [41]. For the benefit of healthcare stakeholders, these sources are combined to quickly and affordably offer prescriptive, descriptive, and diagnostic information [42]. To accurately diagnose breast cancer and get rid of mistakes brought on by radiologists' manual diagnosis, a hybrid machine learning-based classification model including Multilayer Perceptron (MLP), Decision Tree (j48), and K-Means is utilized in [43]. To determine the patient's medical status, a big data platform for linked healthcare that takes emotion into account [44] is developed. Based on the patient's emotion, their state is determined by analyzing their speech and appearance. SVM classifier is used for voice and video processing using Fourier transformation. This framework achieves accuracy of 99.87%. A method for visualizing health data is suggested to track the patient's health [45]. The medical state of a person is determined using colored circles, which is simple to use. Because lung cancer has a greater mortality rate than other diseases, it is more important to predict diseases. For improved sensitivity and accuracy, some often used methods include hybrid SVMs, K means, deep learning, supervised learning, and fusion models [11].

4 Technologies for Big Data Analytics

To carry out big data analytics in ITS, smart cities, smart grids etc., a variety of approaches and techniques have been employed. We examine these from two different viewpoints:

- Technologies for modeling, storage, transfer, and management of data
- Technologies for managing security and privacy of data.

4.1 Technologies for Modeling, Storage, Transfer, and Management of Data

Due to immense growth of data, existing storage systems are not capable of storing and managing data efficiently. There is a vital need for development of more efficient tools for storage of rapidly growing data. The tendency to shift from structural to unstructured data makes traditional RDBMS unsuitable for storage. This inadequacy of RDBMS motivates the development of efficient distributed storage systems [5].

During last few decades relational databases have been used as a data management technology. Datasets such as IMDB movies dataset and MovieLens are available for manipulations using relational operations. Current technologies allow data to be taken from variety of sources having different formats. Structured format of relational database does not support changing formats [46]. Now, data comes with new challenges such as getting real time insights, fast retrieval and processing of huge volumes. Hence the continuous growing data with variety of formats need a different solution. In below section we will discuss some present-day technologies employed for data storage and management.

Database Technologies: With the rapid growth of data, database technology is evolving over years. Traditional database systems are not capable of meeting challenges that are brought up by big data. NoSQL databases (non-traditional relational databases) are more accepted for big data storage. We will discuss NoSQL databases as follows:

(a) **NoSQL Database:** It is a schema-less data model. It does not strictly follow transactional ACID (atomicity, consistency, isolation, durability) properties. NoSQL addresses volume, velocity and variety characteristics of big data. Following are the four main NoSQL databases:

- *Key-value stores*: Here the data is stored as key-value pairs. It allows schema-less storage of data. Compared to relational databases, these databases are easier to expand, have a simpler structure, and respond faster. Many key-value databases have evolved recently including Amazon's Dynamo, LinkedIn's Voldemort. Various other key-value stores include Scalaris, Tokyo Cabinet, Kyoko Cabinet, Memcache DB etc.
- *Document stores:* In comparison to key-value stores, document stores are non-relational databases that allows storage of semi-structured data as documents. Stored values can be queried using a unique key. As documents don't have strict modes, so there is no need for mode migration. Various document-oriented stores include MongoDB, SimpleDB, CouchDB.
- *Columnar stores:* Here data is stored as RDBMS. Related columns are grouped together in a row resulting in column families. Columns and rows are divided into many nodes to ensure sclability. Column oriented data can be highly compressed. Columnar stores are basically motivated by Google's BigTable. Various other column-oriented databases include Cassandra, Hbase, Redshift, HyperTable, etc.
- *Graph stores:* Graph databases store data in graph structures. It is used to store data that is highly associative such as social network graph. High frequency trading systems and recommendation system use graph stores to achieve low latency. Here queries are graph traversals. A flavor of graph stores is triple stores or RDF (Resource Description Framework) triples. It mainly stores semi-structured data.

(b) **In-Memory Storage Technology:** In this technology, datasets are stored in main memory. Cluster based memory allows storage of volumes of data that can be accessed much faster as compared to on-disk storage devices [47]. It can be implemented as:

- *In-memory data grids:* In-memory grids' primary function is to guarantee data is always accessible by storing it in RAM. Key-value pairs (can be any

Table 2 Comparison of various storage technologies

Storage technology	Consistency	Scalability	Schema
NoSQL	Low	Horizontally	No
In-Memory	High	Horizontally	Yes
Cloud	Strong	Horizontally/vertically	Yes

business object) are stored across multiple nodes. It allows schema-less data storage. Data access is provided through API's.

- *In- memory databases:* Here databases are stored in main memory so as to provide faster response time. It is a type of analytic database that stores data on metrics for business applications particularly as a part of data warehouse. Relational datasets are stored in databases, so it stores structured data. Examples of some commercial in-memory relational databases are SAP HANA, VoltDB, Oracle TimesTen, SolidDB and MemSQL.

(c) **Cloud Storage:** It is a mechanism of storing big data online with the aim of providing greater accessibility and mobility to data. Examples include Google Drive, Dropbox, Xdrive, etc. Cloud storage allows both enterprises and end users to store their data. End users store data on cloud in order to access data everywhere and from every device in a reliable way or they can use it to simply backup their data. Enterprises use cloud storage so as to access data from multiple locations, for backup and for disaster recovery. Among the best-known cloud storage providers are Amazon, Google, and Microsoft. CloudView is a Map/Reduce-based framework that offers real-time response for the storing, processing, and analysis of machine-generated data in a cloud computing environment. Table 2 presents the comparison of various storage technologies in terms of consistency, scalability and presence or absence of schema.

4.2 Security Technologies

(a) **Provenance Technology:** In conventional computer systems, the labelled methodology was employed to use the data provenance method to identify the origin of the data in the data warehouse. When acquiring big data, different data sources from the Internet, cloud, social, and IoT networks are used. Despite the innovative encryption methods used for huge sensing data streams, assaults are still conceivable even during the data acquisition stage [48]. In order to make it easier to mine the information during data streaming itself, metadata regarding these data sources, like the origin of data, the method utilized for transmission, and any intermediate calculations, could be stored. To effectively use data analytics techniques for identifying anomalies in the data acquisition stage of big data, it is crucial to adapt data provenance technologies.

(b) **Authentication Technology:** Authenticating involves making sure that claims made by or about a subject are accurate and legal [49, 50]. It accomplishes three crucial objectives for every business: maintaining user identities, controlling access to corporate networks, and confirming that a user is who he professes to be. The majority of cryptographic systems have some type of endpoint authentication expressly to thwart man-in-the-middle (MITM) attacks. Examples of cryptographic protocols that offer security for communications over networks like the Internet incorporate Transport Layer Security (TLS) and Secure Sockets Layer (SSL), which are both versions of the TLS protocol. End-to-end Transport Layer encryption is performed by TLS and SSL for network communications. In applications consisting of Internet faxing, email, voice over IP, instant messaging and web browsing, several versions of the protocols are frequently used (VoIP). Utilizing a mutually trusted certification authority, one can utilize SSL or TLS to authenticate the server. The Bull Eye algorithm can also be utilized to continuously monitor all sensitive data from any angle. In addition to ensuring data security, this technique has been used to preserve relationships between original and copied data. Furthermore, only authorized users are allowed to read or write sensitive data [51].

(c) **Encryption and Access Control Technology:** Modern techniques including homomorphic, attribute-based, and picture encryption are being researched to make sure that private information is protected in cloud storage and other big data service platforms. Although homomorphic encryption permits some operations on encrypted data without first decrypting it, there is still room for improvement in terms of processing effectiveness and scalability in order to handle huge data. The attribute-based encryption method, on the other hand, is thought to be better suitable for end-to-end security of private data in the cloud environment since the encrypted data can only be decrypted if a set of qualities of the user's secret key coincide with those of the encrypted data [52].

(d) **Blockchain Technology:** Blockchain technology, often known as a "distributed ledger," is a decentralized method of storing different transactions and activities in a chain of blocks without the requirement for an unbiased third party. It has shown to be the new reliable method of interaction and exchange of information by removing mediators and enabling two parties to connect directly in a network using information replication and validation processes. The immutability of blockchain has been demonstrated, which aids in maintaining integrity, accountability, and, to some extent, confidentiality thanks to a set of public and private keys. After the successful rise of Bitcoin, blockchain has gained attention [53].

The four main characteristics of Blockchain technology—"decentralization", "immutability", "integrity", and "anonymity"—allow it to be used in non-financial domains like smart contracts, the IoT, security services, wireless network virtualization and other applications [54]. "Distributed Ledger", "Smart Contracts", and "Distributed Applications" are the three main parts of a typical blockchain.

- Digital distributed ledger technology-based distributed ledger (DLT).
- Smart Contracts that give users a mechanism to communicate transactions kept on the distributed ledger.
- Applications created for users that are distributed.

A digital signature is required in a blockchain network to establish a safe route of communication between participants. Each user has a public key and a private key assigned to them. A user signs a transaction with the private key before broadcasting it to the network. The recipients later validate the transaction using the user's public key. This type of encrypted communication maintains the transaction's integrity within the network. When a hashed or encrypted block is rehashed to adhere to difficulty level requirements, a nonce (one-time use number) is a number that is added to the block. Finding nonce, which is difficult in blockchain as it is in general cryptography, is considered a filter to remove the unconnected or incompetent miners [55].

In a Big Data environment, each piece of information, or record, is a block that carries transactional information, such as the date of the transaction and a connection to the block before it. Using blockchain technology, data integrity is preserved. This is due to ongoing checks by the many computers on the network being made in real-time, which prevent corrupted data from being inserted into the blockchain. Blockchain also enables more responsible data exchange as agreed upon by users, reducing the risk of data leakage and cybercrime. Predictive analytics can be performed with blockchain data, which may potentially offer insightful information about patterns and behavior.

Blockchain technology has the potential to be crucial for IoT security. With millions of devices interconnected and interacting, it is crucial to assure that the data flowing is securely through IoT and the accountability of the participants is ensured. Smart energy grid systems are evolving into sophisticated cyber-physical systems (CPS) that allow for complex, bidirectional interactions between power production, distribution, utility offices, and end customers. The key aspects of Blockchain technology offer an environment that is secure and verifiable to allow interactions in the energy CPS. In case of healthcare, personal health information is sensitive and requires strict security handling [56]. Blockchain technology can be employed to encrypt and store such personal details, and it can also give a private key that would only permit certain people to view the records. Blockchain enables the creation of numerous new interaction models that could not be created inside a centralized architecture due to its durable and concurrently distributed storage. This paradigm assures authentication with a high level of security and does away with the need for passwords [57].

Blockchain can also help ITS tremendously by protecting data as digital proof. An ITS infrastructure built on peer-to-peer networks can be implemented with the help of blockchain technology. ITS apps built on blockchain allow users and drivers to securely communicate and update information about the state of roads and infrastructure without the help of third parties [58]. Table 3 provides an idea about how these security technologies can be useful for smart city, smart grids, ITS, smart healthcare.

Table 3 Use of security technologies in smart environments

Application area	Provenance technology	Authentication technology	Encryption and access control technology	Blockchain technology
Smart City	To track the origin and history of data related to city infrastructure, such as energy usage, traffic patterns, and environmental data.	To prevent unauthorized access, data breaches, and cyber-attacks. To authenticate users accessing public services, as well as devices accessing city networks.	To safeguard confidential information including social security numbers, bank account numbers, and medical records.	To store and share data related to public services, such as transportation, energy, and waste management.
ITS & Smart Grid	To track the history of network events and identify the source of security breaches or system failures.	To guarantee that only approved devices are able to access network resources and control systems.	To restrict access to network resources and control systems based on the user's identity, role, and permissions.	To securely store network data and ensure the integrity of network events.
Smart Healthcare	To track the history of patient data, including who accessed it	To ensure that only authorized personnel can access patient data.	To ensure that only authorized personnel can access patient data.	To securely store patient data and ensure its authenticity.

5 Challenges in Big Data Analytics

The use of big data analytics has become increasingly popular in various smart ecosystems such as smart cities, smart healthcare, ITS, and smart grids. However, there are several challenges associated with big data analytics in these areas.

(a) **Privacy and Security Challenges:** Security and privacy are the most difficult issues with big data. A sub field of data mining referred to as privacy preserving data mining (PPDM) deals with privacy issues. Performing analytics on data can reveal confidential data. For companies the privacy issue is relevant to the sensitive data that they work with. Companies can store their valuable information on cloud systems, in-house and a hybrid systems. By storing data on cloud leads to decrease in cost and increase in processing speed of requested operations. The data security still remains issue. The data generated in various smart environments like smart grids, smart city, smart healthcare and ITS, often contain sensitive and personal information, such as medical records, location data, and energy consumption patterns. It is crucial to ensure that the data is protected from unauthorized access and breaches of privacy. Table 4 highlights some of the security issues in these areas.

- *Smart grids:* Smart grid data mostly consists of financial transactions, business secrets, and consumer privacy information. Data security is therefore essential, including privacy, integrity, and authentication. Data from the smart grid needs to be authenticated in order to distinguish between real and fraudulent identities. Data authentication is required to protect user privacy as well as to guarantee data integrity. Authentication, encryption, trust management, and intrusion detection are among the most important security methods because they may prevent, detect, and mitigate network assaults. A variety of methods, including data encryption and signature creation, are generally utilised in smart grids in order to accomplish the tasks of data authentication and security management [59].

Table 4 Security issues in smart infrastructures

Application area	Security issues
Smart City	Cyber-attacks, such as hacking or data breaches, Malware and ransomware attacks on city systems.
Smart Healthcare	Unauthorized access to patient records, data breaches, and identity theft, cyber-attacks on medical devices, Malware attacks on healthcare systems
Smart Grid	Cyber-attacks leading to power outages, Unauthorized access to sensitive data such as energy usage patterns and consumer information. Malware attacks on smart grid systems causing system downtime and blackouts.
ITS	Data breaches, and unauthorized access to critical infrastructure, cyber-attacks on vehicle-to-vehicle communication networks, GPS spoofing, and the use of malware to disrupt traffic signals.

- *Smart city:* Securing data privacy and security is a significant difficulty in developing a smart city. In a smart city, data typically go through the steps of data collection, transmission, storage, and processing before being used. Data can, unfortunately, be intentionally destroyed at any point. For instance, during the data gathering stage, attackers may break into the equipment to steal data, exposing and using sensitive data without authorization. Criminals may start attacks on the data stream during the data transmission stage, which could result in data being distorted while being transmitted. Furthermore, the dataset may experience information leakage due to the heterogeneous, multi-source, and linked nature of big data transfer. Numerous researchers have suggested various solutions, such as anonymous processing, encryption processing, access acquisition control, and the efficient administration of data perception frameworks, to address these data security concerns. Existing data protection techniques may largely ensure data security and privacy. Establishing a more sophisticated and trustworthy data protection system is necessary to make sure that the vast amounts of data in a smart city are not misused or abused [60].
- *Smart Healthcare:* Despite the many advantages and opportunities that come with the implementation of big data technology in the healthcare industry, there are also certain obstacles and difficulties. Indeed, concerns over the safety and privacy of sensitive information continue to mount each year as a result of a number of recent innovations in the medical field. These innovations include clinical mobility and wireless networking, the exchange of health information, cloud computing, and others. Security and privacy concerns are one of the main obstacles to the integration of various healthcare data sources into big data. Data in the healthcare industry are susceptible to a variety of security threats, including the inappropriate sharing of patient information, the unauthorized use of patient data, and the unauthorized loss of patient data. As a direct consequence of this, medical personnel are dissuaded from making use of electronic healthcare systems in order to exchange patient information [51].
- *ITS:* Confidentiality is unquestionably one of the essential security services that must be taken into account in ITS while handling confidential data in a communication network. In ITS confidentiality is important, as the devices and parties interact with each other in a secure and private manner which prevents information from being revealed to other parties [61]. Traveler safety relies on the operability and connectivity of devices to other ITS elements. Denial-of-service attacks are the primary threats to the availability of ITS parts and services. In most of the attack surfaces that are available within ITS, denial-of-service attacks are frequent. Attacks on availability pose a particular threat to ITS since so many ITS components demand real-time functioning. Current protections against denial-of-service attacks include proof-of-work and signature-based authentication [50]. Big data security involves ensuring that data repositories and networks are secured

via authentication and authorization. Big data security techniques include authentication, data provenance, Blockchain as discussed in Sect. 4.2.

(b) **Data Quality and Management Challenges:** One of the major challenges of big data analytics in smart cities, smart healthcare, ITS, and smart grids is the quality and management of data [62]. In these domains, data is often generated from various sources such as sensors, devices, and social media, which can result in data inconsistencies and inaccuracies. The data needs to be cleaned, transformed, and structured before it can be analyzed. Data management challenges also arise in terms of data storage and access, as large amounts of data need to be stored and accessed in real-time. For example: In smart healthcare, patient data is collected from various sources such as medical records, wearables, and health monitoring devices. The data needs to be cleaned, transformed, and structured before it can be analyzed to derive insights. However, data quality issues such as missing or inconsistent data can affect the accuracy of the analysis.

(c) **Integration and Interoperability Challenges:** Big data analytics in smart ecosystem's require the integration of data from various sources and systems. This can pose a significant challenge as the data may be in different formats and may not be compatible with each other. The lack of interoperability can lead to data silos, which can hinder the effectiveness of the analysis [63].

For example: In smart grids, data is collected from various sources such as smart meters, weather sensors, and power grids. However, the data may be in different formats and may not be compatible with each other, making it difficult to integrate and analyze.

(d) **Scalability Challenges**: Big data analytics in smart ecosystems generate vast amounts of data that need to be analyzed in real-time. The scalability of the infrastructure and systems used for data processing and analysis can pose a significant challenge. For example: In ITS, real-time traffic data is collected from various sources such as sensors and GPS devices. However, the volume of data generated can be massive, and the infrastructure used for data processing and analysis needs to be scalable to handle the workload [64].

(e) **Cost Challenge:** Implementing big data analytics can be expensive [65], and the cost challenge is especially relevant for the development and deployment of big data analytics solutions in smart city, smart healthcare, ITS, and smart grids. Some of the factors that contribute to the cost of big data analytics in these areas include:

- Infrastructure: For example, in a smart city, the sensors, cameras, and other devices required to collect data can be costly. Similarly, in a smart healthcare system, the electronic health record (EHR) system and other health IT infrastructure required to store and process health data can be costly.
- Hardware and software: Hardware and software required to store, process, and analyze large data sets can be expensive. In a smart city, a data center may be required to store the data generated from various sources, and

high-performance computing systems may be needed to analyze the data. Similarly, in a smart healthcare system, advanced analytics software may be required to analyze the data from EHR systems.
- Integration with existing systems: In many cases, big data analytics solutions need to be integrated with existing systems, such as EHR systems, transportation systems, or power grids. Integrating these systems can be expensive, as it requires changes to the existing infrastructure and systems.

6 Conclusion

This chapter presents the foundational ideas of big data. It presents the generalized big data platform and explains different storage and processing technologies used. It further defines the applications of big data in various areas such as smart cities, smart grids and smart healthcare systems. It further provides the insights of future research by discussing privacy and security challenges related to big data. This chapter would be helpful for emerging researchers in this area as it briefly explains the fundamentals of big data.

References

1. World Population Clock: 8 Billion People (LIVE, 2023) – Worldometer, https://www.worldometers.info/world-population/, last accessed 2023/03/04
2. Daily social media usage worldwide | Statista, https://www.statista.com/statistics/433871/daily-social-media-usage-worldwide/, last accessed 2021/10/06
3. Number of mobile devices worldwide 2020–2025 | Statista, https://www.statista.com/statistics/245501/multiple-mobile-device-ownership-worldwide/, last accessed 2023/03/04
4. Saleem S, Mehrotra M (2021) Data analytics and mining: platforms for real-time applications. In Data driven decision making using analytics. pp 61–80
5. Gandomi A, Haider M (2015) Beyond the hype: Big data concepts, methods, and analytics. Int J Inf Manag 35:137–144. https://doi.org/10.1016/j.ijinfomgt.2014.10.007
6. Levin N, Salek RM, Steinbeck C (2016) From databases to big data. Metab Phenotyping Pers Public Healthc:317–331. https://doi.org/10.1016/B978-0-12-800344-2.00011-2
7. Laney D (2001) 3D data management: controlling data volume, velocity and variety. META Gr Res note 6:1
8. IBM (2013) Analytics: the real-world use of big data How innovative enterprises in the midmarket extract value from uncertain data
9. Number of social media users 2025 | Statista, https://www.statista.com/statistics/278414/number-of-worldwide-social-network-users/, last accessed 2021/05/30
10. Global Social Media Stats — DataReportal – Global Digital Insights, https://datareportal.com/social-media-users, last accessed 2021/10/06
11. Tsikala Vafea M, Atalla E, Georgakas J, Shehadeh F, Mylona EK, Kalligeros M, Mylonakis E (2020) Emerging technologies for use in the study, diagnosis, and treatment of patients with COVID-19. Cell Mol Bioeng 13:249–257. https://doi.org/10.1007/s12195-020-00629-w
12. Sun Z (2018) 10 Bigs: Big data and its ten big characteristics. Manag Perspect Intell Big Data Anal:1–14

13. Sindhu K, Kumar DSR (2014) Influence of risk perception of investors on investment decisions: an empirical analysis. J Financ Bank Manag 2:15–25
14. Kumar A, Tyagi AK, Tyagi SK (2014) Data mining: various issues and challenges for future a short discussion on data mining issues for future work. Int J Emerg Technol Adv Eng 4:1–8
15. Hurrah NN, Loan NA, Parah SA, Sheikh JA, Muhammad K, de Macedo ARL, de Albuquerque VHC (2021) INDFORG: industrial forgery detection using automatic rotation angle detection and correction. IEEE Trans Industr Inform 17:3630–3639. https://doi.org/10.1109/TII.2020.3014158
16. Abaker I, Hashem T, Yaqoob I, Badrul N, Mokhtar S, Gani A, Ullah S (2015) The rise of " big data " on cloud computing: review and open research issues. Inf Syst 47:98–115. https://doi.org/10.1016/j.is.2014.07.006
17. Zhou H, Sun G, Fu S, Fan X, Jiang W, Hu S, Li N (2020) A distributed approach of big data mining for financial fraud detection in a supply chain. Comput Mater Continua 64:1091–1105
18. Qader WA, Ameen MM, Ahmed BI (2020) Big data characteristics, architecture, technologies and applications. J Comput Sci 16:817–824. https://doi.org/10.3844/JCSSP.2020.817.824
19. Lee KH, Lee YJ, Choi H, Chung YD, Moon B (2011) Parallel data processing with MapReduce: a survey. SIGMOD Rec 40:11–20. https://doi.org/10.1145/2094114.2094118
20. Xianya J, Mo H, Haifeng L (2019) Stock classification prediction based on spark. Procedia Comput Sci 162:243–250. https://doi.org/10.1016/j.procs.2019.11.281
21. Shvachko K, Kuang H, Radia S, Chansler R (2010) The Hadoop distributed file system, pp 1–10
22. Splunk | The Key to Enterprise Resilience, https://www.splunk.com/, last accessed 2023/03/05
23. Apache Cassandra | Apache Cassandra Documentation, https://cassandra.apache.org/_/index.html, last accessed 2023/03/05
24. Rouf N, Malik MB, Arif T, Sharma S, Singh S, Aich S, Kim H-C (2021) Stock market prediction using machine learning techniques: a decade survey on methodologies, recent developments, and future directions. Electronics 10(21):2717
25. Rouf N, Malik MB, Arif T (2021) A regression based approach to predict the Indian stock market trend amid COVID-19. In: 2021 3rd International Conference on Advances in Computing, Communication Control and Networking (ICAC3N). pp 2014–2020
26. Al Nuaimi E, Al Neyadi H, Mohamed N, Al-Jaroodi J (2015) Applications of big data to smart cities. J Internet Serv Appl 6:1–15
27. Zhu L, Yu FR, Wang Y, Ning B, Tang T (2018) Big data analytics in intelligent transportation systems: a survey. IEEE Trans Intell Transp Syst 20:383–398
28. Montoya-Torres JR, Moreno S, Guerrero WJ, Mejía G (2021) Big data analytics and intelligent transportation systems. IFAC-PapersOnLine 54:216–220
29. Abberley L, Gould N, Crockett K, Cheng J (2017) Modelling road congestion using ontologies for big data analytics in smart cities. In: 2017 international smart cities conference (ISC2). IEEE, pp 1–6
30. Rizwan P, Suresh K, Babu MR (2016) Real-time smart traffic management system for smart cities by using Internet of Things and big data. In: 2016 international conference on emerging technological trends (ICETT). IEEE, pp 1–7
31. Sharif A, Li J, Khalil M, Kumar R, Sharif MI, Sharif A (2017) Internet of things—smart traffic management system for smart cities using big data analytics. In: 2017 14th international computer conference on wavelet active media technology and information processing (ICCWAMTIP). IEEE, pp 281–284
32. Zhang Y, Huang T, Bompard EF (2018) Big data analytics in smart grids: a review. Energy inform 1:1–24
33. Hashemi F, Mohammadi M, Kargarian A (2017) Islanding detection method for microgrid based on extracted features from differential transient rate of change of frequency. IET Gener Transm Distrib 11:891–904
34. Alam MR, Muttaqi KM, Bouzerdoum A (2017) Evaluating the effectiveness of a machine learning approach based on response time and reliability for islanding detection of distributed generation. IET Renew power Gener 11:1392–1400

35. Saleem S, Mehrotra M (2022) Emergent use of artificial intelligence and social media for disaster management. In: International conference on data science and applications, pp 195–210. https://doi.org/10.1007/978-981-16-5348-3_15
36. Sun H, Wang Z, Wang J, Huang Z, Carrington N, Liao J (2016) Data-driven power outage detection by social sensors. IEEE Trans Smart Grid 7:2516–2524
37. Hurrah NN, Parah SA, Sheikh JA, Al-Turjman F, Muhammad K (2019) Secure data transmission framework for confidentiality in IoTs. Ad Hoc Netw 95:101989
38. Parah SA, Rashid M, Vijaykumar V (2022) Artificial intelligence for innovative healthcare informatics, Springer, ISBN: 978-3-030-96568-6
39. Hurrah NN, Parah SA, Sheikh JA (2020) Embedding in medical images: an efficient scheme for authentication and tamper localization. Multimed Tools Appl 79:21441–21470
40. Afzal I, Parah SA, Hurrah NN, Song OY (2020) Secure patient data transmission on resource constrained platform. Multimed Tools Appl. https://doi.org/10.1007/s11042-020-09139-3
41. Parsa S, Parah SA, Bhat GM, Khan M (2021) A security management framework for big data in smart healthcare. Big Data Research 25:100225
42. Ahad F, Parah SA, Sheikh JA, Bhat GM (2015) On the realization of robust watermarking system for medical images. In: 2015 annual IEEE India conference (INDICON), New Delhi, India, pp 1–5, https://doi.org/10.1109/INDICON.2015.7443363
43. Kaur A, Rashid M, Bashir AK, Parah SA (2022) Detection of breast cancer masses in mammogram images with watershed segmentation and machine learning approach BT – artificial intelligence for innovative healthcare informatics. Presented at the https://doi.org/10.1007/978-3-030-96569-3_2
44. Hossain MS, Muhammad G (2017) Emotion-aware connected healthcare big data towards 5G. IEEE Internet Things J 5:2399–2406
45. Rouf N, Bashir Malik M, Sharma S, Ra I-H, Singh S, Meena A (2022) Impact of healthcare on stock market volatility and its predictive solution using improved neural network. Comput Intell Neurosci 2022:7097044. https://doi.org/10.1155/2022/7097044
46. Li B, Wang M, Zhao Y, Pu G, Zhu H, Song F (2015) Modeling and verifying Google file system. In: 2015 IEEE 16th international symposium on high assurance systems engineering. IEEE, pp 207–214
47. Yadranjiaghdam B, Pool N, Tabrizi N (2016) A survey on real-time big data analytics: applications and tools. In: 2016 international conference on computational science and computational intelligence (CSCI). IEEE, pp 404–409
48. Hurrah NN, Khan E, Khan U (2023) CADEN: cellular automata and DNA based secure framework for privacy preserving in IoT based healthcare. J Ambient Intell Humaniz Comput 14:2631–2643. https://doi.org/10.1007/s12652-022-04510-8
49. Parah SA, Sheikh JA, Bhat GM (2014) A secure and efficient spatial domain data hiding technique based on pixel adjustment. Am J Eng Technol Res 14(2):33
50. Parah SA, Sheikh JA, Loan NA, Ahad F, Bhat GM (2018) Utilizing neighborhood coefficient correlation: a new image watermarking technique robust to singular and hybrid attacks. Multidim Syst Sign Process 29:1095–1117
51. Rashid M, Singh H, Goyal V, Parah SA, Wani AR (2021) Big data based hybrid machine learning model for improving performance of medical Internet of Things data in healthcare systems. In: Healthcare paradigms in the internet of things ecosystem. Academic Press, pp 47–62
52. Hurrah NN, Parah SA, Sheikh JA (2019) A secure medical image watermarking technique for e-healthcare applications. In: Handbook of multimedia information security: techniques and applications, pp 119–141
53. Rawat D B, Chaudhary V, Doku R (2020) Blockchain technology: emerging applications and use cases for secure and trustworthy smart systems. J Cybersecur Priv 1:4–18
54. Bhardwaj A, Narayan Y, Vanraj P, Dutta M (2015) Sentiment analysis for Indian stock market prediction using sensex and nifty. Procedia Comput Sci 70:85–91
55. Malomo OO, Rawat DB, Garuba M (2018) Next-generation cybersecurity through a blockchain-enabled federated cloud framework. J Supercomput 74:5099–5126

56. Hurrah NN, Loan NA, Parah SA, Sheikh JA (2017) A transform domain based robust color image watermarking scheme for single and dual attacks. In: 2017 fourth international conference on image information processing (ICIIP). IEEE, pp 1–5
57. Rotună C, Gheorghiță A, Zamfiroiu A, Smada Anagrama D (2019) Smart city ecosystem using blockchain technology. Inform Econ 23:41–50
58. Jabbar R, Dhib E, ben Said A, Krichen M, Fetais N, Zaidan E, Barkaoui K (2022) Blockchain technology for intelligent transportation systems: a systematic literature review. IEEE Access
59. Qi J, Hahn A, Lu X, Wang J, Liu C (2016) Cybersecurity for distributed energy resources and smart inverters. IET Cyber-Phys Syst Theory Appl 1:28–39
60. Parah SA, Sheikh JA, Ahad F, Bhat GM (2018) High capacity and secure electronic patient record (EPR) embedding in color images for IoT driven healthcare systems. In: Internet of things and big data analytics toward next-generation intelligence. Springer, Cham, pp 409–437
61. Rouf N, Malik MB, Arif T (2021) Predicting the stock market trend: an ensemble approach using impactful exploratory data analysis. In: International conference on information, communication and computing technology. Springer, pp 223–234
62. Ramasamy A, Chowdhury S (2020) Big data quality dimensions: a systematic literature review. JISTEM-J Inf Syst Technol Manag. Springer
63. Kadadi A, Agrawal R, Nyamful C, Atiq R (2014) Challenges of data integration and interoperability in big data. In: 2014 IEEE international conference on big data (big data). IEEE, pp 38–40
64. Naeem M, Jamal T, Diaz-Martinez J, Butt SA, Montesano N, Tariq MI, De-la-Hoz-Franco E, De-La-Hoz-Valdiris E (2022) Trends and future perspective challenges in big data. In: Advances in intelligent data analysis and applications: proceeding of the sixth Euro-China conference on intelligent data analysis and applications. Springer, Singapore, pp 309–325
65. Balachandran BM, Prasad S (2017) Challenges and benefits of deploying big data analytics in the cloud for business intelligence. In: Procedia Computer Science, pp 1112–1122

Breakthroughs and Challenges in Multimedia Privacy and Security in the Internet of Things (IoT)

Rohun Nisa and Asifa Baba

1 Introduction

The IoT is attaining popularity in fields such as remote sensing, smart manufacturing, and wireless communication. According to Karie et al. [1], "the number of linked IoT devices will climb to 38.6 billion by 2025 and an estimated 50 billion by 2030." The main concept behind the IoT is to supply us with a range of things or devices, including sensors, smartphones, and smart cameras, that can connect and collaborate to accomplish communication, computation, and service functions. As the need for developing applications like Video on Demand (VoD) and real-time video monitoring has expanded dramatically, so does the potential for big data application areas as suggested by Zhou and Chao [2]. With the development of multimedia and network technologies for the IoT, multimedia content such as photos, audio, and videos is becoming more common in industrial and social situations. Because the IoT is designed to extensively run unauthorized user-implemented applications, both programs and humans potentially pose security risks to the IoT. As a result, security is crucial for multimedia content in the IoT.

Multimedia content may be created by a wide range of IoT applications, including smart cities, smart grids, smart homes, and smart transportation. Furthermore, the IoT affects healthcare, surveillance, agriculture, and a variety of other industries. The IoT touches every aspect of a person's life. This indicates that the multimedia information at risk might be financial, medical, personal, or sensitive in nature. Most multimedia content exchanged in open contexts such as wireless networks can be easily unlawfully duplicated and redeployed, resulting in major implications such as privacy leaks and reputational issues [3]. As a result, the security and privacy of

R. Nisa (✉) · A. Baba
Department of Electronics and Communication Engineering, Islamic University of Science and Technology, Awantipora Pulwama, Jammu & Kashmir, India
e-mail: rohunnisa@islamicuniversity.edu.in; asifa.baba@islamicuniversity.edu.in

S. A. Parah et al. (eds.), *Intelligent Multimedia Signal Processing for Smart Ecosystems*, https://doi.org/10.1007/978-3-031-34873-0_3

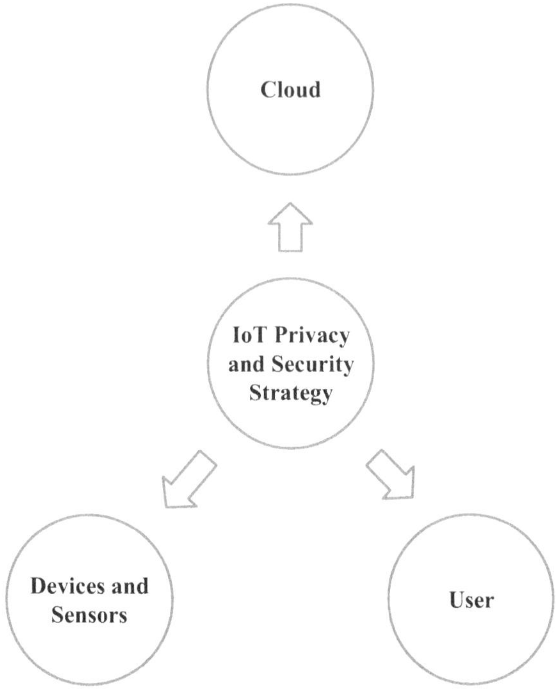

Fig. 1 A comprehensive strategy for IoT security

multimedia data are critical components of a reliable IoT system. If multimedia data security is compromised, it can result in illegal access to IoT applications or devices, the leakage of highly secret information, invasion of privacy, and identity theft. As a result, multimedia information security in the IoT has been a research emphasis in the last few years, and new methods are continually being developed [4]. Figure 1 depicts a comprehensive strategy for IoT security [5].

One of the most crucial aspect of information and communication technology has been the security of information that is being conveyed due to the emergence of digitized media, including audio, images, and videos, and an upsurge in internet usage globally. However, cryptography is being employed as a prospective means of information security. Because the disguised appearance of encrypted information attracts the attention of adversaries, the development of innovative techniques for improving information security has become a focal point for researchers all over the world [6]. Data hiding has received a lot of interest from the research community in recent years as an alternative choice for sharing information securely and without allowing the adversary to understand exactly what communication is underway [7–9]. The driving force behind the entire process is the fact that concealing information in the cover medium, such as a family picnic picture, is less sceptical than revealing an encrypted image.

As multimedia information protection becomes increasingly critical in IoT-enabled contexts, concerns about privacy and security must be addressed immediately. Many IoT devices are restricted by constraints such as computational capability and energy, making them vulnerable to attacks. As a result, putting in place strong security and privacy controls decreases data security threats and dangers in the IoT. Conventional encryption algorithms that use static structures, including Data Encryption Standard (DES), Advanced Encryption Standard (AES), and the Rivest, Shamir, and Adleman (RSA) algorithm for binary or text data, require numerous rounds of calculations, experiencing large costs in computational overhead and execution time, and so do not appear to be appropriate for multimedia-related IoT applications. Furthermore, when multimedia data is managed in the IoT, the problem becomes more difficult since the algorithms in question must fulfil rigorous Quality of Service (QoS) criteria [10].

2 Overview of Previous Work

Several academic publications give overviews of multimedia content privacy and security. To begin, Nahrstedt et al. [11] presented a number of cryptography-and digital watermarking-based security solutions to address multimedia's growing security needs. According to Tzelepi et al. [12], the confidentiality of multimedia medical information is also reliant on security architectures, in addition to encryption and digital watermarking. In this study, they establish an authorization mechanism for controlling access to multimedia medical data. Furthermore, a database management system (DBMS) architecture has been developed to assure the privacy of multimedia medical data. Furht and Socek [13] conducted a survey of multimedia data security which included the following cryptosystem types: symmetric cryptosystems vs. asymmetric cryptosystems; and Digital Rights Management (DRM) vs. communication cryptography, encryption levels, main multimedia content formats such as MPEG and JPEG, as well as communication encryption approaches such as the naive approach, scrambling, and selective encryption for audio and video data. Voloshynovskyy et al. [14] identified some important security concerns with multimedia information and examined watermarking-based solutions to solve those concerns. Furthermore, the authors offered a unified theoretical tool for assessing various data concealing approaches across a wide range of implementations.

Su et al. [9] suggested four multimedia information security evaluation metrics: time assessment, error resilience, compression ratio, and security assessment. The authors then used the four measures to assess chaos-based multimedia encryption methods. After reviewing several typical threats (eavesdropping and port scanning), Moravčík et al. [15] showed how to protect real-time multimedia signalling using asymmetric cryptography and digital signatures. Ashour and Dey [16] provided a brief overview of multimedia information concealing approaches (for example, typical data hiding processes) and applications. In addition, the authors summarised problems and future research in hidden data. Li et al. [17] discuss the privacy and security concerns associated with crowd-sensing multimedia content acquired through a diversity of applications such as air quality data retrieval, healthcare monitoring, and traffic patterns. The authors proposed ways for safeguarding crowd-sensing multimedia content, such as improving data dependability and safeguarding participant privacy and unintentional data.

Alvi et al. [18] presented an Internet of Multimedia Things (IoMT) architecture to allow multimedia-based support to customers in the IoT age. The suggested design is divided into four parts, with technological approaches for each level discussed in the work. Furthermore, the difficulties and practicality of the approaches for each level are explored. Wang et al. [19] conducted a literature review on mobile IoT-based multimedia systems. The writers of this review study concentrated on topics including mobility management, video traffic prioritizing, and security. Kumar et al. [20] carried out a survey on multimedia *big data computing* for the IoT. Furthermore, the writers outlined the obstacles involved as well as applicable solutions, with advantages and cons discussed.

The work presented by Hurrah et al. [21] aims at preserving data confidentiality and privacy in the unprotected surroundings of multimedia communication between two IoT hops. The proposed approach relies on the Random Coefficient Selection and Mean Modification Approach (RCSMMA). To prevent an adversary and maintain confidentiality of data, an effective multi-level security framework relying on information hiding and chaotic theory is put forth, which results in very high reliability of the data exchanged for all signal processing and region-specific attacks, working concurrently or independently. Parah et al. [22] proposed a high-capacity, computationally efficient, and secure Electronic Health Record (EHR) hiding approach in medical images in an IoT-oriented healthcare system. The Pixel Repetition Method (PRM) and modular arithmetic are used in the method. For cover image generation, the PRM was employed to scale up the input medical image, and modular arithmetic was utilized to incorporate the hidden EHR into the scaled-up images. The proposed approach has been thoroughly tested for a variety of frequently encountered medical and test images, as well as a collection of randomly selected images from the UCID repository.

Nauman et al. [23] gave a thorough examination of multimedia IoT. Initially, a summary of possible multimedia IoT architectures is provided. Secondly, multimedia IoT applications that are revolutionizing people's lives are being investigated. Lastly, the limitations and open research concerns of multimedia IoT are found. Verma et al. [24] investigated the IoT multimedia, big data computing technologies and architectures in precision agricultural applications. The authors of this study highlighted the issues created by the huge volumes of data gathered by IoT devices, from storage to transfer. The authors also examined precision agricultural multimedia data gathering, communication, and storage methods. Nath et al. [25] discussed cryptography and automated watermarking strategies for multimedia data security in the IoT era. The work described various types of attacks as well as the corresponding security procedures. Furthermore, the authors proposed that cryptographic techniques be used for digital watermarking as an extra security layer to increase system security.

Given that one of the key concerns about big data is privacy, Yu [26] investigated big data privacy and performed a wide study on relevant privacy studies. This study divides large data-oriented privacy into two groups: data grouping and privacy frameworks. Qu et al. [27] provided an overview of privacy limits and major assaults in the IoT, illustrating their findings with three case studies. The investigators also classified current privacy safeguards into four kinds: clustering-based systems; Differential Privacy (DP); encryption; and machine learning. The authors identified three significant issues, depending on their assessment: the lack of a theoretical framework; the optimization of the trade-off between information value and privacy; and the high complexity and scalability of system isomorphism.

Alhirabi et al. [28] examined the evolution of design notations, models, and languages for addressing IoT privacy and security concerns. The authors also investigated potential risk evaluation models and how they may be used in IoT systems and applications. The authors discussed why it is critical to incorporate privacy into system development at a preliminary phase. According to their findings, whereas the majority of research publications address security in some form, they rarely address data privacy. The authors of this study highlighted the possible obstacles and possibilities for proactive design products that enhance IoT privacy. Furthermore, the authors highlighted six research problems concerning privacy in IoT systems, as well as their consequences for the IoT scientific community on how to address such challenges.

DP has evolved as one of the privacy safeguards for IoT applications. Husnoo et al. [29] offered a comprehensive and in-depth review of the use and implementation of DP in four domains of IoT applications: intelligent transportation systems; healthcare and medical systems; the smart grid; and Industrial Internet of Things (IIoT). A thorough examination of unique privacy potential and uses in the IIoT was offered by Jiang et al. [30]. Firstly, relevant IIoT and privacy protection research is reported. Following that, the industrial data privacy metrics are emphasized. The issue of deep model data use and individual privacy protection is then examined. Lastly, numerous interesting questions are raised, and new study avenues are proposed.

3 Prerequisites for IoT-Based Multimedia Information Privacy and Security

Texts, photos, audio data, animation, and videos are all part of multimedia (Furht and Socek [13]). According to Nahrstedt et al. [11] and Furht and Socek [13], the following are some fundamental privacy and security considerations for multimedia data:

- Authentication: Authentication is the process of determining if a communication is truly from where it purports to be or what it purports to be. Communication partners should be able to be authenticated by the authentication system.

There are several types of entity authentication, such as digital signatures and watermarking.

- Confidentiality: It implies that communication is treated privately and only those who need to communicate have access to it. It is among the most desirable characteristics for IoT networks, as well as crucial for critical social and industrial purposes. Cryptographic systems, for example, can protect information from unauthorized parties.
- Data Integrity: To maintain data integrity in the IoT, communication receivers must be able to confirm that received messages have not been altered during transit. Data integrity indicates that communicated data cannot be manipulated in any way (Oracevic et al. [31]). All communication parties can identify tampering with a message. Various methods, such as one-way hash functions, message authentication codes, digital signatures, and watermarks, can be used to identify data manipulation.
- Non-Repudiation: Non-repudiation establishes the occurrence of a given event or action. An event or action might be the creation, transmission, or receiving of a message, for example. Non-repudiation mechanisms include digital signatures and message authentication codes.
- Privacy: Privacy is about protecting information, ensuring that no one's private or sensitive data is exposed to prying eyes. Because privacy is a human right, policymakers, researchers, and legal experts should collaborate to develop suitable multimedia information mining policies and strategies to ensure privacy and correct use of multimedia information [32].

In addition, the following additional needs for multimedia content in the IoT might be considered:

- Individualized privacy: Generally, a consistent privacy level is applied to a wide variety of audio-visual information in the IoT. This is both unrealistic and results in wastage of resources. However, individualised privacy may be handled by assigning specific privacy levels depending on individual criteria (Qu et al. [27]).
- Privacy evaluation: Generally, measurement serves as the foundation for scientific endeavour. Thus far, privacy measurement has not yet been well-defined. Nevertheless, the authors feel that in the future, privacy measurement should be a prerequisite for multimedia content in the IoT. In certain circumstances, direct measurements are impossible to acquire, but relative measures such as Kullback-Leibler distance, and Kolmogorov-Smirnov distance can be used instead (Yu [26]).
- Privacy versus data utility trade-off: Because IoT users can post multimedia data anonymously to avoid disclosure of sensitive information and data utility degradation, the purpose of multimedia information privacy protection in the IoT is to achieve an ideal trade-off between data utility and privacy (Qu et al. [27]).

4 Data Classification, Privacy, and Security Aspects for Multimedia Content

In the IoT, multimedia things outfitted with network connections that communicate with other things in the absence of human participation provide tremendous prospects for enhancing people's everyday lives (Nauman et al. [23]). Because multimedia content contains a great deal of information, some information is considered normal and non-sensitive and does not require protection, whereas other information, such as *biometric data* and *healthcare data*, is considered *private* and *sensitive*, and thus requires protection to prevent *data leaks* or *exposure*. In this part, multimedia content based on application categories is classified, as indicated in Table 1, to establish levels of security depending on different types of multimedia information. This classification yields fewer overlapping categories than *Zikria* et al.'s classification [34].

4.1 Agriculture Multimedia Content

Agricultural multimedia content gathered by IoT sensors may be utilized to develop an intelligent system in which environmental data such as *temperature*, CO_2, *water level*, and *humidity* as well as *images* can be seen to perform *data analysis*. The data collection methods covered in Precision Agriculture are listed in Table 2. Collecting and evaluating *agricultural multimedia* content collected by *IoT sensors* is critical to increasing productivity and improving *crop quality* (Verma et al. [24]). For example, Khattab et al. [35] created a tailored IoT architecture for *precision agriculture* applications, for example, *crop monitoring* and *irrigation control*, in which IoT sensors gather agricultural information including *air temperature*, *leaf wetness*, *rain gauge*, *solar radiation*, and *wind speed*. The suggested *three-tier* architecture collects the necessary data and sends it to a *cloud-based backend* for processing and analysis. Instructions can be sent back to the *front-end sensors* based on data analysis input. Mohanraj et al. [36] proposed an *e-agriculture* application built on a *knowledge base* and a *monitoring module* framework. The *knowledge base*

Table 1 Classification of multimedia content and associated privacy and security levels in the Internet of Things

Type of content	Level of privacy and security
Agriculture multimedia content	Low
Biometric multimedia content	High
Healthcare multimedia content	High
Social multimedia content	Medium
Surveillance multimedia content	Medium
Transport multimedia content	Low

Table 2 Gathering methods for multimedia content

Data gathering and communicating devices	Sensors available	Multimedia content category
IoT sensors	Air humidity and temperature sensor Electric conductivity sensor Gas sensor Leaf wetness sensor Light sensor Liquid-level sensor Passive InfraRed (PIR) sensor pH sensor PPFD sensor Pressure sensor Rain volume sensor Smoke sensor Soil humidity and temperature sensor Thermocouple sensor URD sensor Water-level indicator sensor Wind direction/ speed sensor	Audio Text
Remote sensing and imaging	Depth cameras Digital cameras Hyperspectral cameras IR cameras Multispectral cameras	Audio, Images (spatial and spectral) Video
Microcontrollers	Arduino Atmega 128 L, S3C2440, CC2420, STM85103F3 Raspberry Pi	–
Wireless technologies	Multi-hop SMS RFID UART WiFi ZigBees	–

is organized around crop specifics such as *geolocation data* and *weather forecasts* acquired by the *monitoring module* made up of IoT sensors.

4.2 *Biometric Multimedia Content*

In biometric systems, including *face* and/or *fingerprint recognition*, biometric multimedia content is acquired from an individual's *behavioural* or *physical* features (Yang et al. [37]). While biometrics offers inherent advantages over traditional personal recognition technologies—for example, passwords—ensuring the *confidentiality* and *integrity* of biometric multimedia data is critical. Biometric

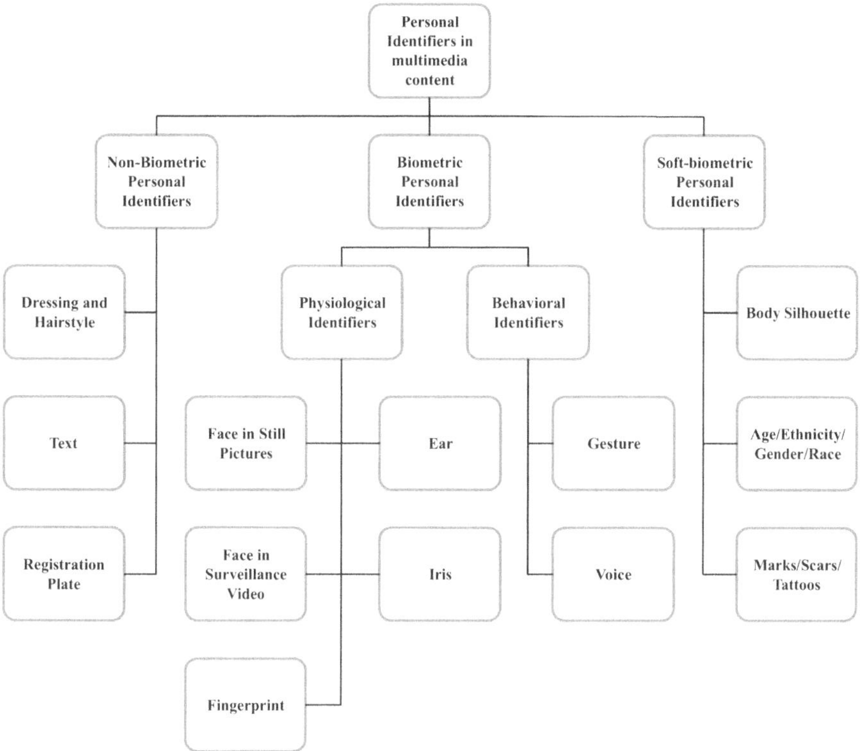

Fig. 2 Personal Identifier taxonomy in multimedia content

data (such as facial photos) cannot be republished or substituted in the same way that a *stolen password* or *token* can. Individual identifiers should be *de-identified* or *biometric* multimedia content protection should be strengthened to ensure *user privacy* (Ribaric et al. [38]). Figure 2 illustrates the taxonomy of personal identifiers of multimedia content employed in Biometric applications. Given the benefits of *high accuracy*, *resistance to variations in illumination*, and rapid implementation of *finger vein detection*, Lu et al. [39] created a quick local classifier for finger vein feature extraction in IoT applications known as the Histogram of Competing Orientation and Magnitude (HCOM). From finger vein pictures, two kinds of local histograms are produced and integrated to effectively and fully capture the competing data. Thilagavathi and Suthendran [40] developed a novel approach for automated real-time face identification from IoT device recordings. Instead of static photos, the suggested technique makes use of video-based biometric information, which contains more relevant information than just a single image. For feature extraction and training, the suggested technique employs techniques including the Haar cascade classifier and the local binary pattern histogram. Yang et al. [41] demonstrated a lightweight biometric system designed for resource-constrained

IoT devices in order to reduce memory utilization and processing expenses. The suggested technique employs simple yet effective block-based logic operations to reduce biometric feature sizes. The results of the experiments reveal that the suggested lightweight biometric system achieves good recognition accuracy while using a much smaller feature size.

4.3 Healthcare Multimedia Content

Healthcare experts, hospitals, patients, insurers, and pharmacies, among others, have easy access to healthcare multimedia content. Healthcare information is frequently fragmented, sensitive, and confidential, making it difficult to transfer between organisations (Xu et al. [42]). Body sensor networks are important IoT applications that record and handle numerous crucial bio-signals (BK and Muralidhara [43]; Muhammad et al. [44]). BK and Muralidhara [43] suggested an IoT-based *smart health monitoring system* wherein *IoT sensors* act as a catalyst in healthcare application areas. A microcontroller is employed as a gateway in the proposed system to communicate with various sensors, including *pulse oximetry sensors* and *temperature sensors*. The *microcontroller* gathers sensor data and transmits it to the network through *Wi-Fi*, allowing clinicians to monitor health data in real time. Muhammad et al. [44] developed a speech pathology detection system and highlighted the idea of utilizing *IoT-cloud computing* to examine people for *speech pathology*. To diagnose disease, the proposed system employs *local binary patterns* from a *mel-spectrum* representation of speech sounds and an *Extreme Learning Machine* (*ELM*) classifier.

In order to effectively preserve the patient data and authenticate the collected information, Parah et al. [45] have presented a new high-payload and reversible EHR embedding schema. The approach relies on *Left Data Mapping* (*LDM*), *PRM*, RC4 encryption, and checksum computation techniques. In addition to EHR, a fragile logo is incorporated into the cover images to aid in initial tamper detection. The average *Peak-Signal-to-Noise-Ratio* (*PSNR*) of the obtained stego images is 41.95 dB, corresponding to an extremely high embedding capacity of 2.25 bits per pixel. All of the characteristics indicate that the proposed framework is a viable candidate for offering enhanced authentication and security options for IoMT-based smart health. Rashid et al. [46] proposed a novel big data pipeline solution for storage and analysis of IoT medical data. The presented Big Data processing application makes use of *Apache Flume* to collect and transfer massive volumes of IoT data from a cloud-based server into the *Hadoop Distributed File System* for storage of IoT-based sensor medical information. *Recursive Feature Elimination with Cross Validation* (*RFECV*) is employed to eliminate less important features. The authors then proposed a hybrid prediction system of *Density-Based Spatial Clustering of Applications with Noise* (*DBSCAN*) to eliminate sensor data anomalies and improve diabetes disease detection accuracy by utilizing the Random Forest machine learning classification method. Because of the accelerating rise of the

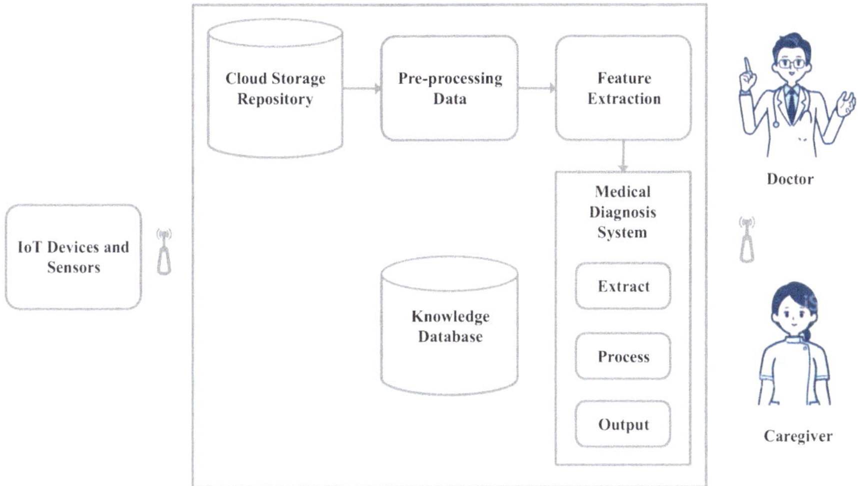

Fig. 3 IoT and Big Data functions in healthcare systems

Internet of Health Things (*IoHT*) in health services and infrastructure, high volume of health-related information is accumulating in *Personal Health Records* (*PHR*) and being handled in storage data centres. Figure 3 represents the functioning of IoT and Big Data in healthcare systems. Sarosh et al. [47] assessed a distributed security feature for clinical images, which accounts for 80% of all healthcare information. For distributed storage of clinical images, authors use the *Rivest Cipher 6* (RC6) encryption algorithm in conjunction with a *Computational Secret Sharing Scheme* (*CSIS*). The generated shares have an entropy greater than 7.99, and the *Structural Similarity* (*SSIM*) values are insignificant.

4.4 Social Multimedia Content

People engage with one another, and sustain social interactions more conveniently and effectively than before, thanks to the IoT. We may communicate with friends by getting and uploading real-time updates via *social media* such as *Facebook* or *Applications* like *TikTok*. Individual *data and social ties* created by IoT must be safeguarded with proper privacy and security technology (Obaidat et al. [48]). In a social IoT ecosystem, each item may use its relationships to find a specific service. Hsu and Tung's [49] work shows that multimedia IoT devices communicate using *Peer-to-Peer* (*P2P*) networks. Multimedia IoT devices that communicate with users, represented by Fig. 4, depict peers. Because users have *friend lists*, *social linkages* are prioritised depending on *social relationships*, including *family*, *friends*, and *others*. The authors suggested a *P2P video transmission* approach based on *weighted fair queues* with variable *queuing* priority. The suggested technique, by

Fig. 4 The flowchart for distributing user-generated content (UGC) to social networking platforms and video-sharing webpages using IoT multimedia devices

leveraging the inherent trust in social relationships, can reduce the effect of free riders while providing users with a *positive video sharing* and *viewing experience* on multimedia IoT devices.

4.5 Surveillance Multimedia Content

Because many internet-connected devices can capture movies, video streaming has become more common in IoT-based networks. Surveillance multimedia data is often obtained from *video surveillance* applications, for example, *intelligent traffic surveillance* and *house monitors*, where surveillance is used to monitor human activities such as *public safety*, *incident detection*, and *transportation control*. In practice, *video surveillance data* is viewed as a continuous stream of *high-dimensional*, *unstructured pictures* (Yu et al. [50]). Feng et al. [51] investigated security flaws in IoT-related *video surveillance* from the standpoint of hardware systems. The authors began by creating a concrete evidence device that mimicked a video replay assault. To overcome the security challenge, they created a hardware-based IoT security architecture. The suggested design creates a trusted *computing context* and physically isolates *security-sensitive* aspects, for example, the *motion detection module* from the rest of the system. Yu et al. [50] introduced an online prediction system based on multilayer ELM for video surveillance applications. Temporal and geographical properties in this system provide a *dynamic semantic* representation among adjacent frames. The suggested method not only recognises several objects moving in different directions, but it also recognises fine-tuned semantic properties.

4.6 Transport Multimedia Content

There are several IoT uses, including *intelligent transportation systems* and *self-driving cars*. Buses, cars, and trains may now be outfitted with *smart sensors* that generate huge volumes of multimedia information on road conditions, traffic congestion, temperature, and so on, providing users with information to help them drive safely (Obaidat et al. [48]). Using photos acquired by multi-view visual traffic surveillance sensors, Liu et al. [52] developed a novel *vehicle type classification*. *Deep neural networks* were integrated with balanced sampling by the authors. The suggested categorization is divided into two stages. To address the issue of imbalanced datasets, data augmentation with balanced sampling is used in the first step. The parameters obtained from the expanded training dataset are used in the second step to build a variety of *convolutional neural network models* with various designs to acquire *better accuracy* for classification purposes.

4.7 Chaos-Based Encryption Techniques

The major advantage of *chaos-based encryption* is that *chaotic signals* seem to be noise for *unauthorised users*, regardless of how they are generated. Furthermore, the cost of producing *chaotic signals* is often inexpensive, making it suited for *encrypting* huge amounts of data (Su et al. [9]). Many *chaos-based picture*, *audio*, and/or *video encryption* algorithms have been developed so far. Maggo and Chhillar [60] created a shuffling procedure that was performed on several photos to reduce unnecessary data. The suggested architecture for lightweight devices, when combined with the Henon chaotic system algorithm, uses minimal computer resources while providing great security for multimedia information, particularly photos. Dhall et al. [61] presented a *cryptosystem* based on *chaotic sequence* generation for replacement and diffusion to fulfil specific resource efficiency criteria and to satisfy the intrinsic qualities of bulkiness and redundancy of audio-visual in *real-time* and *resource-constrained* environments. Wang et al. [62] investigated a chaotic system devoid of equilibrium as well as its multimedia security applications. The suggested approach encrypts and hides 128 Kbs of data in the sound file. This work adds to our understanding of practical applications involving systems with no *hidden attractors*. Mekki et al. [63] investigated the characteristics of *chaotic systems* in the context of multimedia information encryption. The authors created a safe *surveillance* method for the *healthcare application* by integrating an effective chaotic map to extract information from the watched scene by using a *visual sensor in IoT*. Nie et al. [64] presented *window-based fountain codes* to ensure multimedia information security in edge computing *wireless transmission*. Data security for wireless transmission is ensured, thanks to fountain codes, only when the target recipient gathers enough *coded packets* ahead of *eavesdroppers*. Furthermore, the researchers used *constellation rotational technology* and *interfering noise* in the wireless signal to disturb the eavesdropper's *signal quality* and increase the eavesdropper's *packet loss*.

5 Levels of Multimedia Content Privacy and Security

Application areas largely determine multimedia content privacy and security. As a result, various types of multimedia content necessitate varying levels of protection. This section defines three degrees of privacy and security for multimedia content: *high-privacy and security*; *medium-privacy and security*; and *low-privacy and security*. The reasoning behind categorising multimedia content privacy and security into three different levels is that it can be expensive to install a *protection measure*, particularly due to the constraints of many IoT devices, so applying a high degree of protection to *sensitive*, *private*, or *confidential* multimedia content and low or no protection to publicly available data would be more *cost-effective*. When choosing a protective method, the level of privacy and security of the multimedia content handled by IoT devices must be considered. In other words, whether multimedia content is extremely private-restricted, or public, determines its privacy and security degree.

- *Level 1-Privacy and Security (Low)*
  This is the most basic degree of privacy and security. Its installation is simple and inexpensive. Agriculture multimedia content and transport multimedia content are two examples of multimedia content that may be classified as private and secure. This information is not necessarily private.
- *Level 2- Privacy and Security (Medium)*
  Level 2 privacy and security enhance the level 1 protective methods. Its execution is more costly and difficult than Level 1. Social multimedia content and surveillance multimedia content are examples of Level 2 privacy and security content.
- *Level 3-Privacy and Security (High)*
  Level 3 privacy and security ensures that multimedia content is well protected. It is the greatest degree of privacy and security of the three, and consequently more expensive and hard to achieve than Levels 1 and 2 privacy and security. Human bio-signal privacy and security, for example, are controlled and must be properly secured (Zhao et al. [53]; Pandey et al. [54]).

6 Multimedia Privacy and Security Protecting Solutions for the Internet of Things

In practice, multimedia privacy and security solutions are designed to protect multimedia material. These methods rely heavily on cryptographic techniques to

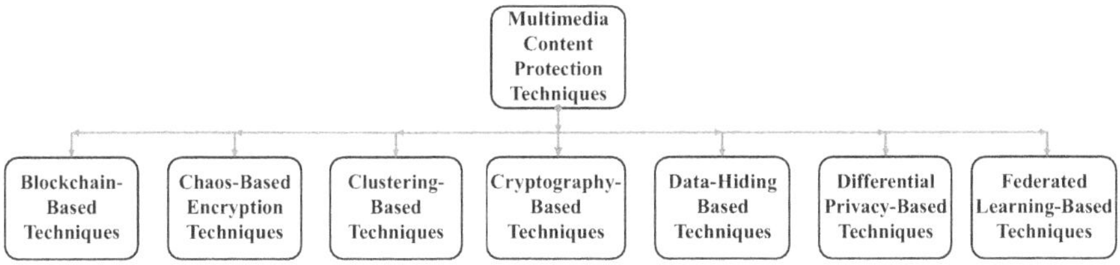

Fig. 5 Multimedia Content Protection Techniques in the IoT

provide secure communication (Furht and Socek [13]). As seen in Fig. 5, there are several ways and strategies specifically intended for multimedia privacy and security in the IoT. A detailed summary of various multimedia protection techniques is given in Table 3.

6.1 Blockchain-Based Techniques

Blockchain evolved as a contemporary decentralised architecture and model for distributed computing as a result of the rising popularity of digital currency such as Bitcoin. Its major characteristics include *decentralisation*, *time-series data*, *communal maintenance*, and *security* (Ren et al. [55]). Vishwa and Hussain [56] suggested decentralised information management based on *blockchain technology* to facilitate user data while protecting *user privacy*. The suggested protocol gives consumers total control over personal multimedia content without requiring them to rely on a third party. Li et al. [57] investigated blockchain's privacy protection method and secured data privacy in order to improve blockchain's data search and analysis capabilities. The authors created a *data security module* for *multimedia blockchains* by combining distributed, trustworthy communication networks with rapid information retrieval using *Bloom filters*. Abdur Rahman et al. [58] proved the feasibility of *blockchain-based* secured IoT services, wherein IoT devices may be rented from *service providers* while being safe and private. By concealing the participants' true identities, the suggested system allows many parties to securely share multimedia material and conduct financial transactions. Rathee et al. [59] used blockchain technology to secure healthcare multimedia information in order to prevent IoT devices from attacks launched by various types of hackers. The authors created a *hash* for every *data stream* so that any modification or change in the data, or the destruction of drugs, would be mirrored across the entire *blockchain network*.

6.2 Clustering-Based Techniques

Yu [26] claims that data clustering-based approaches evolved from the *k-anonymity* approach, then from the *l-diversity* technique, and ultimately from the *t-closeness*

Table 3 An assessment of various multimedia protection techniques

Classifications	Strategies	Methodology	Specifications Met	
			Privacy	Security
Blockchain-based techniques	Vishwa and Hussain [56]	A decentralised data management approach based on blockchain technology that allows complete control over a user's multimedia information while also ensuring privacy, auditing, and trust management in multiple areas	✓	✓
	Li et al. [57]	A blockchain security system built using distributed communicable network infrastructure and multimedia content security technology to analyze and evaluate data and protect information	✓	✓
	Abdur Rahman et al. [58]	An IoT ecosystem with IoT devices rented as a service in a diverse array of shared economic situations in order to display live visual content throughout varied IoT-based business transactions	✓	✓
	Rathee et al. [59]	Multimedia content processing in an IoT healthcare service and a secure healthcare structure based on blockchain technology	✓	✓

Chaos-based encryption techniques	Maggo and Chhillar [60]	A shuffling procedure combined with the Henon chaotic system algorithm was performed on several photos to reduce unnecessary data, using minimal computer resources while providing great security for multimedia information, particularly photos	✓	✓
	Dhall et al. [61]	A cryptosystem based on chaotic sequence generation for replacement and diffusion to fulfil specific resource efficiency criteria and to satisfy the intrinsic qualities of bulkiness and redundancy of audio-visual in real-time and resource-constrained environments	✓	✓
	Wang et al. [62]	A chaotic system devoid of equilibrium as well as its multimedia security applications, involving systems for practical applications with no hidden attractors	✓	✓
	Mekki et al. [63]	Investigated a safe surveillance method for the healthcare application by integrating an effective chaotic map to extract information from the watched scene by using a visual sensor in IoT	✓	✓
	Nie et al. [64]	Presented a window-based fountain code to ensure multimedia information security in edge computing wireless transmission, together with constellation rotational technology and interfering noise in the wireless signal to disturb the eavesdropper's signal quality and increase the eavesdropper's packet loss	✓	✓

(continued)

Table 3 (continued)

Classifications	Strategies	Methodology	Specifications Met	
			Privacy	Security
Clustering-based techniques	Liu and Li [65]	A clustering-based k-anonymity strategy to safeguard the data privacy of wearable IoT devices while also ensuring data usability	✓	×
	Xiong et al. [66]	A privacy and availability data clustering approach that strengthens the original centroid selection as well as other point-to-centroid distance computation methods, using the k-means algorithm and DP	✓	×
	Guo et al. [67]	A reciprocal privacy-preserving k-means approach based on homomorphic encryption that does not disclose either the participants' privacy or the cluster center's private information	✓	×
	Onesimu et al. [68]	A privacy-protecting data collection method with a clustering-based k-anonymization model to meet privacy requirements and protect the healthcare IoT from a range of privacy assaults	✓	×
	Tian et al. [69]	A structural entropy-based graph clustering privacy protection approach that combines data mining and structural information theory with a two-dimensional structural data solution technique and an entropy minimization principle node module division technique	✓	×

Cryptography-based techniques	Xia et al. [70]	A ciphertext-policy attribute-based encryption delegation approach that allows the data owner to encrypt the data by setting an access policy for the attribute	✓	✓
	Noura et al. [10]	A lightweight cryptographic technique that targets multimedia IoT and focused on a single-round dynamic structure with basic operations	✓	✓
	Aljawarneh et al. [71]	A powerful GPU encryption method to safeguard multimedia data against real-time assaults such as DoS attacks and tampering attacks	✓	✓
	BD et al. [74]	A hash-based secure radio-frequency IDentification (RFID) authentication technique for context-aware sensor management systems to withstand assaults such as replay attacks and man-in-the-middle attacks, based on hash modification of synchronised secret session keys	✓	✓
	Omrani et al. [75]	A lightweight encryption for picture data implemented on low-resource systems based on an "outer-inner structure" designed to manage picture correlation and redundancy while also ensuring the maximum level of concealment and diffusion features	✓	✓
	Ma et al. [76]	A lightweight, privacy-preserving data acquisition incorporated with virtual aggregation zones to balance multimedia data and batch verification used to increase system performance	✓	✓

(continued)

Table 3 (continued)

Classifications	Strategies	Methodology	Specifications Met	
			Privacy	Security
Data-hiding based techniques	Hurrah et al. [78]	A technique for copyright protection, data security, and multimedia picture content authentication incorporating a robust watermark based on an effective inter-block coefficient difference technique embedded to protect copyright and authenticate the media content in the spatial domain	✓	✓
	Rani et al. [80]	A method that combines steganography and visual cryptography to improve the security and dependability of multimedia data	✓	✓
	Huang et al. [77]	Employed a vector quantisation (VQ) transformation and least significant bits (LSB) to disguise secret data encoded in a cover picture to transmit confidential information across an IoT network	✓	✓
	Wu et al. [79]	Suggested a reversible data hiding strategy capable of achieving contrast enhancement (CE) of magnetic resonance (MR) brain pictures while also being capable of losslessly retrieving the concealed raw medical image using the secret retrieval key encoded in the cover image	✓	✓

Differential privacy-based techniques	Xu et al. [81]	A methodology for local differential privacy obfuscation (LDPO) at the edge for IoT data analysis without disclosing sensitive personal information	✓	×
	Qiao et al. [82]	A wavelet-based partitioned histogram data publishing technique with a greedy algorithm-based segmentation strategy used to provide a better partitioning structure	✓	×
	Xue et al. [84]	A privacy-preserving solution for IoT device categorization in an edge computing context to perturb the feature extraction component when machine learning models are transferred from the cloud to the edge nodes to alleviate the privacy problem	✓	×

(continued)

Table 3 (continued)

Classifications	Strategies	Methodology	Specifications Met	
			Privacy	Security
Federated learning-based techniques	Imteaj and Amini [86]	Employed a token recognition system to activate dispersed end-user devices to communicate data to the cloud and preserve it on a cloud server in the right format when needed	✓	✓
	Lu et al. [87]	Sharing data in IoT networks was defined as a machine learning problem, and privacy-preserving FL was added to it, integrated into the permission blockchain's consensus process, allowing the computational labour for consensus to be applied to FL	✓	✓
	Wu et al. [88]	Examined emerging customised FLs that may minimise the negative impacts of heterogeneity in several aspects to solve the problem of heterogeneity in IoT settings, giving a case study of IoT-based human activity identification such as human activity recognition	✓	✓

approach. Liu and Li [65] suggested a *clustering-based k-anonymity* strategy to safeguard the data privacy of *wearable IoT devices* while also ensuring data usability. Data holders can use k-anonymity to separate *identities* from sensitive information before sharing it, where in this case, *identity information* cannot be reliably correlated with *sensitive data information*. The authors of the proposed strategy first examined the possible weaknesses of current *privacy-preserving* techniques for data exchange on wearable devices. The authors then advocated adjusting the record split so that the identities inside each comparable set are indistinguishable, hence ensuring privacy. Using the k-means algorithm and DP, Xiong et al. [66] created a privacy and availability *data clustering* approach that strengthens the original centroid selection as well as other *point-to-centroid distance* computation methods. Furthermore, the suggested technique attempts to reduce the outlier impact by recognizing outliers during the *clustering phase*. The security analysis demonstrates that the suggested technique satisfies the DP goal and prevents the leakage of private information.

Guo et al. [67] introduced a reciprocal *privacy-preserving k-means* approach based on *homomorphic encryption* that does not disclose either the participants' privacy or the cluster center's private information. Onesimu et al. [68] proposed a privacy-protecting *data collection* method for an IoT-based health care system. In the suggested approach, the authors employed a *clustering-based* anonymization model to meet privacy requirements and protect the healthcare IoT from a range of privacy assaults. An enhanced clustering-based *k-anonymity* approach is used on the client side to anonymize data generated by *IoT nodes*. The authors used a bottom-up clustering technique to produce groupings of records depending on privacy constraints. Tian et al. [69] suggested a structural entropy-based *graph clustering* privacy protection approach that combines *data mining* and *structural information theory*. In particular, in the *social IoT*, user private information is *encrypted* using a *homomorphic technique* to build a graph structure in the *ciphertext state*. To divide the ciphertext network structure into distinct modules, the authors used a *two-dimensional* structural data solution technique and an entropy minimization *principle node module* division technique. Using the *k-dimensional* structural data solution technique, *the internal nodes* of the partitioned modules are further grouped.

6.3 Cryptography-Based Techniques

Cryptography is a method of converting data into an unreadable format and spreading it over *networked media*. It is generally performed by a combination of *computing functions* (Nath et al. [25]). Xia et al. [70] developed a *ciphertext-policy attribute-based* encryption delegation approach that allows the data owner to encrypt the data by setting an *access policy* for the *attribute*, so that only users associated with the attribute who meet the policy may decode the data. Furthermore, the suggested technique is activated *adaptively* depending on the estimated *decryption overhead* for distinct vehicles. Noura et al. [10] created a *lightweight cryptographic* technique that targets multimedia IoT and is focused on a *single-round* dynamic structure with basic operations. This approach generates a *dynamic key*, which is then used to generate two *robotic substitution tables*, a *dynamic permutation table*, and two *pseudo-random matrices*. The number of rounds in this dynamic cryptographic framework is reduced to one while preserving a high level of *unpredictability* and *security*. Aljawarneh et al. [71] created a powerful GPU encryption method to safeguard multimedia data against *real-time* assaults such as *DoS attacks* and *tampering attacks*. The suggested method takes advantage of the benefits of *Feistel encryption* and evolutionary algorithms. Furthermore, the devised encryption method combines the processes of *bit scrambling* and *box replacement* to achieve a strong avalanche impact. Mishra et al. [72] investigated the authentication schemes of Kumari and Om [73], revealing a number of design flaws and providing a safe mutual authentication scheme with key establishment approaches to address the flaws discovered in the Kumari and Om [73] procedures.

To reduce sensor system complexity, BD et al. [74] introduced a *hash*-based *secure Radio-Frequency Identification* (*RFID*) authentication technique for *context-aware* sensor management systems. To withstand assaults such as *replay attacks* and *man-in-the-middle attacks*, the described protocol is based on *hash modification* of synchronised *secret session keys*. Omrani et al. [75] created a *lightweight encryption* for picture data that can be implemented on *low-resource* systems, making it the first effort to handle multimedia protection in the context of the IoT. The suggested cypher is based on an "*outer-inner structure*", which is designed to manage picture *correlation* and *redundancy* while also ensuring the maximum level of concealment and diffusion features. Ma et al. [76] presented a *lightweight, privacy-preserving data acquisition method for mobile multimedia*. Most present data privacy protection solutions rely on a trusted third party, which creates a bottleneck. However, there is no reliable third party in the suggested solution, and the endpoint processing is lightweight. Furthermore, virtual aggregation zones are used to balance multimedia data, and batch verification is used to increase system performance.

6.4 Data-Hiding Based Techniques

Data-hiding strategies offer a distinct benefit in which an opponent is unaware of the presence of a secret, making them an effective tool for multimedia *privacy* protection. They are mostly used in *image* and *video-based* applications, as demonstrated by Nath et al. [25], Nauman et al. [23], Huang et al. [77], Hurrah et al. [78], and Wu et al. [79]. The basic goal behind these techniques is to conceal a secret in the *image* or *video* cover material while retaining the *image-visual video quality*. *Watermarking*, *Steganography*, and *Visual Cryptography* (*VC*) are traditional *data-hiding* techniques.

Watermarking is primarily used to identify the source of multimedia information, trace illicit distribution, and discourage unauthorized access by putting distinctive *watermarks* into multimedia information (Nauman et al. [23]). Due to its maturity and long history, watermarking continues to be the technology that is most frequently used in commercial applications. Most watermarking schemes, however, have been defeated if the multimedia item carrying the secret is examined. As a result, *steganography* and/or *VC* have been used. Hurrah et al. [78] suggested a technique for *copyright protection*, *data security*, and *multimedia picture content authentication*. A robust *watermark* based on an effective inter-block coefficient difference technique is embedded to protect copyright, while a fragile watermark is inserted to authenticate the media content in the *spatial domain*. To confuse the *adversary* and guarantee that he or she cannot simply retrieve the actual embedded information, the authors encrypt the data before embedding it using an encryption method and the Arnold transform. Rani et al. [80] used *steganography* and *VC* to improve the security and dependability of multimedia data. *Steganographic* pictures are buried in VC shares that are illogical and cannot be recognised by *Steganalysis* tools in the suggested technique. Data security relies heavily on VC and steganography. The suggested approach is tested with *text* and *picture files*, and the findings show that the quality of the recovered information and images is not degraded when compared to the original files. The disadvantage of such designs is inefficiency, as many cover pictures are required.

As per Huang et al. [77] and Wu et al. [79], *reversible data hiding* is an innovative data-hiding approach that may smoothly incorporate the benefits of *watermarking* and *cryptography*. Huang et al. [77] employed *Vector Quantisation* (*VQ*) transformation and *Least Significant Bits* (*LSB*) to disguise secret data encoded in a cover picture to transmit confidential information across an IoT network. The authors used a technique known as *two-level coding* to remap the *VQ-compressed* picture to the *VQ-transformed* image, and then utilized the hidden data to adjust the new image to be more like the original. Wu et al. [79] recently suggested a *reversible data hiding* strategy capable of achieving *Contrast Enhancement* (*CE*) of *Magnetic Resonance* (*MR*) brain pictures while also being capable of losslessly retrieving the concealed *raw medical image* using the *secret retrieval key* encoded in the *cover image*. This is an intriguing conclusion since it can achieve various purposes in addition to the traditional goal of data concealment.

6.5 Differential Privacy-Based Techniques

The main idea of DP is to introduce random noise to the data before releasing it. This random noise addition may be accomplished in a variety of ways. Xu et al. [81] suggested a methodology for *Local Differential Privacy Obfuscation* (*LDPO*) at the edge for IoT data analysis without disclosing sensitive personal information. The authors originally introduced the *LDPO framework's* design and advantages. They then explored some technical issues and demonstrated an early version of the *LDPO framework*. The authors also validated the proposed framework's effectiveness in terms of privacy protection and data usefulness using practical applications and datasets. Qiao et al. [82] created a *wavelet-based partitioned histogram* data publishing technique. To begin with, a *greedy* algorithm-based segmentation strategy is used to provide a better partitioning structure. The *wavelet transform* is then used to introduce noise. Lastly, a simplified *histogram structure* is constructed to confirm the *histogram's legitimacy* and use. The suggested approach can reduce the complexity of *wavelet trees* generated by the wavelet transform while increasing the accuracy of *histogram counting* searches. Zia et al. [83] conducted a DP *case study*. The authors investigated the effect of noise on the original information as well as the link between noise in the data and its usefulness. The impact of data leaks on privacy infractions is also examined. In Xue et al. [84], the authors introduced a *privacy-preserving* solution for IoT device categorization in an *edge computing* context. Because direct disruption to the input information in the training set has a significant influence on *classification precision*, the suggested approach is optimized to perturb the feature extraction component when *machine learning models* are transferred from the *cloud* to the *edge nodes* to alleviate the privacy problem. Experiment findings on several datasets show that the suggested approach achieves a fair balance of privacy preservation and usefulness.

6.6 Federated Learning-Based Techniques

A promising approach for user privacy protection in the IoT is training models on massive libraries of decentralised multimedia information (Pang et al. [85]). Imteaj and Amini [86] used existing sensors incorporated into smartphones to collect and distribute audio-visual data for various IoT *decision-making* objectives. The goal of this distributed sensing technique is to employ *token recognition* systems to activate dispersed end-user devices such that data may be communicated to the cloud and preserved on a *cloud server* in the right format when needed. To overcome privacy issues, the authors employ FL to exploit *sensing platforms* in IoT networks for distributed *decision-making* and learning. The problem of sharing data in IoT networks was defined as a machine learning problem by Lu et al. [87], and privacy-preserving FL was added to it. Privacy is properly safeguarded by releasing the data model rather than disclosing the actual data. Furthermore, the authors integrated FL into the permission *blockchain's consensus* process, allowing the computational labour for consensus to be applied to FL. Wu et al. [88] examined emerging customised FLs that may minimise the negative impacts of heterogeneity in several aspects to solve the problem of heterogeneity in IoT settings. The power of edge computing can also meet the demand for rapid processing and low latency in smart multimedia IoT applications. To highlight the usefulness of personalised FL in smart multimedia IoT systems, the authors gave a case study of *IoT-based human activity identification* such as *human activity recognition*.

7 Open Issues in Multimedia Privacy and Security Protection in the IoT

The new *IoT-based multimedia applications* provide new issues for multimedia content protection. In this part, we will discuss some of the open problems of multimedia privacy and security in the IoT.

- To begin with, determining what level of privacy and security is suitable for multimedia content in IoT-related services is a difficult challenge. A sufficient degree of security leads to more efficient usage and protection of multimedia content. In the *decision-making phase*, the value of the multimedia data to be secured should be thoroughly examined against the expense of the protection itself. If the multimedia information to be safeguarded—for example, *agricultural multimedia data*—is not inherently *valuable* or *sensitive*, then a reasonably *low-cost* and *basic encryption* solution is adequate. If the multimedia content, for example, *biometric* or *healthcare* multimedia information, is personal and/or secret, a *high level* of protection must be used (Furht and Socek [13]).
- Secondly, the variety of IoT devices, data formats, and systems poses problems for IoT multimedia security.

  IoT device heterogeneity: In the rapidly expanding IoT, diverse items, both *physical* and *virtual*, are connected to the internet and communicate with other entities using *unique addressing* methods to provide or request multiple services (Hamoudy et al. [89]).

  Data heterogeneity in terms of data kinds: When compared to traditional data, multimedia data collected, created, and/or transmitted by IoT devices has varying characteristics owing to different forms of data such as *photos*, *audio*, and *videos*, *unstructured features*, *noise*, and so forth (Kumar et al. [20]).

  System heterogeneity: IoT systems differ in terms of *computing capacity*, *memory storage*, and *communication range*. IoT systems must be: a) able to accept heterogeneous devices, b) able to build valid models with limited *device involvement*, and c) *fault-tolerant* in the situation of distributed or inactive devices in the network, since IoT devices can be dispersed or inactive or just a limited number of devices may be operational to continue connection with the server (Imteaj and Amini [86]).
- Thirdly, practicality and computing costs are significant considerations for IoT multimedia privacy and security. While certain encryption methods provide good security, their computing costs rise as the amount of data increases. As a result, new privacy and security requirements increase the energy consumption and computational capacity of IoT devices. To manage such instances, *lightweight privacy* and *security* approaches are needed for IoT devices with *limited resources*, including *memory*, *computing power*, and *energy*, together with insufficient cyber security measures (Hamoudy et al. [89]).
- Fourth, DP is a primary method of protecting the privacy of IoT multimedia content. The execution of DP on IoT devices necessitates a significant processing

burden. In the IoT, developing a dependable DP with minimum *computing overhead* is difficult. Furthermore, as the IoT expands, a tremendous amount of audio-visual information is created in real time. As a result, it is difficult to manage high-dimensional multimedia information and perform efficient computation of DP composition in large multimedia data analytics when applying DP to big multimedia data in the IoT (Husnoo et al. [29]).

- Lastly, in addition to using *privacy-preserving* measures, a lack of transparency and oversight about the usage of IoT multimedia content is a barrier to safeguarding user privacy while reaping the benefits of IoT. Individuals are seldom told that their personal information is being collected. Users of IoT devices in public places might not be aware that their personal information is being taken because these devices can automatically acquire information. Transparency and regulation of IoT devices are thus required to preserve users' privacy by protecting the way IoT devices acquire or utilise *personal information* (OVIC [90]).

8 Conclusion

With the increasing growth of IoT devices and the advancement of IoT multimedia technologies in modern environments, multimedia content privacy and security concerns should be taken seriously, notably in data-sensitive IoT systems. In response to these issues, investigators have developed a number of solutions for multimedia content privacy and security in the IoT. We examined recent research breakthroughs in IoT-related multimedia security, focusing on upcoming technologies such as blockchain. We have also categorized multimedia information into several sorts and levels of protection. Furthermore, we have highlighted critical issues in multimedia privacy and security protection in the IoT.

We suggest the following future directions for research premised on the comprehensive examination and thorough discussion in this review article:

1. *A systematic security model to deal with the IoT's heterogeneity:* Because no specific content protection system can resolve all security concerns, a systematic security architecture that integrates multiple encryption methods, authentication procedures, and defence strategies to encourage multimedia content protection in the heterogeneous IoT is worthwhile to consider.
2. *Adaptable protection procedures depending on multimedia content privacy and security levels:* In terms of privacy and security, data classification is an effective approach for providing a suitable and adaptive privacy and security outcome based on the type, privacy and security level of multimedia content. An adversary finds a high-cost attack frustrating, so a simple encryption process may suffice for conveying MPEG videos. DRM or selective encryption is more appealing for such applications (Furht and Socek [10]). The basic idea behind selective encryption is to encrypt only a portion of the compressed bitstream. For example,

only the most significant coefficients from the final or intermediate phases of the compression stage could be encrypted. Significant coefficients are either lightly scrambled or not at all encrypted. (Furht and Socek [10]).

3. *Cost versus privacy/security trade-off:* The privacy and security of multimedia information are crucial in the IoT, but it is difficult to achieve. Whereas the standard trade-off between high privacy and security and the accessibility of multimedia content for the IoT is more pressing than before, privacy and security are regarded as a hugely expensive characteristic, as several IoT systems possess finite resources such as computing power and energy. As a result, developers of IoT devices require not only methods to assess potential privacy and security threats and investigate solutions, but also techniques to evaluate expenses and strike an acceptable balance between privacy and security and cost (Bodei et al. [91]).
4. *Lightweight solution advancement:* Multimedia transmission security over IoT networks has received insufficient attention. With regard to energy consumption and power computation, IoT devices are typically limited. As a result, secure communication in the IoT necessitates computationally effective methods. Lightweight multimedia content protection options are available in a variety. Shifa et al. [92], for instance, suggested a lightweight video encryption algorithm wherein video information is not entirely encrypted. To protect individual privacy and location, varying levels of encryption are employed to protect multimedia information from attacks.
5. *Utilization of new technologies:* Throughout this study, we looked at several emerging technologies, including blockchain and FL. All such newer technologies provide enhanced security for multimedia content. We recommend that investigators weigh the benefits and drawbacks of evolving technologies including artificial intelligence (AI) and fog/edge computing in order to more effectively apply them to multimedia content protection in IoT applications. Even if AI can support information security, utilising AI algorithms on resource-constrained IoT systems is probably impossible due to their extensive computations, which consume a lot of energy. For IoT devices, fog/edge computing is superior and quicker. However, the security of this new technique has not been thoroughly investigated.

References

1. Karie NM, Sahri NM, Haskell-Dowland P (2020) IoT threat detection advances, challenges and future directions. In: 2020 workshop on emerging technologies for security in IoT (ETSecIoT), April 2020. IEEE, pp 22–29
2. Zhou L, Chao HC (2011) Multimedia traffic security architecture for the internet of things. IEEE Netw 25(3):35–40
3. Yongliang L, Gao W, Liu S (2004) Multimedia security in the distributed environment. In: APCC/MDMC'04. The 2004 joint conference of the 10th Asia-Pacific conference on communications and the 5th international symposium on multi-dimensional mobile communications proceeding, August 2004, vol 2. IEEE, pp 639–642

4. Punia A, Gupta D, Jaiswal S (2017) A perspective on available security techniques in IoT. In: 2017 2nd IEEE international conference on recent trends in electronics, information & communication technology (RTEICT), May 2017. IEEE, pp 1553–1559
5. Riahi A, Challal Y, Natalizio E, Chtourou Z, Bouabdallah A (2013) A systemic approach for IoT security. In: 2013 IEEE international conference on distributed computing in sensor systems, May 2013. IEEE, pp 351–355
6. Parah SA, Sheikh JA, Hafiz AM, Bhat GM (2014) Data hiding in scrambled images: a new double layer security data hiding technique. Comput Electr Eng 40(1):70–82
7. Parah SA, Sheikh JA, Bhat GM (2012) Data hiding in intermediate significant bit planes, a high capacity blind steganographic technique. In: 2012 international conference on emerging trends in science, engineering and technology (INCOSET), December 2012. IEEE, pp 192–197
8. Parah SA, Sheikh JA, Bhat GM (2014) A secure and efficient spatial domain data hiding technique based on pixel adjustment. Am J Eng Technol Res 14(2):33
9. Parah SA, Sheikh JA, Bhat GM (2012) On the realization of a secure, high capacity data embedding technique using joint top-down and down-top embedding approach. Comput Sci Eng 49:10141
10. Noura H, Chehab A, Sleem L, Noura M, Couturier R, Mansour MM (2018) One round cipher algorithm for multimedia IoT devices. Multimed Tools Appl 77(14):18383–18413
11. Nahrstedt K, Dittmann J, Wohlmacher P (2000) Approaches to multimedia and security. In: 2000 IEEE international conference on multimedia and Expo. ICME2000. Proceedings. Latest advances in the fast changing world of multimedia (Cat. No. 00TH8532), July 2000, vol 3. IEEE, pp 1275–1278
12. Tzelepi S, Pangalos G, Nikolacopoulou G (2002) Security of medical multimedia. Med Inform Internet Med 27(3):169–184
13. Furht B, Socek D (2003) A survey of multimedia security. Comprehensive report.
14. Voloshynovskyy S, Koval O, Deguillaume F, Pun T (2005) Multimedia security: open problems and solutions. In: Proceedings of NATO-advanced study institute: security through science program 2005
15. Moravčík M, Segeč P, Hrabovský J, Papán J, Uramová J (2016) Survey of real-time multimedia security mechanisms. In: 2016 international conference on emerging eLearning technologies and applications (ICETA), November 2016. IEEE, pp 233–238
16. Ashour AS, Dey N (2017) Security of multimedia contents: a brief. In: Intelligent techniques in signal processing for multimedia security. Springer, Cham, pp 3–14
17. Li Y, Jeong YS, Shin BS, Park JH (2017) Crowdsensing multimedia data: security and privacy issues. IEEE Multimed 24(4):58–66
18. Alvi SA, Afzal B, Shah GA, Atzori L, Mahmood W (2015) Internet of multimedia things: vision and challenges. Ad Hoc Netw 33:87–111
19. Wang Q, Zhao Y, Wang W, Minoli D, Sohraby K, Zhu H, Occhiogrosso B (2017) Multimedia IoT systems and applications. In: 2017 global internet of things summit (GIoTS), June 2017. IEEE, pp 1–6
20. Kumar D, Kumar P, Ashok A (2020) Introduction to multimedia big data computing for IoT. In: Multimedia big data computing for IoT applications. Springer, Singapore, pp 3–36
21. Hurrah NN, Parah SA, Sheikh JA, Al-Turjman F, Muhammad K (2019) Secure data transmission framework for confidentiality in IoTs. Ad Hoc Netw 95:101989
22. Parah SA, Sheikh JA, Akhoon JA, Loan NA (2020) Electronic health record hiding in images for smart city applications: a computationally efficient and reversible information hiding technique for secure communication. Futur Gener Comput Syst 108:935–949
23. Nauman A, Qadri YA, Amjad M, Zikria YB, Afzal MK, Kim SW (2020) Multimedia internet of things: a comprehensive survey. IEEE Access 8:8202–8250
24. Verma S, Bhatia A, Chug A, Singh AP (2020) Recent advancements in multimedia big data computing for IoT applications in precision agriculture: opportunities, issues, and challenges. In: Multimedia big data computing for IoT applications. Springer, Singapore, pp 391–416
25. Nath MP, Priyadarshini SBB, Ray M, Das DS (2022) An overview of multimedia technologies in current era of internet of things (IoT). In: Multimedia technologies in the internet of things environment, vol 2, pp 1–23

26. Yu S (2016) Big privacy: challenges and opportunities of privacy study in the age of big data. IEEE Access 4:2751–2763
27. Qu Y, Yu S, Zhou W, Peng S, Wang G, Xiao K (2018) Privacy of things: emerging challenges and opportunities in wireless internet of things. IEEE Wirel Commun 25(6):91–97
28. Alhirabi N, Rana O, Perera C (2021) Security and privacy requirements for the internet of things: a survey. ACM Trans Internet Things 2(1):1–37
29. Husnoo MA, Anwar A, Chakrabortty RK, Doss R, Ryan MJ (2021) Differential privacy for IoT-enabled critical infrastructure: a comprehensive survey. IEEE Access 9:153276–153304
30. Jiang B, Li J, Yue G, Song H (2021) Differential privacy for industrial internet of things: opportunities, applications, and challenges. IEEE Internet Things J 8(13):10430–10451
31. Oracevic A, Dilek S, Ozdemir S (2017) Security in internet of things: a survey. In: 2017 international symposium on networks, computers and communications (ISNCC), May 2017. IEEE, pp 1–6
32. Bhat GM, Mustafa M, Parah SA, Ahmad J (2010) Field programmable gate array (FPGA) implementation of novel complex PN-code-generator-based data scrambler and descrambler. Maejo Int J Sci Technol 4(1):125–135
33. Qu Y, Yu S, Gao L, Zhou W, Peng S (2018) A hybrid privacy protection scheme in cyber-physical social networks. IEEE Trans Comput Soc Syst 5(3):773–784
34. Zikria YB, Afzal MK, Kim SW (2020) Internet of multimedia things (IoMT): opportunities, challenges and solutions. Sensors 20(8):2334
35. Khattab A, Abdelgawad A, Yelmarthi K (2016) Design and implementation of a cloud-based IoT scheme for precision agriculture. In: 2016 28th international conference on microelectronics (ICM), December 2016. IEEE, pp 201–204
36. Mohanraj I, Ashokumar K, Naren J (2016) Field monitoring and automation using IOT in agriculture domain. Procedia Comput Sci 93:931–939
37. Yang W, Wang S, Zheng G, Chaudhry J, Valli C (2018) ECB4CI: an enhanced cancelable biometric system for securing critical infrastructures. J Supercomput 74(10):4893–4909
38. Ribaric S, Ariyaeeinia A, Pavesic N (2016) De-identification for privacy protection in multimedia content: a survey. Signal Process Image Commun 47:131–151
39. Lu Y, Wu S, Fang Z, Xiong N, Yoon S, Park DS (2017) Exploring finger vein based personal authentication for secure IoT. Futur Gener Comput Syst 77:149–160
40. Thilagavathi B, Suthendran K (2018) Boosting based implementation of biometric authentication in IoT. J Cyber Secur Mobil:131–144
41. Yang W, Wang S, Zheng G, Yang J, Valli C (2019) A privacy-preserving lightweight biometric system for internet of things security. IEEE Commun Mag 57(3):84–89
42. Xu J, Glicksberg BS, Su C, Walker P, Bian J, Wang F (2021) Federated learning for healthcare informatics. J Healthc Inform Res 5(1):1–19
43. Bhoomika BK, Muralidhara KN (2015) Secured smart healthcare monitoring system based on Iot. Int J Recent Innov Trends Comput Commun 3(7):4958–4961
44. Muhammad G, Rahman SMM, Alelaiwi A, Alamri A (2017) Smart health solution integrating IoT and cloud: a case study of voice pathology monitoring. IEEE Commun Mag 55(1):69–73
45. Parah SA, Kaw JA, Bellavista P, Loan NA, Bhat GM, Muhammad K, de Albuquerque VHC (2020) Efficient security and authentication for edge-based internet of medical things. IEEE Internet Things J 8(21):15652–15662
46. Rashid M, Singh H, Goyal V, Parah SA, Wani AR (2021) Big data based hybrid machine learning model for improving performance of medical internet of things data in healthcare systems. In: Healthcare paradigms in the internet of things ecosystem. Academic Press, pp 47–62
47. Sarosh P, Parah SA, Bhat GM, Heidari AA, Muhammad K (2021) Secret sharing-based personal health records management for the internet of health things. Sustain Cities Soc 74:103129
48. Obaidat MS, Rana SP, Maitra T, Giri D, Dutta S (2019) Biometric security and internet of things (IoT). In: Biometric-based physical and cybersecurity systems. Springer, Cham, pp 477–509

49. Hsu TH, Tung YM (2020) A social-aware P2P video transmission strategy for multimedia IoT devices. IEEE Access 8:95574–95584
50. Yu H, Wang J, Sun X (2019) Surveillance video online prediction using multilayer ELM with object principal trajectory. SIViP 13(6):1243–1251
51. Feng X, Ye M, Swaminathan V, Wei S (2017) Towards the security of motion detection-based video surveillance on IoT devices. In: Proceedings of the on thematic workshops of ACM multimedia, October 2017. Association for Computing Machinery, New York, NY, pp 228–235
52. Liu W, Zhang M, Luo Z, Cai Y (2017) An ensemble deep learning method for vehicle type classification on visual traffic surveillance sensors. IEEE Access 5:24417–24425
53. Zhao H, Qin J, Hu J (2012) An energy efficient key management scheme for body sensor networks. IEEE Trans Parallel Distrib Syst 24(11):2202–2210
54. Pandey C, Sharma S, Matta P (2021) Privacy techniques for body sensor network in healthcare internet of things (HIoT)-a critical survey. In: 2021 5th international conference on computing methodologies and communication (ICCMC), April 2021. IEEE, pp 385–389
55. Ren Y, Zhu F, Zhu K, Sharma PK, Wang J (2021) Blockchain-based trust establishment mechanism in the internet of multimedia things. Multimed Tools Appl 80(20):30653–30676
56. Vishwa A, Hussain FK (2018) A blockchain based approach for multimedia privacy protection and provenance. In: 2018 IEEE symposium series on computational intelligence (SSCI), November 2018. IEEE, pp 1941–1945
57. Li D, Liu W, Deng L, Qin B (2021) Design of multimedia blockchain privacy protection system based on distributed trusted communication. Trans Emerg Telecommun Technol 32(2):e3938
58. Abdur Rahman M, Loukas G, Maruf Abdullah S, Abdu A, Sadiqur Rahman S, Hassanain E, Arafa Y (2019) Blockchain and IoT-based secure multimedia retrieval system for a massive crowd: sharing economy perspective. In: Proceedings of the 2019 on international conference on multimedia retrieval, June 2019, pp 404–407
59. Rathee G, Sharma A, Saini H, Kumar R, Iqbal R (2020) A hybrid framework for multimedia data processing in IoT-healthcare using blockchain technology. Multimed Tools Appl 79(15):9711–9733
60. Maggo P, Chhillar RS (2013) Lightweight image encryption scheme for multimedia security. Int J Comput Appl 71(13)
61. Dhall S, Pal SK, Sharma K (2014) Cryptographic primitives for multimedia security. INROADS-Int J Jaipur National Univ 3(1s2):335–339
62. Wang X, Akgul A, Kacar S, Pham VT (2017) Multimedia security application of a ten-term chaotic system without equilibrium. Complexity 2017
63. Mekki N, Hamdi M, Aguili T, Kim TH (2018) A real-time chaotic encryption for multimedia data and application to secure surveillance framework for IoT system. In: 2018 international conference on advanced communication technologies and networking (CommNet), April 2018. IEEE, pp 1–10
64. Nie H, Jiang X, Tang W, Zhang S, Dou W (2020) Data security over wireless transmission for enterprise multimedia security with fountain codes. Multimed Tools Appl 79(15):10781–10803
65. Liu F, Li T (2018) A clustering-anonymity privacy-preserving method for wearable iot devices. Secur Commun Netw 2018
66. Xiong J, Ren J, Chen L, Yao Z, Lin M, Wu D, Niu B (2018) Enhancing privacy and availability for data clustering in intelligent electrical service of IoT. IEEE Internet Things J 6(2):1530–1540
67. Guo X, Lin H, Wu Y, Peng M (2020) A new data clustering strategy for enhancing mutual privacy in healthcare IoT systems. Futur Gener Comput Syst 113:407–417
68. Onesimu JA, Karthikeyan J, Sei Y (2021) An efficient clustering-based anonymization scheme for privacy-preserving data collection in IoT based healthcare services. Peer Peer Netw Appl 14(3):1629–1649
69. Tian Y, Zhang Z, Xiong J, Chen L, Ma J, Peng C (2021) Achieving graph clustering privacy preservation based on structure entropy in social IoT. IEEE Internet Things J 9(4):2761–2777

70. Xia Y, Chen W, Liu X, Zhang L, Li X, Xiang Y (2017) Adaptive multimedia data forwarding for privacy preservation in vehicular ad-hoc networks. IEEE Trans Intell Transp Syst 18(10):2629–2641
71. Aljawarneh S, Yassein MB, Talafha WAA (2017) A resource-efficient encryption algorithm for multimedia big data. Multimed Tools Appl 76(21):22703–22724
72. Mishra D, Vijayakumar P, Sureshkumar V, Amin R, Islam SK, Gope P (2018) Efficient authentication protocol for secure multimedia communications in IoT-enabled wireless sensor networks. Multimed Tools Appl 77(14):18295–18325
73. Kumari S, Om H (2017) Cryptanalysis and improvement of an anonymous multi-server authenticated key agreement scheme. Wirel Pers Commun 96(2):2513–2537
74. Deebak BD, Al-Turjman F, Mostarda L (2019) A hash-based RFID authentication mechanism for context-aware management in IoT-based multimedia systems. Sensors 19(18):3821
75. Omrani T, Rhouma R, Becheikh R (2019) LICID: a lightweight image cryptosystem for IoT devices. Cryptologia 43(4):313–343
76. Ma S, Zhang T, Wu A, Zhao X (2019) Lightweight and privacy-preserving data aggregation for mobile multimedia security. IEEE Access 7:114131–114140
77. Huang CT, Tsai MY, Lin LC, Wang WJ, Wang SJ (2018) VQ-based data hiding in IoT networks using two-level encoding with adaptive pixel replacements. J Supercomput 74(9):4295–4314
78. Hurrah NN, Parah SA, Loan NA, Sheikh JA, Elhoseny M, Muhammad K (2019) Dual watermarking framework for privacy protection and content authentication of multimedia. Futur Gener Comput Syst 94:654–673
79. Wu HT, Zheng K, Huang Q, Hu J (2021) Contrast enhancement of multiple tissues in MR brain images with reversibility. IEEE Signal Processing Lett 28:160–164
80. Rani M, Germine Mary G, Rosemary Euphrasia K (2016) Multilevel multimedia security by integrating visual cryptography and steganography techniques. In: Computational intelligence, cyber security and computational models. Springer, Singapore, pp 403–412
81. Xu C, Ren J, Zhang D, Zhang Y (2018) Distilling at the edge: a local differential privacy obfuscation framework for IoT data analytics. IEEE Commun Mag 56(8):20–25
82. Qiao Y, Liu Z, Lv H, Li M, Huang Z, Li Z, Liu W (2019) An effective data privacy protection algorithm based on differential privacy in edge computing. IEEE Access 7:136203–136213
83. Zia MT, Khan MA, El-Sayed H (2020) Application of differential privacy approach in healthcare data–a case study. In: 2020 14th international conference on innovations in information technology (IIT), November 2020. IEEE, pp 35–39
84. Xue W, Shen Y, Luo C, Xu W, Hu W, Seneviratne A (2022) A differential privacy-based classification system for edge computing in IoT. Comput Commun 182:117–128
85. Pang J, Huang Y, Xie Z, Han Q, Cai Z (2020) Realizing the heterogeneity: a self-organized federated learning framework for IoT. IEEE Internet Things J 8(5):3088–3098
86. Imteaj A, Amini MH (2019) Distributed sensing using smart end-user devices: pathway to federated learning for autonomous IoT. In: 2019 international conference on computational science and computational intelligence (CSCI), December 2019. IEEE, pp 1156–1161
87. Lu Y, Huang X, Dai Y, Maharjan S, Zhang Y (2019) Blockchain and federated learning for privacy-preserved data sharing in industrial IoT. IEEE Trans Industr Inform 16(6):4177–4186
88. Wu Q, He K, Chen X (2020) Personalized federated learning for intelligent IoT applications: a cloud-edge based framework. IEEE Open J Comput Soc 1:35–44
89. Hamoudy MA, Qutqut MH, Almasalha F (2017) Video security in internet of things: an overview. IJCSNS 17(8):199
90. OVIC (2021) Internet of things and privacy – issues and challenges [Available Online] https://ovic.vic.gov.au/privacy/internet-of-things-and-privacy-issues-and-challenges/ (Accessed 30 July 2022)
91. Bodei C, Chessa S, Galletta L (2019) Measuring security in IoT communications. Theor Comput Sci 764:100–124
92. Shifa A, Asghar MN, Fleury M (2016) Multimedia security perspectives in IoT. In: 2016 sixth international conference on innovative computing technology (INTECH), August 2016. IEEE, pp 550–555

Analysis of Ensemble Methods for Phishing Detection

Deepak Gupta, Ekta Gandotra, Yogesh Mohan, and Sukhvir Singh

1 Introduction

The recent boom in multimedia technology and Internet is providing overabundance of benefits in everyday life. Due to their extensive capabilities in carrying out a variety of activities, the employment of powerful processors and sensors deployed in several sectors is proving to be advantageous. Digital platforms are being used widely for a variety of purposes, which has drawn the attention of attackers to launch more and more cyber attacks. The cyber attackers attempt to steal the credit card details, passwords or pass on malware to the victims' device through fake webpages. Such illegal actions of endeavoring to seek valuable or sensitive information from the user are called phishing. It involves the use of fake webpages or websites and e-mail spoofing. Web phishing is one of the persistent security threats on the Internet and is responsible for around 90% of data breaches [1].

During coronavirus (COVID-19) pandemic [2], many offices were forced to shut down and the employees were made to work from home where they didn't have

The original version of the chapter has been revised. A correction to this chapter can be found at https://doi.org/10.1007/978-3-031-34873-0_16

D. Gupta · E. Gandotra (✉)
Department of Computer Science & Engineering and Information Technology, Jaypee University of Information Technology Waknaghat, Solan, Himachal Pradesh, India

Y. Mohan
Department of Computer Science and Applications, Himachal Pradesh University, Shimla, Himachal Pradesh, India

S. Singh
Department of Computer Science, Himachal Pradesh University Regional Centre, Dharamshala, Himachal Pradesh, India

S. A. Parah et al. (eds.), *Intelligent Multimedia Signal Processing for Smart Ecosystems*, https://doi.org/10.1007/978-3-031-34873-0_4

security checks which were typically there in the offices earlier. Many organizations have experienced changes in their working environment and remote activities are increased. Further, the dependency on email for communication has created the perfect conditions for email fraud and phishing. Cybercriminals are utilizing it to deceive users into disclosing personal information or opening malicious links or attachments, which results in the unintentional download of malware onto their systems. These cybercriminals may even pretend to be government agencies or influential people in order to breach trustworthy sources.

Attackers are taking benefit of the aforementioned facts and performing phishing attacks through fake websites or emails by allowing users to visit fake websites to steal their confidential information. Therefore, the number of phishing attacks keeps on increasing day by day. According to a report by APWG [3], the number of phishing attacks has almost doubled from the year 2020. A total of 2,22,127 unique phishing attacks were reported in June 2021. According to another security report [4], phishing attacks are increased by 161% in the year 2021 as compared to the second half of 2020. The rise in phishing attacks demands big data framework based solutions for detecting phishing pages [5, 6]. In addition, the motive of phishing attacks is not only to get the sensitive information of the victim, it has now become one of the most popular methods of malware dissemination [7–11]. There are various tools [12–14] that are used by attackers to create duplicate websites for performing phishing attacks. Such duplicate websites are sometimes detected by anti-phishing tools. There are other methods for the detection of phishing websites. One of the most well-liked methods is based on signatures in which a list of existing blacklisted webpages or websites is maintained in a database. The requested webpage is matched in the database to check whether it is phished. Google Safe Browsing (GSB) [15] and PhishTank [16] are the two most popular databases of blacklisted webpages/websites. The main disadvantage of this method is that it is not able to detect the newly created phishing websites as they are not still added to the database. Heuristic approach analyzes to find the features to discriminate between phished pages from the benign ones. The visual based approach creates a database of logos, images, tags, etc. to identify if a webpage is phished or not. However, this approach is time intensive. In recent times, researchers are focusing on machine learning, deep learning and computational approaches [17] for the detection of phishing webpages. These methods first extract the features related to benign and phished websites and then use these features for training to build the classification model. The drawback of these methods is that these provide high false negative and false positive rate [18]. Ensemble methods are used to improve the accuracy of detection and classification. This chapter aims to study and carry out a comparative analysis of various base classifiers and ensemble methods for the detection and classification of phishing webpages.

The chapter is organized as follows: Section 2 provides the research work related to the classification of phishing webpages. Section 3 discusses the approach used to carry out the comparative analysis. Section 4 provides the experimental results and discussion followed by Sect. 5 which presents the conclusion.

2 Related work

This section discusses the state of the art research carried out to perform the classification of phishing websites using machine and deep learning methods.

Tan et al. [19] presented an approach for the classification and detection of phishing webpages. The approach comprised of three steps. In the first step, keywords were mined from the webpages using the N-gram method. In the second step, a search engine was used to find the target domain name followed by detecting the webpage legality in the third step. Another image based approach was proposed in [20] for the detection and identification of fake webpages. The logo on the webpage has been used to find the identity of the webpage using Google image search engine. The domain name returned by Google and the webpage query are compared to distinguish between phishing and benign webpages. Jain and Gupta [21] used the 19 important features of phishing webpages and used various machine learning methods for the classification of phishing webpages. Gandotra and Gupta [22] proposed a different set of features from the various categories like webpage, URL, and HTML. They first used the features under an individual category to classify webpages. Then an integrated set of features from all categories are used for the purpose of classification. The experimental results revealed that the classification accuracy improved significantly using the proposed approach. In another work [23], authors carried out a comparative study of machine learning algorithms with or without using feature selection methods. They conducted the experiments on a dataset with or without features and concluded that utilizing a feature selection method improved the classification model building time for phishing detection and classification without compromising the accuracy.

Sahingoz et al. [24] proposed a real-time system relying on natural language processing (NLP) based features. Seven machine learning algorithms were used for the classification of phishing webpages. The proposed system was tested and validated on a self-created dataset. The random forest (RF) provided the best classification results. Afzal et al. [25] proposed a secure data transmission method for providing security to the medical images. They used two encryption techniques, i.e., chaotic encryption and bitplane encryption on the images. The experimental results validate the value of the proposed methods in terms of security, authenticity, and encryption time. In an unsecure context of multimedia sharing between two IoT hops, reference [26] concentrated on protecting data privacy and confidentiality. It was based on random coefficient selection and mean modification method which used several discrete cosine transform coefficients chosen at random from two non-neighbourhood blocks. The suggested method produced extremely high robustness of the data sent between two nodes, for all signal processing and the region-focused attacks, functioning simultaneously or independently. The experimental results showed that the proposed strategy outperforms other cutting-edge methods. In [27], the authors proposed Hessenberg transform and Chaotic encryption based image encryption framework to improve the security and reducing the computational cost. It was concluded from the experimental results that the proposed method offered less computational cost and proved to be more robust to attacks as compared to the other methods. In [28], the framework for securing the information of a patient and authenticating the received content is presented. The suggested method was based on new left data mapping, pixel repetition method, checksum computing method, and RC4 encryption. The experimental results revealed that the suggested method provides a better authentication and security solution for smart health based on the Internet of Medical Things (IoMT). A cybersecurity framework created for medical imaging in a smart healthcare system was presented in [29]. The authors suggested two brand-new 2D chaotic maps with enormous chaotic ranges and compound chaotic features named the logistic regulated quadratic map and quadratic regulated quadratic map. They described a method of encryption based on these maps and demonstrated the extremely erratic pseudo-random number sequence produced by the maps yields a very strong cypher picture. NLP methods were used to extract some of the features from a system that was given in [30]. Before opening them with the help of extracted features, the system was implemented by looking over the URLs used in phishing attacks. It was determined from the experimental results that RF algorithm, which provided an overall success rate of 89.9%, is the best algorithm of those examined. Reference [31] proposed a technique for phishing detection, where machine learning and image checking methods were used. The proposed method used third party, hyperlink and URL obfuscation based features.

The "Cantina+" approach was put forward in [32] which was an extension of "Cantina" [33]. It used an additional features for the detection of phishing webpages. A total of eight features were used in Cantina+ method. In this work, document object model, search engines and machine learning techniques were used for the classification of webpages. In [34], the authors presented a feature selection and machine learning based study to improve the efficiency of phishing detection. A dataset consisting of 10,000 phishing webpages (5000 benign and 5000 phishing) with 48 features was used. The experimental results revealed that an RF with 20 features provided the best classification accuracy. In [35], the authors proposed a deep learning based approach for the detection and classification of phishing webpages. A Convolution Neural Networks (CNN) model was used for this purpose and a dataset of 4898 phishing and 6157 genuine websites were used for evaluating the proposed approach. The experimental results were compared with the traditional machine learning methods to claim that their approach performed better. A nonlinear regression based method for detection and classification of a phishing webpages was proposed in [36]. They used a Support Vector Machine (SVM) meta-heuristic algorithm and harmony search for training the model. It was concluded that the harmony search method provides better accuracy results. In [37], the authors proposed a URL lexical analysis based phishing detection method. The proposed approach was tested on a dataset of 1,056,937 instances (both phishing and legitimate) containing 22 features. The experimental results showed that SVM provided the highest accuracy of 99.89%. Jain et al. [38] proposed a search engine based method to match the domain name of the site under consideration with the sites that appear as a search query. They first examined the performance of traditional "Term Frequency-Inverse Document Frequency" (TF-IDF) method on the classification of websites as phishing or legitimate. Afterward, they assigned weights to different tag data and modified the TF-IDF score to improve the performance of the proposed method. Experiment results show that the proposed search-based method with weighted TF-IDF outperforms conventional TF-IDF.

Reference [39] proposed a hybrid ensemble feature selection method for the detection of phishing webpages. In the first step, they used a Cumulative Distribution Function (CDF) gradient algorithm to generate a subset of primary features, which are then fed into a data perturbation ensemble to generate secondary feature subsets. The experimental results revealed that the proposed approach is highly advantageous for the detection of phishing webpages using machine learning techniques. A comparison of some of the phishing detection systems is given in Table 1.

From the literature, it is found that sufficient work is done on the classification and detection of phishing pages using machine learning algorithms. However, very less focus is given to the exploration of ensemble methods. This chapter presents a study to compare the traditional machine learning techniques and ensemble methods for the classification and detection of phishing pages.

3 Methodology Used

This section presents the methodology used for the classification and detection of phishing webpages. After acquiring the suitable dataset, five machine learning models named as Naïve Bayes (NB), logistic regression (LR), k-nearest neighbor (k-NN), decision table (DT), RF, and four ensemble methods, i.e., bagging, boosting, stacking, and voting are employed to classify phishing webpages. LR is employed as a meta-classifier in case of the stacking ensemble method. The performance of all these methods is measured using different evaluation metrics and a comparative analysis is carried out. Figure 1 demonstrates the workflow of the methodology used, and a detailed description of all these steps is given below:

3.1 Data Acquisition

The dataset used in this work is acquired from [40, 41]. The reason for using this dataset is that it is the most recently updated and is available freely in the data repository. It contains a total of 11,055 instances including 4898 phishing and 6157 legitimate webpages. Every instance of the dataset has an index value, 30 features, and a class label. After pre-processing the data, five base classifiers (discussed in the sub-section 3.2) are employed.

Table 1 Comparison of phishing detection systems

Technique used	Features used	Dataset size	Accuracy	Ref.
Machine Learning (RF, SVM, Adaboost, k-NN, NB, DT)	20 features (webpage, URL and HTML based)	5223 instances (Phishing: 2500 Benign: 2723)	99.5% using RF	Gandotra and Gupta [22]
Machine Learning (IG for feature selection. RF, IB1, J48, DT, Bagging, Adaboost, NB)	15 features (like SFH, Web_Traffic, oogle_index, Age of domain, page rank etc.)	11,055 instances (Phishing: 4898 Benign: 6157)	96.5% using RF	Gandotra and Gupta [23]
Machine learning (RF, SVM, NB, DT)	Word Vectors, NLP-based and hybrid features	73,575 instances (Phishing: 36,400 Benign: 37,175)	97.98%	Sahingoz et al. [24]
Machine learning	278 features (238 word and 40 NLP)	7357 instances (Phishing:3717 Benign: 3640)	89.9%	Buber et al. [30]
Machine Learning (PCA along with RF, J48, SMO, SVM, LR)	16 features (hyperlink, third-party URL features)	3526 instances (Phishing: 1407 Benign: 2119)	99.5%	Rao and Pais [31]
Content and Machine Learning	15 features (6 URL, 4 HTML, 5 webpage)	13,001 instances (Phishing: 8118 Benign: 4883)	92%	Xiang et al. [32]
Traditional machine learning and Convolutional neural network	30 features (SFH, web traffic, page rank, Google index, Age of domain etc.)	11,055 instances Phishing: 4898 Benign: 6157)	98.2%	Yerima et al. [35]
Meta-heuristics (HS, SVM) feature selection methods: decision tree and wrapper	20 features (Having IP address, URL length, Shortening Service etc.)	11,055 instances	92.80%	Babagoli et al. [36]
Machine learning (RF, Gradient Boosting, ANN, and SVM)	22 features (URL length, dot number, host name, URL tokens count)	10,56,937 instances	99.89% using SVM	Abutaha et al. [37]
Content based and Modified TF-IDF	Domain name, keywords (comprising of title, body text, meta data etc.)	–	89%	Jain et al. [38]
Machine Learning (Cumulative distribution function gradient, RF, SVM, NB, C4.5, JRip, and PART)	–	11,055 instances	94.6%	Chiew et al. [39]

IG Information Gain, *PCA* Principal Component Analysis, *ANN* Artificial Neural Network, *SVM* Support Vector Machine

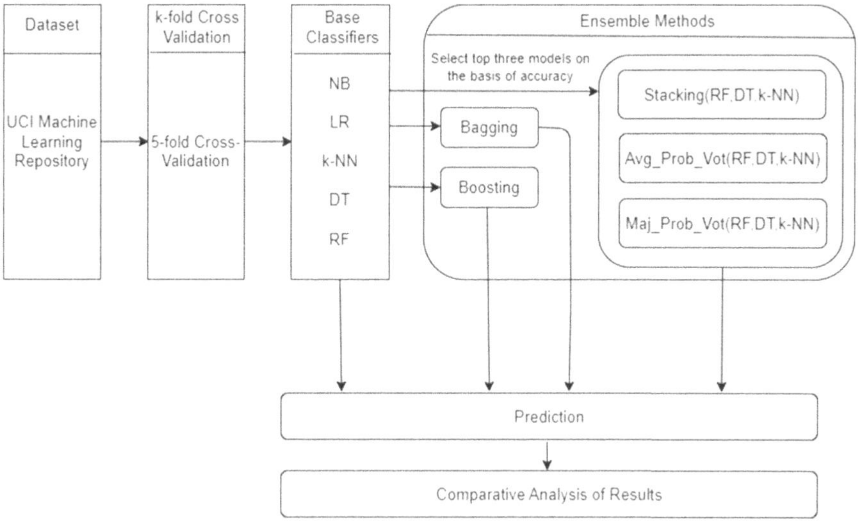

Fig. 1 Workflow of methodology used

3.2 Base Classifiers Used

A brief description of the base classifiers used in this work is as follow:

- **Naïve Bayes**
  This machine learning method is based on Bayes' theorem [42]. It assumes that all the features being used for classification are independent of each other. It is one of the most effective algorithms to solve classification problems. It helps in building the model fast. For a dataset with a feature vector $f = (f_1, f_2, \ldots\ldots\ldots f_n)$ the posterior probability of class *C* can be computed using Eq. (1).

$$P\left(C|f\right) = \frac{P\left(f|C\right)P(C)}{P(f)} \tag{1}$$

Where $P(C|f)$ is the probability of class *C* given predictor attribute *f*. $P(f)$ is the prior probability of the predictor attribute *f* and $P(f|C)$ is the probability of *f* given class *C*.

- **Logistic Regression**
  It is a machine learning algorithm used to solve classification problems and is based on the concept of probability. Sigmoid function (also known as logistic function) is used as an activation function for LR. It maps the predicted (any real) values to probabilities (values lying between 0 and 1). Equation (2) is used to describe the logistic function.

$$f(x) = \frac{1}{1 + e^{-x}} \tag{2}$$

where x is the real numeric value to be transformed to probabilistic value.

- **K-Nearest Neighbor**
  It is a lazy machine learning algorithm that makes use of k closest training data instances for classification purposes. The nearest k neighbors are chosen using distance functions like Euclidean and Manhattan, etc. In this work, we have considered the distance function as Euclidian with a value of k as 5. The Euclidian distance between $P1(x, y)$ and $P2(a, b)$ points is calculated using Eq. (3).

$$Dist\left((x, y), (a, b)\right) = \sqrt{(x - a)^2 + (y - b)^2} \tag{3}$$

- **Decision Table**
  DT [42] makes use of some specific features for performing the training. Cross validation performance is done for different sub-set of features to select the best set of features. We have used the best-first search algorithm to search the feature space.

- **Random Forest**
  RF [42] is an ensemble machine learning method that considers the aggregated result of various decision trees. Several decision trees are trained independently by considering the random samples of the original dataset and their average result is used to improve the predictive accuracy. The large number of decision trees in RF helps to improve the classification results and prevent the problem of overfitting.

3.3 Ensemble Methods Used

Ensemble learning methods combine basic machine learning techniques to improve the prediction and classification results. Different types of ensemble learning techniques are discussed in this sub-section.

- **Bagging**
  It is also known as bootstrap aggregation. This technique is useful for both regression and classification. It helps in stabilizing the models by improving their accuracy and reducing overfitting. Multiple models are created using a sub-dataset of training instances using random sampling with replacement. The results of these multiple models are then combined together using majority voting [43].

- **Boosting**
  It is an ensemble learning technique that combines a set of weak learners to create a strong learner. It minimizes training errors and improves prediction accuracy. In this method, a model is trained using a training dataset sequentially and every time it compensates the weaknesses of its predecessor. It selects the misclassified instances and modifies their weights. The weighted mean obtained from all weak learners is used for the final prediction. There are three different types of boosting algorithms, namely, AdaBoost, Gradient Tree Boosting and XGBoost [44].
- **Stacking**
  Unlike bagging and boosting ensemble techniques which use homogenous weak learners, stacking makes use of heterogeneous learners. The process of stacking gets completed in two steps. In the first step, heterogeneous weak learners are trained in parallel, and in the second step, their prediction results are combined together by training a meta-learner to produce the final prediction.

3.4 Validation and Evaluation Parameters Used

In this work, a five-fold cross-validation method is used for performing experiments. This method divides the original dataset into five equal sub-datasets. Four sub-datasets are used for training the model and the remaining one is used for testing. This process is repeated five times. Average results are generated by considering the different combinations of training and testing sub-datasets. False positive rate (FPR), true positive rate (TPR), Precision, Accuracy, and F-Measure are used to compare the performance of different machine learning techniques [42]. Their values are computed using True Negative (TN), True Positive (TP), False Negative (FN), and False Positive (FP) fields of confusion matrix. TN denotes the number of benign webpages predicted correctly as benign. TP denotes the number of phishing webpages predicted correctly as phishing. FN is the number of phishing webpages predicted incorrectly as benign, and FP is the number of benign webpages predicted incorrectly as phishing. The different evaluation metrics used in this work are described as follows:

- **FPR:** The rate of incorrectly recognized benign webpages is called FPR. It is computed using Eq. (4).

$$FPR = \frac{FP}{FP + TN} \tag{4}$$

- **TPR:** The rate of correctly recognized phishing webpages is known as TPR (also known as recall). It is computed using Eq. (5).

$$TPR = \frac{TP}{TP + FN} \tag{5}$$

- **Precision:** It is the degree of exactness and is computed using Eq. (6).

$$Precision = \frac{TP}{TP + FP} \tag{6}$$

- **Accuracy (%):** It is defined as the percentage of correctly identified benign and phishing webpages and is computed using Eq. (7).

$$Accuracy\ (\%) = \frac{TP + TN}{TP + FN + TN + FP} \times 100 \tag{7}$$

- **F-Measure:** The harmonic mean of recall and precision is known as F-measure. It is computed using Eq. (8).

$$F\text{-}Measure = \frac{2 \times Precision \times Recall}{Precision + Recall} = \frac{2 \times TP}{2 \times TP + FN + FP} \tag{8}$$

4 Experimental Results

This section presents the experimental results and carries out a comparative analysis of results obtained. All five base classifiers, their bagging and boosting version and stacking models are built using five-fold cross-validation approach and these are compared using the various evaluation parameters like FPR, TPR, Precision, Accuracy, and F-measure.

Table 2 presents the experimental results for the classification and detection of phishing webpages for five base classifiers. It shows that RF provides the best classification results with an accuracy of 96.924% followed by k-NN which provides an accuracy of 95.785%. The value of TPR for RF and k-NN are 0.969 and 0.958 respectively. It is also found that the NB model based on probabilistic approach provides a minimum accuracy of 90.701%. Figure 2 visualizes the comparison of these base classifiers based on Accuracy and F-measure. It shows that RF classifier suits best (among five base classifiers) to our dataset.

Tables 3 and 4 present the classification results of five base models using bagging and boosting ensemble methods. Table 3 shows that NB, LR, k-NN, DT, and RF with bagging (named as NB_Bagging, LR_Bagging, k-NN_Bagging, DT_Bagging, and RF_Bagging) provide almost similar results as that of base models. DT with bagging

Table 2 Classification results for base classifiers

Classifier	TPR	FPR	Accuracy (%)	Precision	F-measure
NB	0.907	0.098	90.701	0.907	0.907
LR	0.927	0.077	92.745	0.927	0.927
k-NN	0.958	0.044	95.785	0.958	0.958
DT	0.934	0.074	93.405	0.935	0.934
RF	0.969	0.033	96.924	0.969	0.969

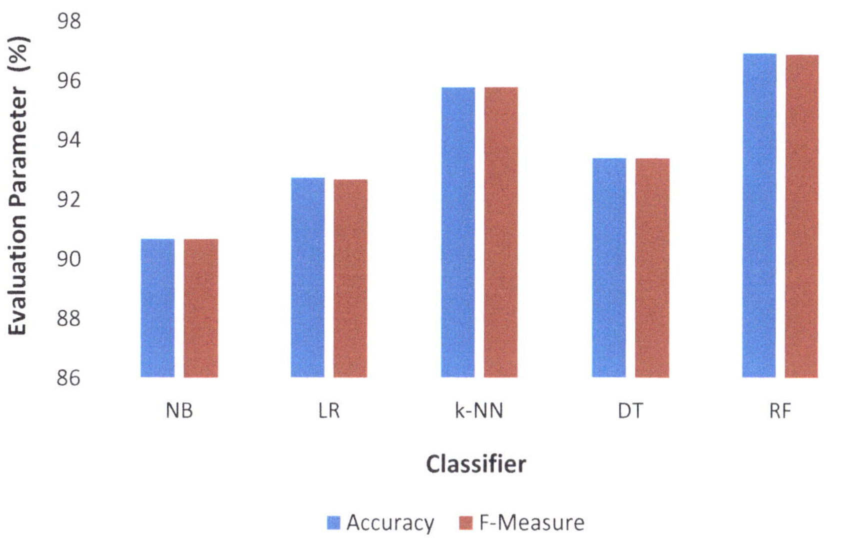

Fig. 2 Comparison of base classifiers using Accuracy and F-Measure

Table 3 Classification results using bagging ensemble technique

Classifier	TPR	FPR	Accuracy (%)	Precision	F-measure
NB_Bagging	0.907	0.098	90.719	0.907	0.907
LR_Bagging	0.927	0.078	92.673	0.927	0.927
k-NN_Bagging	0.955	0.046	95.550	0.955	0.955
DT_Bagging	0.943	0.064	94.265	0.943	0.942
RF_Bagging	0.967	0.035	96.735	0.967	0.967

(DT_Bagging) shows an improvement in the classification results as compared to the basic DT. The values of TPR, FPR, and Accuracy for DT_Bagging are 0.943, 0.064 and 94.265% respectively whereas for base DT, these values are 0.934, 0.074, and 93.405% respectively. For all other classifiers with bagging, the values of TPR and FPR are almost similar to base classifiers.

Table 4 shows that NB, LR, k-NN, DT, and RF with boosting (named asNB_Boosting, LR_Boosting, k-NN_Boosting, DT_Boosting, and RF_Boosting)

Table 4 Classification results using boosting ensemble technique

Classifier	TPR	FPR	Accuracy (%)	Precision	F-measure
NB_Boosting	0.906	0.100	90.565	0.906	0.905
LR_Boosting	0.927	0.077	92.745	0.927	0.927
k-NN_Boosting	0.958	0.044	95.785	0.958	0.958
DT_Boosting	0.952	0.050	95.242	0.952	0.952
RF_Boosting	0.969	0.033	96.861	0.969	0.969

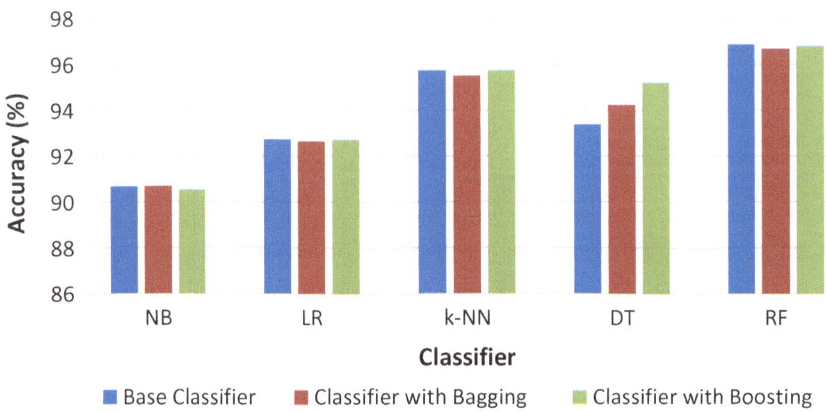

Fig. 3 Comparison of base classifiers with their bagging and boosting versions using Accuracy

and provide almost similar classification results as that of base models. DT with boosting (DT_Boosting) shows an improvement in the classification results as compared to the basic DT. The values of TPR, FPR, and Accuracy for DT_Boosting are 0.952, 0.050 and 95.242% respectively whereas, for base DT, these values are 0.934, 0.074, and 93.405% respectively. Further, it is found that DT_Boosting has performed better than DT_Bagging.

Figures 3 and 4 visualize the comparison of five base classifiers with their bagging and boosting versions using Accuracy and F-Measure respectively. Figure 3 shows that the maximum accuracy is given by RF and its boosting version followed by k-NN and its boosting version. Bagging and boosting versions of DT are performing better than base DT. Similar is the case for F-Measure (shown in Fig. 4). RF and its boosting version provide the highest F-Measure. The minimum F-Measure value is provided by the bagging version of NB.

Table 5 presents the webpage classification results for other ensemble methods, i.e., stacking, average probability voting (Avg_Prob_Vot), and majority probability voting (Maj_Prob_Vot). The base classifiers are ranked based on their accuracy to implement these ensemble methods.

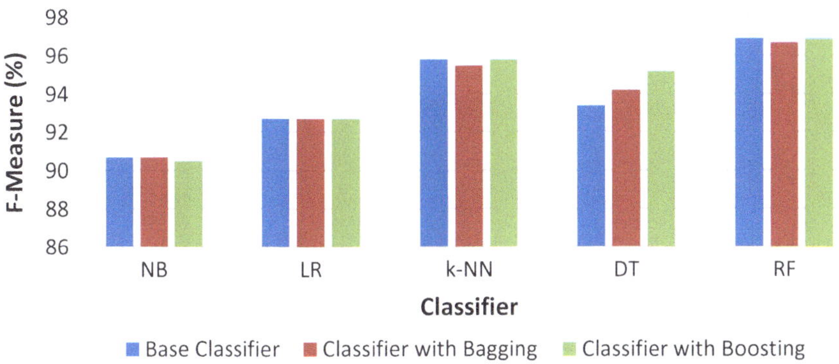

Fig. 4 Comparison of the base classifiers with their bagging and boosting versions using F-Measure

Table 5 Classification results using other ensemble techniques

Classifier	TPR	FPR	Accuracy (%)	Precision	F-measure
Stacking (RF,DT,k-NN)	0.97	0.032	96.988	0.97	0.97
Avg_Prob_Vot (RF,DT,k-NN)	0.969	0.033	96.870	0.969	0.969
Maj_Prob_Vot (RF,DT,k-NN)	0.968	0.035	96.753	0.968	0.967

In our results, the top three base models obtained are RF, DT, and k-NN. Three ensemble models, i.e., Stacking (RF,DT,k-NN), Avg_Prob_Vot (RF,DT,k-NN), and Maj_Prob_Vot (RF,DT,k-NN) are trained using these top three base models. In stacking ensemble technique, LR is used as a meta classifier. The results of Table 5 show that the stacking ensemble technique performs better than the base classifiers and other ensemble methods. The accuracy obtained by the stacking of RF, DT, and k-NN provides an accuracy of 96.988%. The accuracy provided by Avg_Prob_Vot (RF,DT,k-NN) and Maj_Prob_Vot (RF,DT,k-NN) is 96.87% and 96.753% respectively.

Figure 5 illustrates the comparison of top three base classifiers with the ensemble methods, i.e., Stacking (RF,DT,k-NN), Avg_Prob_Vot (RF,DT,k-NN), and Maj_Prob_Vot (RF,DT,k-NN) on the basis of Accuracy and F-Measure. It shows that the stacking ensemble method provides the maximum accuracy while classifying phishing webpages. Further, it is evident that the stacking ensemble model shows improvement over base classifiers as well as their bagging and boosting versions.

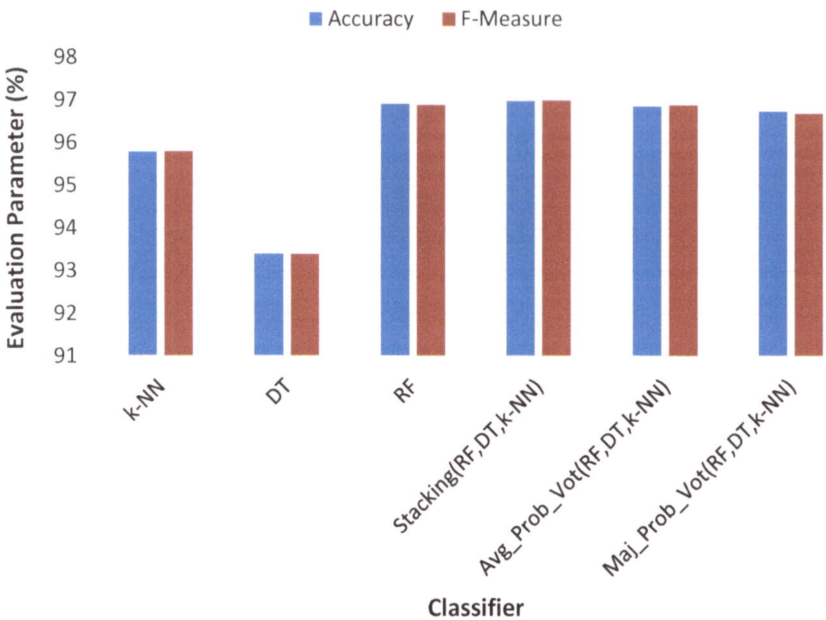

Fig. 5 Comparison of top three base classifiers and other ensemble techniques using Accuracy and F-Measure

5 Conclusion and Future Scope

Nowadays, the phishing attacks are becoming more complex and sophisticated as attackers are using novel techniques for launching the attacks. This chapter presented a comparative study of various machine and ensemble learning methods for classification and detection of phishing webpages. A comparative analysis of traditional machine learning algorithms and ensemble methods such as bagging, boosting, stacking and voting is carried out based on various evaluation metrics. A dataset consisting of 6157 benign and 4898 phishing webpages has been used to carry out the comparative analysis. Experimental results revealed that for the considered dataset, ensemble methods performed better than base classifiers. Further, it is concluded that stacking (of RF, DT, and k-NN) outperformed as compared to other ensemble methods and provided an accuracy of 96.988%.

In future, we intend to use deep learning methods and an extended set of features for improving the detection of phishing webpages. The experimental work may be extended to evaluate the performance of deep learning models. The study will be carried out on large datasets using big data tools and technologies.

References

1. Peng P, Xu C, Quinn L, Hu H, Viswanath B, Wang G (2019) What happens after you leak your password: Understanding credential sharing on phishing sites. In: Proceedings of the 2019 ACM Asia conference on computer and communications security, pp 181–192
2. Threat Analysis Group, Findings on COVID-19 and online security threats. Available online at: https://blog.google/technology/safety-security/threat-analysis-group/findings-covid-19-and-online-security-threats/
3. Phishing Activity Trends Report, APWG (2021) Available online at: https://docs.apwg.org/reports/apwg_trends_report_q2_2021.pdf
4. Security. Available online at: https://www.securitymagazine.com/articles/96430-mobile-phishing-threats-surged-161-in-2021
5. Gupta D, Rani R (2019) A study of big data evolution and research challenges. J Inf Sci 45:322–340
6. Gupta D, Rani R (2020) Improving malware detection using big data and ensemble learning. Comput Electr Eng 106729
7. Gupta D, Rani R (2018) Big Data Framework for Zero-Day Malware Detection. *Cybernetics and Systems 49*:103–121
8. Dhalaria M, Gandotra E (2021) A hybrid approach for android malware detection and family classification. Int J Interact Multimed AI:174–188
9. Gandotra E, Singla S, Bansal D, Sofat S (2018) Clustering morphed malware using opcode sequence pattern matching. *Recent Patents on Engineering*, 12:30–36
10. Gandotra E, Bansal D, Sofat S (2017) Malware threat assessment using fuzzy logic paradigm. Cybern Syst 48:29–48
11. Sharma A, Gandotra E, Bansal D, Gupta D (2019) Malware capability assessment using fuzzy logic. *Cybernetics and Systems 50*: 323–338
12. Selenium (2021) Available online at: http://docs.seleniumhq.org/download/
13. Gandotra E, Bansal D, Sofat S (2016) Tools & techniques for malware analysis and classification. Int J Next-Gener Comput:176–197
14. Jsoup Java HTML Parser, with best of DOM, CSS, and jquery. Available online at: https://jsoup.org/
15. GSB. Available online at: http://code.google.com/apis/safebrowsing/
16. Phishtank. Available online at: http://www.phishtank.com/
17. Gandotra E, Bansal D, Sofat S (2015) Computational techniques for predicting cyber threats. In: Intelligent computing, communication and devices. Springer, New Delhi, pp 247–253
18. Gandotra E, Bansal D, Sofat S (2014) Malware analysis and classification: a survey. J Inf Secur 05:56–65
19. Tan CL, Chiew KL, Wong K (2016) PhishWHO: Phishing webpage detection via identity keywords extraction and target domain name finder. Decis Support Syst 88:18–27
20. Chiew KL, Chang EH, Tiong WK (2015) Utilisation of website logo for phishing detection. Comput Secur 54:16–26
21. Jain AK, Gupta BB (2018) Towards detection of phishing websites on client-side using machine learning based approach. Telecommun Syst 68:687–700
22. Gandotra E, Gupta D (2020) Improving spoofed website detection using machine learning. Cybern Syst Int J 52(2):169–190
23. Gandotra E, Gupta D (2021) An efficient approach for phishing detection using machine learning. In: Giri KJ, Parah SA, Bashir R, Muhammad K (eds) Multimedia security: algorithm development, analysis and applications. Springer, Singapore, pp 239–253
24. Sahingoz OK, Buber E, Demir O, Diri B (2019) Machine learning based phishing detection from URLs. Expert Syst Appl 117:345–357
25. Afzal I, Parah SA, Hurrah NN, Song OY (2020) Secure patient data transmission on resource constrained platform. In: Multimedia tools and applications. Springer, pp 1–26

26. Hurrah NN, Parah SA, Sheikh JA, Al-Turjman F, Muhammad K (2019) Secure data transmission framework for confidentiality in IoTs. Ad Hoc Netw 101989:101989
27. Jan A, Parah SA, Malik BA (2022) IEFHAC: image encryption framework based on hessenberg transform and chaotic theory for smart health. Multimed Tools Appl 81:18829–18853
28. Parah SA, Kaw JA, Bellavista P, Loan NA, Bhat GM, Muhammad K, de Albuquerque VHC (2022) Efficient security and authentication for edge-based internet of medical things. IEEE Internet Things J 8:15652–15662
29. Sarosh P, Parah SA, Malik BA, Hijji M, Muhammad K (2022) Real-time medical data security solution for smart healthcare. In: IEEE transactions on industrial informatics, pp 1– 11. https://doi.org/10.1109/TII.2022.3217039
30. Buber E, Dırı B, Sahingoz OK (2017) Detecting phishing attacks from URL by using NLP techniques. In: 2017 international conference on computer science and engineering (UBMK). IEEE, pp 337–342
31. Rao RS, Pais AR (2019) Detection of phishing websites using an efficient feature-based machine learning framework. Neural Comput Appl 31:3851–3873
32. Xiang G, Hong J, Rose CP, Cranor L (2011) Cantina+ a feature-rich machine learning framework for detecting phishing web sites. In: ACM transactions on information and system security (TISSEC), pp 1–28
33. Zhang Y, Hong JI, Cranor LF (2007) Cantina: a content-based approach to detecting phishing web sites. In: Proceedings of the 16th international conference on World Wide Web. ACM, pp 639–648
34. Almseidin M, Zuraiq AA, Al-kasassbeh M, Alnidami N (2019) Phishing detection based on machine learning and feature selection methods. Int J Interact Mob Technol:171–183
35. Yerima SY, Alzaylaee MK (2020) High accuracy phishing detection based on convolutional neural networks. arXiv preprint arXiv:2004.03960
36. Babagoli M, Aghababa MP, Solouk V (2019) Heuristic nonlinear regression strategy for detecting phishing websites. Soft Comput 23:4315–4327
37. Abutaha M, Ababneh M, Mahmoud K, Baddar SAH (2021) URL phishing detection using machine learning techniques based on URLs lexical analysis. In: 2021 12th international conference on information and communication systems (ICICS), IEEE, pp 147–152
38. Jain AK, Parashar S, Katare P, Sharma I (2020) Phishskape: a content based approach to escape phishing attacks. Procedia Comput Sci 171:1102–1109
39. Chiew KL, Tan CL, Wong K, Yong KS, Tiong WK (2019) A new hybrid ensemble feature selection framework for machine learning-based phishing detection system. Inf Sci 484:153–166
40. UCI Machine Learning Repository, Phishing Websites Dataset. Available online at: https://archive.ics.uci.edu/ml/datasets/phishing+websites
41. Mohammad RM, Thabtah F, Mc Cluskey L (2012) An assessment of features related to phishing websites using an automated technique. In: 2012 international conference for internet technology and secured transactions. IEEE, pp 492–497
42. Witten IH, Frank E (2002) Data mining: practical machine learning tools and techniques with Java implementations. ACM SIGMOD Rec 31:76–77
43. Leo B (1996) Bagging predictors. Mach Learn:123–140
44. Quinlan JR (1996) Bagging, boosting, and C4.5. In: AAAI/IAAI, pp 725–730

Part II
Intelligent Signal Processing for Smart Health

Machine Learning Based Diabetic Retinopathy Detection and Classification

Fei Long, Jun Sang, and Mohammad S. Alam

1 Introduction

In this section, at first, the diagnosis principles of diabetic retinopathy are introduced. Since the diagnosis of diabetic retinopathy is usually based on the fundus structures, the fundus imaging methods are discussed. Finally, the need of applying intelligent multimedia signal processing for diabetic retinopathy diagnosis in smart ecosystems is presented.

1.1 Principles and Characteristics of Diabetic Retinopathy

According to the Lancet Commission on diabetes [1], approximately 463 million people worldwide have diabetes, and a third of them suffer from vision impairments. As a microvascular complication of diabetes, diabetic retinopathy accounts for most of those with ocular disease [2, 3]. It is caused by morphological alterations such as pericyte deformation, which leads to the dilatation of capillary walls. As the disease progresses, the retinal capillary wall or microaneurysm becomes vulnerable to rupture, which can lead to intraretinal hemorrhage. Diabetic

F. Long · J. Sang
School of Big Data & Software Engineering, Chongqing University, Chongqing, China

M. S. Alam (✉)
Department of Electrical Engineering and Computer Science, Texas A&M University – Kingsville, Kingsville, TX, USA
e-mail: Mohammad.alam@tamuk.edu

S. A. Parah et al. (eds.), *Intelligent Multimedia Signal Processing for Smart Ecosystems*, https://doi.org/10.1007/978-3-031-34873-0_5

Table 1 Diagnosis criteria in ICDRDSS

Disease	Diagnosis criteria
Without diabetic retinopathy	Do not have any abnormalities
Mild diabetic retinopathy	Only microaneurysms exist
Moderate diabetic retinopathy	It is more severe than microaneurysms but less than non-proliferative diabetic retinopathy
Severe diabetic retinopathy	More than 20 intraretinal haemorrhages may exist in each of four quadrants, or definite venous beading may exist in two or more quadrants, or prominent intraretinal microvascular abnormality may exist in one or more quadrants; however no signs of proliferative retinopathy exists
Proliferate diabetic retinopathy	Neovascularization or/and vitreous/preretinal haemorrhage are present

retinopathy can be divided into two categories: non-proliferative and proliferative. The non-proliferative category includes mild, moderate, and severe, while proliferative diabetic retinopathy can be divided into high-risk and non-high-risk categories. Other features of non-proliferative diabetic retinopathy include soft or hard exudates and microvascular abnormalities within the retina. Non-proliferative diabetic retinopathy is relatively mild and can be asymptomatic. However, if the disease is left untreated, it can progress to proliferative diabetic retinopathy, where neovascularization may occur, primarily due to ischemia and oxidative stress [4]. At this stage, the patient's vision can deteriorate rapidly, eventually leading to complete blindness. The diagnostic criteria for each stage can be found in the International Clinical Diabetic Retinopathy Disease Severity Scale (ICDRDSS) [5], which is listed from low severity to high severity in Table 1.

If a patient contains two or more of the characteristics of severe diabetic retinopathy, he or she may be considered to have highly severe non-proliferative diabetic retinopathy [6].

Therefore, early diagnosis and treatment of diabetic retinopathy are necessary to prevent the progression of diabetic retinopathy to the point where it has irreversible effects on the patient's vision [7]. The diagnosis of diabetic retinopathy currently relies on a comprehensive analysis of fundus images by experienced physicians.

1.2 Fundus Imaging Methods

The diagnosis of diabetic retinopathy is mainly based on the structure and characteristics of the fundus, which requires imaging methods to capture the fundus images. This section introduces a variety of fundus imaging methods, including fundus images, fundus fluoroscopy angiography, and optical coherence tomography angiography.

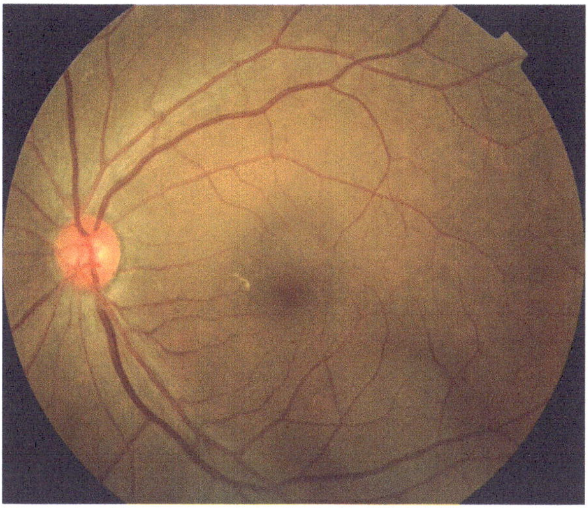

Fig. 1 Fundus image

Fundus Images

Fundus images can assist in the diagnosis of diabetic retinopathy [8]. A dedicated fundus camera is used to capture eye fundus images to provide clear and accurate pictures of the fundus. The imaging system captures various parts of the fundus and combines them into a complete fundus image. With the development of technology, the imaging technology of fundus images has also been improved. Recently, a stereoscopic fundus imaging technology has been developed, which captures features containing depth information through two shots at different angles, and it can assist the computer to perform more accurate feature recognition. Furthermore, the confocal laser scanning technology can capture a wider fundus image, which generally contains abundant edge features. A sample fundus image is shown in Fig. 1.

Fundus Fluorescein Angiography

Fundus fluorescein angiography (FFA) [9] is a method of detecting diabetic retinal microvascular changes using fluorescein sodium imaging. When compared with the diagnosis of diabetic retinopathy relying only on fundus images, it can more clearly detect microaneurysms and neovascularization. Therefore, it can be used to confirm the grading of diabetic retinopathy clinically. However, due to the medicines which may generate adverse reactions and cause specific damage to renal function, it cannot be used for specific groups of people. In particular, fundus fluoroscopy can only generate two-dimensional images, so capillaries of different layers and

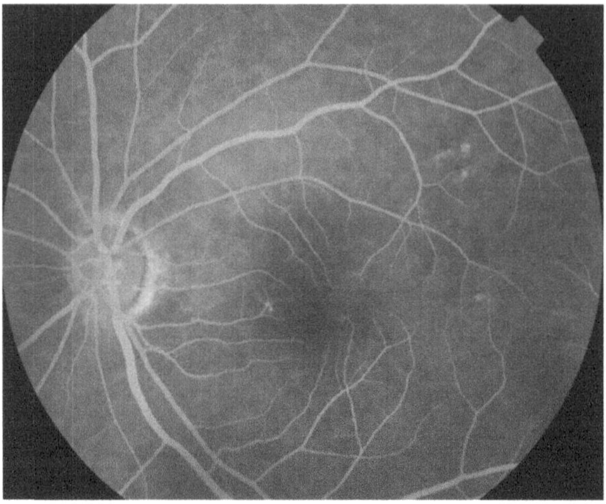

Fig. 2 Fundus fluorescein angiography image

depths may be superimposed on the two-dimensional image, causing interference and making it challenging to observe abnormal blood vessels. A sample of FFA image is shown in Fig. 2.

Optical Coherence Tomography Angiography

Optical coherence tomography (OCT) angiography [10] is a fast and safe method to assess retinal microvasculature. Because of its non-invasive, three-dimensional display and high resolution, OCT angiography can be used to examine fundus diseases. OCT angiography can find blood flow changes in fundus diseases, and can clearly display retinal and choriocapillary blood flow in normal fundus. It can assist physicians to diagnose age-related macular degeneration, diabetic retinopathy, retinal vascular occlusion, and other diseases. A sample of OCT angiography image from Bélanger Nzakimuena's work [11] is shown in Fig. 3.

1.3 Intelligent Multimedia Signal Processing for Diabetic Retinopathy Diagnosis in Smart Ecosystems

A smart ecosystems may incorporate a set of intelligent platforms to help doctors successfully achieve digital transformation. With the development of smart ecosystems, intelligent multimedia signal processing methods are widely used to obtain better results [12–15]. For diabetic retinopathy, specialists can also design smart

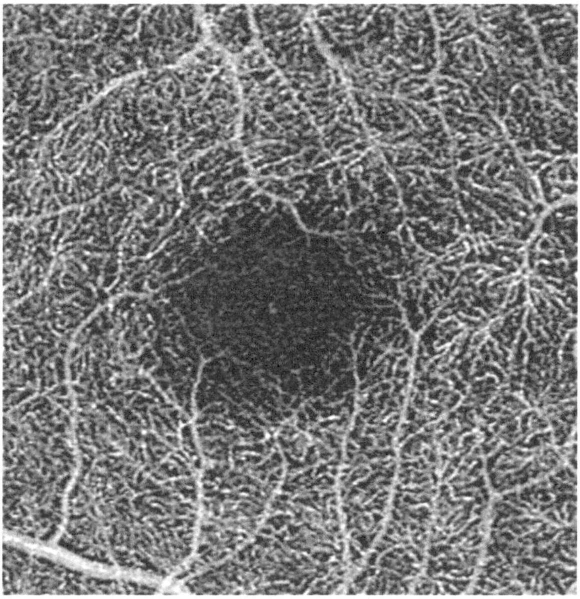

Fig. 3 Optical coherence tomography angiography image [11]

ecosystem with intelligent multimedia signal processing for improved diagnostic purposes. For example, an ophthalmologist can use mobile devices such as a smart phone to capture fundus images [16] and a smart ecosystem can be established for improved diabetic retinopathy diagnosis. The typical workflow of an intelligent diabetic retinopathy diagnosis system is shown in Fig. 4. In general, a physician uses the fundus camera or smart phone (Fig. 4a) to capture the fundus image (Fig. 4b). The captured image is then used in the machine learning algorithm to complete various tasks (Fig. 4c), such as microaneurysms segmentation, haemorrhages segmentation, exudates segmentation, and optic disk segmentation. To satisfy emerging needs, specialists also proposed the next generation diabetic retinopathy screening device with associated algorithms. Pertinent to mention that smart health care scenarios also need to be secure and computationally efficient [17, 18].

The rest of this chapter is organized as follows: Section 2 introduces the machine learning methods commonly used in the diagnosis of diabetic retinopathy. Section 3 presents the existing fundus image datasets for diabetic retinopathy. In Sect. 4, we introduce the machine learning based diabetic retinopathy diagnosis methods, including detection methods, classification methods, and segmentation methods. Section 5 introduces applications involving real-life clinical scenarios. Section 6 discusses future directions.

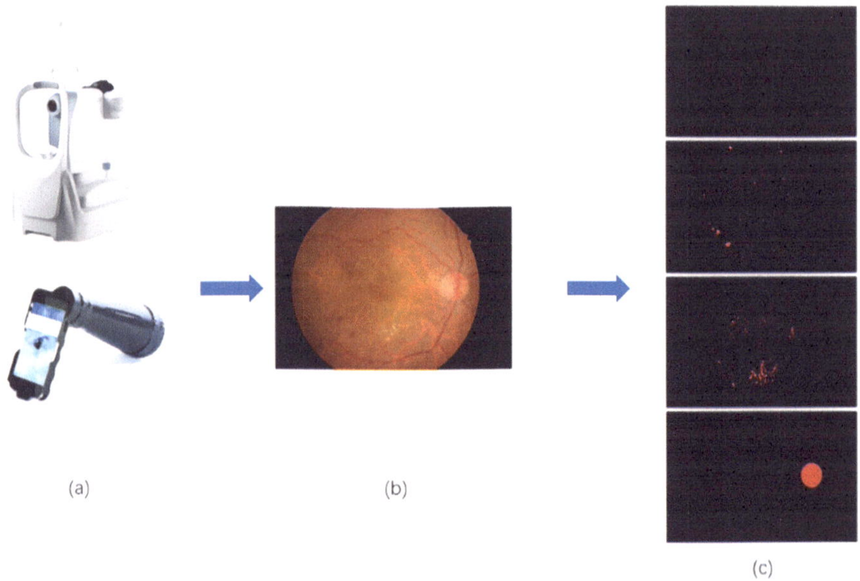

Fig. 4 Workflow of a typical diabetic retinopathy diagnosis system

2 Overview of Commonly Used Machine Learning Methods for Diabetic Retinopathy Diagnosis

This section introduces the machine learning methods commonly used in diabetic retinopathy diagnosis.

2.1 Traditional Methods

Support Vector Machine

Support Vector Machine (SVM) is an attractive technique for traditional machine learning. It is a linear classifier for binary classification of samples and it is extensively used in pattern recognition applications [19]. In SVM, the learning strategy is to maximize the interval and the purpose is to finally transform the task into a convex quadratic programming problem.

Autoencoder

Autoencoder [20] is a widely used method for weakly supervised learning and unsupervised learning environments. Its purpose is to perform efficient and accurate feature representation on the input data. It consists of two parts - an encoder and a decoder. The encoder part reduces the dimension of the data into low-dimensional hidden variables, and the decoder part reconstructs the data from the encoded variable. The goal is to ensure that the output data of the decoder is closer to the original input data, i.e., the distance metric between the two is small or negligible. It is widely used in the field of computer vision.

K-Means Clustering

The K-means [21] algorithm is an unsupervised clustering algorithm. For a given sample set, the data set is divided into K clusters according to the distance among the samples. Assume that the points within the cluster are as close as possible, and the distance among the clusters is as far as possible. In the training process, it selects the initial centers of K clusters, calculates the distance from any sample point to the K cluster centers, and classifies the sample points into the cluster with the smallest distance center, and so on by using an iterative process. In each iteration, the center point (centroid) of each cluster is updated by using parameters such as the mean value. For K cluster centers, after 2 or 3 steps of iterative updating, if the change of the position point is small based on a prescribed threshold, then it is inferred that a stable state is reached and the iteration is over. The principle of K-means clustering is simple and the interpretability is strong, but the selection of K is not easy to grasp, and the final result is closely related to the initial point selection.

2.2 Deep Learning Methods

AlexNet

AlexNet [22] is a deep learning technique, which achieved a first-class error rate of 37.5% and a top-five-class error rate of 17% on ImageNet LSVRC-2010 dataset. The Convolutional Neural Network (CNN) generally uses Sigmoid as the network activation function, while AlexNet uses the Rectified Linear Unit (ReLU) function as its activation function, and solves the gradient dispersion problem of the previous neural network activation function when it encounters a deep model. The model uses data augmentation to process the image, which significantly improves the generalization ability. Also, it uses overlapping max-pooling layers, which significantly improves the richness of features. In addition, it uses Compute Unified Device Architecture (CUDA) to accelerate the speed of deep learning, thus, improving the usefulness of CNN for computer vision applications. It is considered a landmark neural network model when applied to wider and deeper network structures. The architecture of AlexNet is shown in Fig. 5.

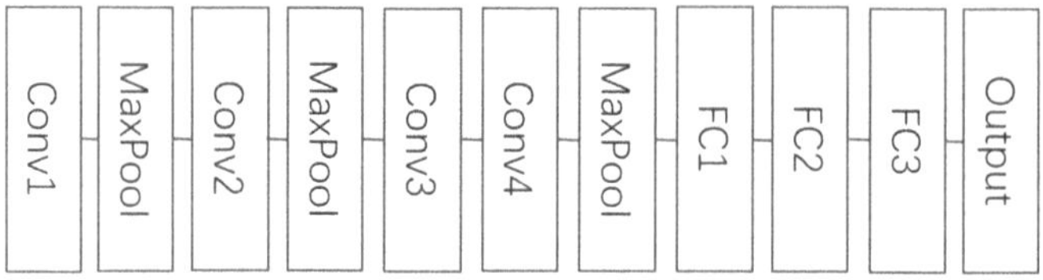

Fig. 5 Architecture of AlexNet. (Conv – Convolution Layer, MaxPool – Max pooling layer, FC – Fully connection layer)

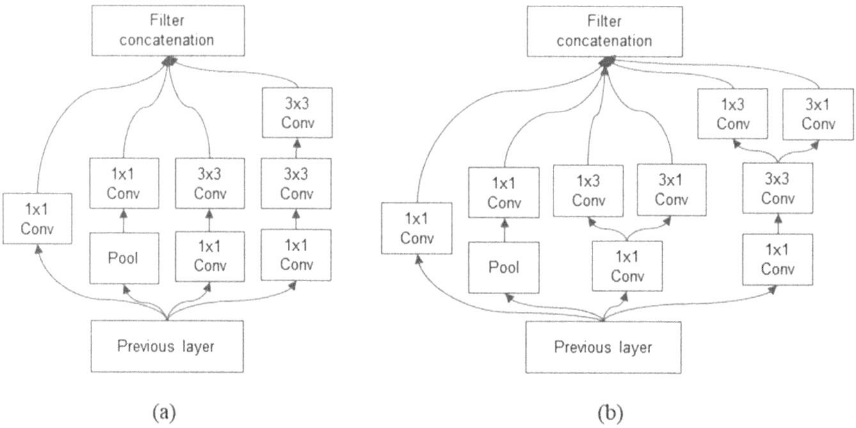

Fig. 6 Architecture of Inception block. (Conv – Convolution layer, Pool – Pooling layer)

Inception

Inception [23] is mainly dedicated to solving the problem of limited performance caused by the increase in depth of the deep neural network. It is also used to ascertain whether the accuracy of the neural network can be improved to the optimal level while keeping it stable. Under the framework of this concept, the Inception-V1 structure was proposed, which reduces the number of channels of the input image by 1×1 and the convolution layers with different convolution kernel sizes, so that features of different scales can be extracted. Finally, the features of different scales are superimposed to generate the final output, i.e., the Inception network (e.g., GoogleNet) is formed by stacking modules. With the development of Inception V1, enhanced structures such as Inception-V2 and Inception-V3 were proposed, which operate in parallel through the convolution layer and the pooling layer. These two layers are then connected, and the parameters required for calculation are reduced by decomposing the size of the convolution kernel while ensuring the extraction of features of different sizes. It is a classic model which yields reasonably good performance with high efficiency. The architecture of the Inception block is shown in Fig. 6.

MobileNet

The performance requirements of today's large-scale neural networks for computing devices gradually increases with the increase of parameters. Although large-scale deep neural networks provide good performance, for practical applications, many embedded devices or mobile devices are needed and it is a challenge to provide such huge computing resources. Therefore, many lightweight network structures are proposed, such as MobileNet [24], which makes a compromise among improving execution speed, reducing model size, and ensuring the accuracy of the task. The separable convolutional structure greatly reduces the amount of computation burden when compared with ordinary convolutional layers. The MobileNet-V2 [25] adds an Inverted Residual Block based on the stacking of convolution modules. This approach retains some low-dimensional modules that can be passed to deep layers, thereby improving the accuracy.

U-Net

The U-Net [26] was originally proposed to solve the task of medical image segmentation. In the beginning, it was designed to solve the problem of cell segmentation. It is one of the most commonly used and simplest deep segmentation models. Its model structure is based on an encoder-decoder structure to realize segmentation. Because the model structure is U-shaped, it is named U-Net. The encoder part of the model consists of two 3×3 convolutional layers (with ReLU) and a 2×2 maxpooling layer to form a down-sampling module. The decoder part of the model consists of an up-sampled convolutional layer i.e., deconvolutional layer with feature concatenation and two 3×3 convolutional layers with ReLU. The model also incorporates a skip-connection structure, which can preserve rich semantic information. The reason is that both low-level features and high-level semantic features are important in medical imaging, so the transfer and retention between semantic features is required. Currently, U-Net is widely used in medical image segmentation applications. The architecture of U-Net is shown in Fig. 7, which was illustrated by ML visuals project [27].

3 Diabetic Retinopathy Fundus Image Datasets

This section introduces common datasets used in machine learning-based diagnosis of diabetic retinopathy.

Kaggle APTOS dataset [28] contains 5590 images, where 3662 images are labeled. This dataset was proposed in APTOS 2019 Blindness Detection Competition. The competition aims to help local medical institutions gain better automated screening capabilities for diabetic retinopathy. This dataset has rich real-world data, but, it also has the quantitative imbalance between the different categories. Also, the

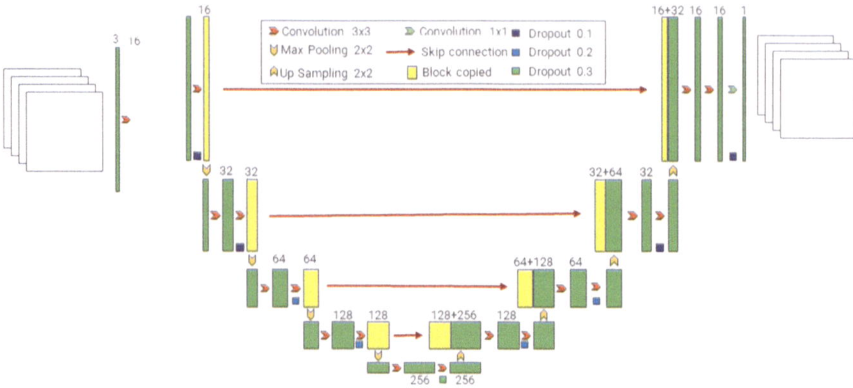

Fig. 7 Architecture of U-Net

quality of the images is not very stable due to the fact that the images are collected from multiple devices involving multiple institutions.

Messidor dataset [29] contains 1748 images collected from 4 ophthalmic medical institutions in France. Every image in this dataset is of exceptionally high quality, with no noise or blurring affecting the quality of the image. The resolution of each image is 2240×1488 pixels.

Kaggle EyePACS dataset [30] is the largest public dataset of diabetic retinopathy available in the open domain. The dataset has 88,702 images, and 35,126 images in this dataset have labels. The quality of images in this dataset varies as some images have noise and blur caused by out-of-focus recording thus affecting the overall image quality.

IDRiD dataset [31] was first proposed in the 2018 IDRiD Fundus Impact Analysis Challenge, which annotated the structures and lesion locations contained in the fundus images at the pixel level. In this dataset, the degree of retinopathy was classified. The dataset contains 516 images collected from an eye clinic and the resolution of each image is 4288×2848 pixels. In this dataset, 81 images are annotated with microaneurysms and hard exudates, 80 images are annotated with hemorrhages, and 40 images are annotated with soft exudates.

DRIVE dataset [32] is a dataset for blood vessel segmentation on fundus images. It contains 40 images, and the resolution of each image is 768×584, collected from patients between the ages of 25 and 90 in Dutch medical institutions. Among the patients, 33 patients had no features of diabetic retinopathy, and 7 patients showed mild diabetic retinopathy features on fundus images. This dataset can be used for fundus vessel segmentation in diabetic retinopathy.

ODIR-5k dataset [33] is a fundus image database containing a variety of ophthalmic diseases involving 5000 images. This dataset contains 8 categories, which are normal, diabetes, glaucoma, cataract, age-related macular degeneration, hypertension, myopia, and other disease abnormalities. This dataset was designed to represent a real patient information set collected by medical companies from different institutions in China.

Chase_DB1 dataset [34] is a small-scale fundus segmentation dataset, proposed by Kinston University, which includes 28 fundus photos, where each photo corresponds to two retinal vessel segmentation images. It is used to identify and automatically classify key retinal features, and to analyze the relationship between retinal vascular morphology and retinal and systemic diseases for the timely detection and treatment of diseases in their early stages.

DIARETDB0/DIARETDB1 are two datasets used for training and evaluating diabetic retinopathy diagnostic models [35, 36]. The DIARETDB0 dataset contains 130 fundus images and the resolution of each image is 1500 × 1152 pixels. Among the 130 fundus images, 110 images are with diabetic retinopathy and the remaining 20 are normal images. The lesion pictures contain at least one microaneurysm, hemorrhage point, hard exudate, soft exudate, and new blood vessels. The DIARETDB1 consists of fundus images with labels manually annotated by experts. The label contains exudates, microaneurysms, and other lesions. DIARETDB1 has 89 images and the resolution of each image is the same as that in DIARETDB0.

STARE dataset [37] was initiated by Goldbaum in 1975. It contains 400 raw fundus images. In particular, it contains 40 hand labeled images for blood vessel segmentation, and 10 images with artery/vein labeling.

4 Machine Learning-Based Diabetic Retinopathy Diagnosis Methods

This section mainly introduces the diagnosis methods for diabetic retinopathy based on machine learning. For clinical diagnosis of diabetic retinopathy, at first, clinical features are detected. Then the preliminary diagnosis is performed according to the clinical features to classify diabetic retinopathy. After classification, basic structural segmentation of the fundus is needed for further diagnosis. To further explore the above-mentioned diagnosis process, this section introduces the detection methods for exudate and microaneurysm, and then analyses the classification methods. Finally, we discuss the segmentation methods of basic structures in the fundus.

4.1 Clinical Features Detection

Exudate Detection

Retinal exudate occurs at the early stage of diabetic retinopathy. In order to make timely diagnosis of diabetic retinopathy, it is necessary to perform exudation detection on fundus images. A sample image for exudate detection from e-Ophtha [38] dataset is shown in Fig. 8.

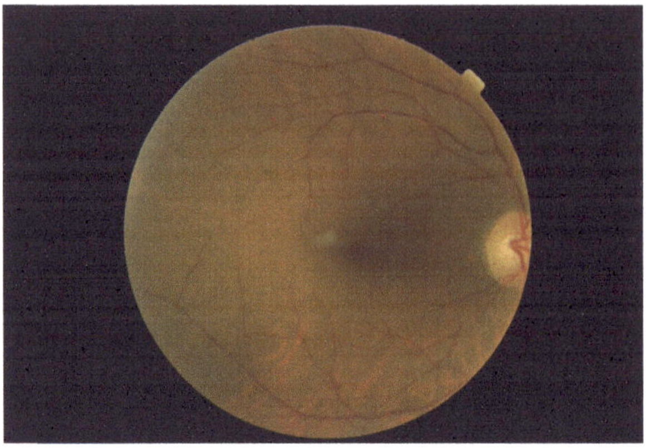

Fig. 8 A sample picture for exudate detection from e-Ophtha dataset [38]

For the traditional machine learning methods, Phillips et al. [39] used the noise removal thresholding method to detect exudates in images and achieved 87% sensitivity. Shilpa et al. [40] used an ensemble method to detect exudates. At first, the original image was converted into a grayscale image. Then, the blood vessel features were removed from the image with morphological opening and closing operations. Thereafter, the image was segemented into subimages by applying a prescribed threshold. Finally, exudates were detected with 100% sensitivity and 89.6% specificity on the DIARETDB1 dataset. Asha et al. [41] used an extreme learning machine to detect diabetic exudates, where the original image was segmented, and the feature values were extracted from the image via CIELUV color space. Then, the features were input into the extreme learning machine to detect exudates. This method achieved 90% accuracy, 100% sensitivity and 87% specificity on DIARETDB0 and DIARETDB1 datasets. Lin et al. [42] used the k-nearest neighbor (k-NN) graph algorithm to detect exudate locations with pixel location and color information. Their method achieved 91.08% sensitivity and 95.42% specificity on the STARE dataset.

For deep learning methods, Khojasteh et al. [43] used a variety of deep learning methods, including CNN, ResNet-50+k-NN, ResNet50+OPF, ResNet-50+SVM, and deep restricted Boltzmann machine (DRBM), to detect exudation. In the comparative experiments involving the above-mentioned methods, ResNet50+SVM achieved the best results with 98.2% accuracy, 99.0% sensitivity and 96% specificity on the DIARETDB1 dataset, while 97.6% accuracy, 98.0% sensitivity and 95.0% specificity on the e-Ophtha dataset. Adem [33] proposed a new approach by combining circular Hough transform and CNN to detect ocular exudates. After performing image augmentation, the circular Hough transform was applied to segment the optic disk so as not to affect the subsequent detection operations. Finally, CNN was used to detect the exudates. On the DIARETDB0 dataset, this

Table 2 Performance measures of exudate detection

Method	Dataset	Accuracy	Sensitivity	Specificity	AUC
Ensemble Method [40]	DIARETDB1	–	100.0%	89.6%	–
Extreme learning machine [41]	DIARETDB0 and DIARETDB1	–	87%	–	–
k-NN [42]	STARE	–	91.08%	95.42%	–
Combining [46]	DIARETDB0	–	100.0%	98.41%	–
Combining [46]	DIARETDB1	–	99.2%	97.97%	–
CNN [44]	DIARETDB1	–	–	–	0.96
CNN [44]	Messidor	–	–	–	0.91
CNN [45]	DRIDB	–	78%	–	–

method achieved 100% sensitivity and 98.41% specificity, and it achieved 99.2% sensitivity and 97.97% specificity on the DIARETDB1 dataset. Li et al. [44] used a CNN to detect exudates with a strategy to accelerate model training. The method achieved 0.96 area under curve (AUC) on the DIARETBDB1 dataset and 0.91 AUC on the Messidor dataset. Prentašić et al. [45] also proposed detecting exudates with CNN, which achieved 78.0% sensitivity on the DRIDB dataset. The performance measures of exudate detection are shown in Table 2.

Microaneurysm Detection

Microaneurysm is an early clinical symptom of diabetic retinopathy and the detection of microaneurysms is important for early diagnosis. Chudzik et al. [47] proposed a patch-based fully convolutional neural network (PB-CNN) with a batch normalization layer. Their technique utilized a dice loss function for precise diagnosis and achieved the highest sensitivity on three datasets. Yadav et al. [48] used a variety of methods to detect microaneurysms. At first, the retinal image was segmented according to the histogram. Then, the features were extracted from the image using the gray-level co-occurrence matrix. Finally, the extracted feature was classified by using a series of traditional machine learning methods, where random forest achieved the highest accuracy of 83.33%. Liao et al. [49] designed an encoder-decoder network to detect microaneurysms. This method detected microaneurysms by comparing differences between skip-connections in a network. Furthermore, they proposed a smooth dice loss to focus on the misclassified microaneurysms and achieved reasonable results on multiple datasets. Murugan et al. [50] used CNN and the majority voting technique to train microaneurysm detection and achieved 92% on the receiver operation characteristic curve (ROC) at the retinopathy online challenge [51]. A sample picture for microaneurysms detection is shown in Fig. 9.

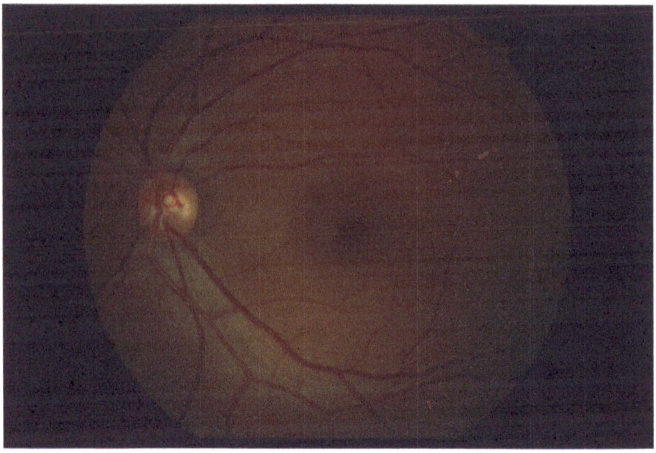

Fig. 9 Example picture for microaneurysms detection from e-Ophtha dataset [38]

4.2 Diabetic Retinopathy Classification

After detecting lesions, classification is applied to clarify the severity level. For traditional machine learning methods, Abramoff et al. [52] used the k-NN classifier to dichotomize diabetic retinopathy (normal, lesion) and achieved an AUC score of 0.839. Noronha et al. [53] used SVM to perform binary classification of the disease and obtained 99.1% accuracy. Acharya et al. [54–56] applied SVM to classify diabetic retinopathy. In [54], the authors used SVM to perform a four-classification diagnosis of the disease. Based on the experiments on a dataset with 238 images, their method achieved 98.9% sensitivity, 89.5% specificity, and 100% accuracy. In [55], they used higher-order spectra to extract non-linear features, then supplied these features into SVM for diabetic retinopathy diagnosis, and achieved 82% accuracy and 82.5% sensitivity for five-classification tasks. In [45], they provided morphological features extracted by morphological image-processing techniques which was fed into an SVM classifier, and achieved 85.9% accuracy, 82% sensitivity and 86% specificity. The dichotomous diagnosis of diabetic retinopathy using Fuzzy C-Means (FCM) was performed by Oasreh et al. [57]. This method pinpointed the exudative lesions and the optic disc which achieved 90.1% dichotomous accuracy on 142 images. Dupas et al. [58] used the k-NN classifier for binary diagnosis of 761 images, which could effectively detect the location of microaneurysms and achieved 83.9% sensitivity and 72.7% specificity. Quellec et al. [59] used the k-NN method for binary diagnosis of diabetic retinopathy. They used this model to detect drusen on a small dataset with 67 images and achieved an AUC score of 0.927. Philip et al. [60] used the Wilson score and kappa statistic methods to classify 527 images, which achieved 99.1% accuracy.

Besides the aforementioned traditional machine learning techniques, neural network based techniques for diabetic retinopathy classification are also reported in the literature. Gardner et al. [61] used neural networks to perform binary diagnosis on 301 pictures and achieved a sensitivity of 88.4%. Sinthanayothin et al. [62] used a neural network-based method to perform binary diagnosis on 767 images and achieved 80.21% sensitivity and 70.66% accuracy. This method is also used in actual scenarios for diagnosis purposes.

Data fusion based strategies using multiple methods were also integrated for diagnosis purposes. Antal et al. [63] used decision tree, k-NN, AdaBoost, and other machine learning methods, as well as the fusion strategy to adjust the model. They achieved 90% accuracy, 90% sensitivity and 91% specificity for the binary classification task on the Messidor dataset.

The recent development of deep learning techniques provide more possibilities for disease diagnosis. Also, the availability of a large number of large-scale datasets and the application of deep learning significantly improved the diagnosis of diabetic retinopathy. Abramoff et al. [64] proposed a hybrid CNN model and classical non-deep algorithms using 10,000–1,250,000 images of different types to train the model and obtained 87.00% specificity and 96.80% sensitivity. Gargeya et al. [65] trained their specific CNN on the Kaggle EyePACS dataset and achieved 0.99 AUC on the Messidor dataset. Islam et al. [66] proposed a novel CNN that can identify microaneurysms, which achieved an AUC score of 0.844 for the five-classification tasks on the Kaggle EyePACS dataset. Ghosh et al. [67] used a custom CNN model to identify features associated with diabetic retinopathy, such as microaneurysms and hemorrhages in fundus images. The model obtained 95% accuracy on binary-classification task and 85% accuracy on five classification tasks with 3000 validation samples from the Kaggle dataset. Xu et al. [68] proposed a custom CNN to classify diabetic retinopathy and achieved an accuracy of 94.5% on the Kaggle EyePACS dataset for binary classification task. Pratt et al. [69] used a custom CNN to classify fundus images, which achieved 75% accuracy and 30% sensitivity for the five-classification task on the Kaggle EyePACS dataset. Lin et al. [70] converted the fundus image into an entropy image and then input it into a custom CNN. They achieved 86.1% accuracy and 73.24% sensitivity on the Kaggle EyePACS dataset for five classification tasks. This method proved that converting an image into an entropy image may contribute to the extraction of features and improve the model's efficiency. Torre et al. [71] proposed a CNN model with several improved strategies, focusing on the interpretability of a diabetic retinopathy classifier that not only classifies images but also generates visual maps to interpret the results. It achieved 91% accuracy and 91.1% sensitivity for the binary classification task on the Messidor dataset.

Inception is also used as a classifier for the diagnosis of diabetic retinopathy. Gulshan et al. [72] used Inception to extract features from a dataset with 128,175 images, which were manually labeled several times by different experts to obtain more accurate ground truth. The model achieved 97.50% sensitivity and 93.40% specificity. Krause et al. [73] trained the Inception-V3 model with 1,670,000 images and validated the model on the EyePACS dataset achieving 97.1% sensitivity and 92.3% specificity. Li et al. [74] collected 58,790 images from the Ophthalmology Center of Sun Yat-sen University to train the Inception-V3 model, and achieved an AUC score of 0.909 on the dataset. Tsighe et al. [75] attempted to solve the problem of insufficient medical imaging data via transfer learning. They used a pre-trained Inception-V3 model and a small number of images on the Kaggle EyePACS dataset for fine-tuning, which obtained 90.9% accuracy on a binary classification task. Wang et al. [76] used Inception-v3 to perform five classifications on the Kaggle EyePACS dataset and obtained 63.23% accuracy.

Rakhlin et al. [77] modified VGG16 network and tested it on two datasets. On the Kaggle EyePACS dataset, it achieved 92% sensitivity and 72% specificity. On the Messidor dataset, the model achieved 99% sensitivity and 71% specificity. Garcia et al. [78] used the EyePACS dataset to train and test on a number of different models, achieving 74.3% binary classification accuracy with VGG16 dataset and 83.68% accuracy and 54.47% sensitivity with VGG16noFC dataset.

Zhang et al. [79] proposed a high-quality diabetic retinopathy dataset and developed an efficient diabetic retinopathy identification system with a sensitivity of 97.5%. Zhou et al. [80] proposed a multi-cell multi-task CNN, which aimed to solve some tiny lesions that were not easily detectable. Using this technique [80], a kappa value of 0.841 was achieved for the five classification tasks on the Kaggle EyePACS dataset. Zhao et al. [81] used BiRA-Net, a bilinear attention net, to classify diabetic retinopathy, which was dedicated to fine-grained feature extraction and classification. They also proposed a new hierarchical loss function that improved task process convergence. On the Kaggle EyePACS dataset, this model achieved an F1-Score of 0.5436. Li et al. [82] proposed a cross-disease attention network, which combined diabetic macular degeneration and diabetic macular edema for joint classification, and used multiple attention modules to capture lesions. It achieved 92.6% accuracy and 92% sensitivity on the Messidor binary-classification task.

Zhou et al. [83] proposed a collaborative learning method that combines the two tasks of disease severity grading and lesion classification. The performance of the two tasks was improved through semi-supervised learning and adversarial architecture. The model achieved an accuracy of 89.1% on the EyePACS five classification tasks and 91.3% on the iDRiD five classification tasks. Vives-Boix et al. [84] summarized the performance of multiple award-winning CNNs on the Kaggle APTOS dataset, including VGG16, InceptionV3, Densenet121 and a series of networks. Liu et al. [85] proposed WP-CNN for the diagnosis of diabetic retinopathy, which incorporated multiple weighted paths. It can quickly converge the network by optimizing multiple path weight coefficients. For the dataset with 60,000 pictures, this method achieved 94.23% accuracy, 90.94% sensitivity, and 95.74% specificity. Chen et al. [86] proposed a method based on the integration of multiscale shallow CNN. This method solved the problem of a deeper CNN, which requires more training time and results in overfitting. The multiple shallow convolutional network's aggregated outputs from softmax function classifies diabetic retinopathy, which achieved 92% accuracy for the five classification tasks on the Kaggle EyePACS dataset. It also achieved better results on smaller datasets when compared to other models with 86% accuracy on the dataset of 3500 images extracted from the EyePACS dataset, and 77% accuracy on the dataset of 1000 images extracted from the EyePACS dataset. Zago et al. [87] proposed a patch-based deep learning approach. This method focused on dividing the lesion area of the fundus image into patches, and then supplying them into the model one at a time. The model inferred whether the lesion will lead to the occurrence of diabetic retinopathy. It achieved a specificity of 91.1% on the Kaggle EyePACS dataset with a AUC score of 0.764. Shaban et al. [88] proposed a low-complexity CNN, which aimed to reduce the complexity of the diabetic retinopathy diagnosis model, and achieved 88.41% accuracy, 88.44% sensitivity and 96.14% specificity for four classifications (normal, mild and moderate, severe, proliferate) on 768 images. The performance measures of diabetic retinopathy classification are shown in Table 3.

4.3 Basic Structure Segmentation

The segmentation of the basic structure can assist doctors to further analyze the disease. Various types of structure segmentation is discussed in this subsection.

Optic Disk Segmentation

Optic disk segmentation assists ophthalmologists to observe optic disc lesions, the width of blood vessels entering the optic disc, and the color of the optic disc. A sample picture for optic disk segmentation is shown in Fig. 10.

For traditional machine learning methods, Mendels et al. [90] segmented nine images using morphological operations and active contours and achieved 100% accuracy. Walter et al. [91] used the watershed transform to segment the optic disk and achieved 90% accuracy on 30 images. Charastek et al. [92] used active contour model and Hough transform to process optic disk, which achieved 72.3% accuracy on 159 images. Lee et al. [93] used morphological operations and spline interpolation for segmentation operations and achieved 92% accuracy. Mahajan et al. [94] corrected the optical distortion when the image was captured, and used the template matching algorithm to detect the optic disc. A precision rate of 96.9% was achieved on the DRIVE dataset. Gopi et al. [95] used Markov random fields to segment the optic disc and achieved 98.49% accuracy on the DRIVE dataset. Bharkad [96] proposed a new algorithm by designing an equi-ripple low pass finite impulse response (FIR) filter to enhance the optic disc region in the fundus image. By setting appropriate thresholds to locate the optic disc region in retinal images, the method achieved 87.07% sensitivity and 99.39% specificity on the DRIVE dataset, and 74.60% sensitivity and 99.61% specificity on the DIARETDB0 dataset. A sensitivity of 75.11% and a specificity of 99.54% were achieved on the DIARETDB1 dataset.

For deep learning methods, Gu et al. [97] proposed a technique, called DeepDisc, which can precisely segment the optic disk. DeepDisc includes two components -

Table 3 Performance measures of diabetic retinopathy classification

Method	Dataset	Accuracy	Sensitivity	Specificity	AUC
k-NN [52]	–		–	–	0.839
SVM [53] (2-class classification)	–	99.1%	–	–	–
SVM [54] (4-class classification)	238 images	100%	98.9%	89.5%	–
SVM [55] (5-class classification)	–	82%	82.5%	88.9%	–
SVM [56] (5-class classification)	–	85.9%	82%	86%	–
FCM [57]	142 images	90.1%	–	–	–
k-NN [58] (2-class classification)	761 images	–	83.9%	72.7%	–
k-NN [59] (2-class classification)	67 images	–	–	–	0.927
Wilson score and kappa statistic methods [60]	527 images	99.1%	–	–	–
Neural network [61] (2-class classification)	301 images	–	88.4%	–	–
Neural network [62] (2-class classification)	767 images	70.66%	80.21%	–	–
Ensemble methods [63]	–	90%	90%	91%	–
Hybrid algorithm [64]	10,000-1,250,000 samples	–	96.8%	78.0%	–
Inception V3 [72]	128,175 images	–	97.5%	93.4%	–
Modified CNN [65]	EyePACS (training), Messidor-2 (testing)	–	–	–	0.99
Inception V3 [73]	1,670,000 images (training), EyePACS (testing)	–	97.1%	92.3%	–
Inception V3 [74]	58,790 images from ophthalmology center of Sun Yat-sen University	–	–	–	0.909
Modified CNN [79]	High quality DR dataset	–	97.5%	–	–

DenseNet-121 [89]	Kaggle dataset	R0:92% R1:70% R2:64% R3:67% R4:69%	–	–	–
Modified CNN [66] (5-class classification)	Kaggle EyePACS	–	–	–	0.844
InceptionV3 [75] (2-class classification)	Kaggle EyePACS	90.9%	–	–	–
Modified CNN [67] (2-class classification)	Kaggle EyePACS	95%	–	–	–
Modified CNN [67] (5-class classification)	Kaggle EyePACS	85%	–	–	–
Modified CNN [68] (2-class classification)	Kaggle EyePACS	94.5%	–	–	–
Modified VGG16 [77] (2-class classification)	Kaggle EyePACS	92%	72%	–	–
Modified VGG16 (2-class classification) [77]	Messidor	99%	71%	–	–
VGG16 [78] (2-class classification)	Kaggle EyePACS	74.3%	–	–	–
VGG16noFC [78] (2-class classification)	Kaggle EyePACs	83.68%	54.47%	96.14%	–
Modified CNN [69] (5-class classification)	Kaggle EyePACs	75%	30%	–	–
InceptionV3 [76] (5-class classification)	Kaggle EyePACs	63.23%		–	–
BiRA-Net [81] (5-class classification)	–	–	–	–	0.5436
Modified CNN [70] (5 4 classification)	Kaggle EyePACS	86.1%	73.24%	–	–
Collaborative learning method [83] (5-class classification)	Kaggle EyePACS	89.1%	–	–	–
Collaborative learning method [83] (5-class classification)	iDRiD	91.3%	–	–	–

(continued)

Table 3 (continued)

Method	Dataset	Accuracy	Sensitivity	Specificity	AUC
Modified CNN [71] (2-class classification)	Messidor	91%	91.1%	–	–
CANET [82] (2-class classification)	Messidor	92.6%	92%	–	–
ResNet50 [84] (2-class classification)	Kaggle APTOS	84.5%	–	–	0.8208
ResNet50 [84] (5-class classification)	Kaggle APTOS	71.35%	–	–	0.8928
AmDenseNet121 [84] (2-class classification)	Kaggle APTOS	91.95%	–	–	0.8685
AmDenseNet121 [84] (5-class classification)	Kaggle APTOS	67.26%	–	–	0.6698
WP-CNN [85] (2-class classification)	60,000 images	94.23%	90.94%	95.74%	0.9823
ShallowNet+PI [86] (5-class classification)	Kaggle EyePACS	92%	–	–	–
PB-CNN [87] (4-class classification)	Kaggle EyePACS	–	91.1%	–	0.764
LCNN [88] (4-class classification)	768 images	88.41%	88.44%	–	–

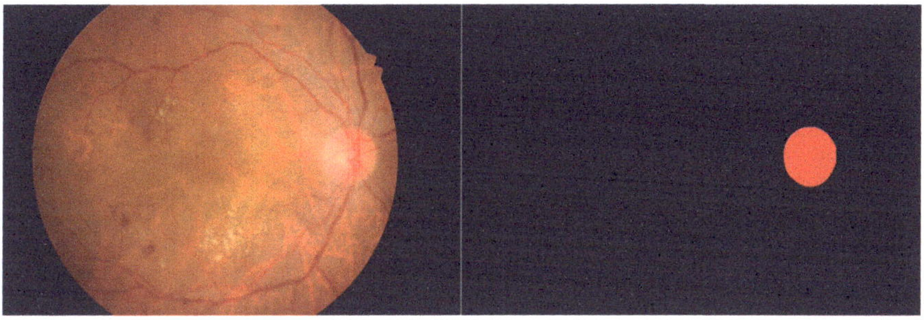

Fig. 10 Example picture for optic disk segmentation from IDRiD dataset

atrous convolution is used to adjust the feature size, and spatial pyramid pooling is used to extract features with multiple scales. This method further improved the accuracy of optic disk segmentation. The overlapping error on the Messidor dataset was 0.64%, and the overlapping error on the ORIGA dataset was 0.69%. Tan et al. [98] proposed a CNN model to segment the optic disc, fovea and blood vessels simultaneously. The model extracted features and generated 4 dimensions through the fully connected layer for representing the background, optic disc, fovea, and blood vessels. This method achieved 87.9% sensitivity and 99.27% specificity on the DRIVE dataset. The performance measures of optic disk segmentation are shown in Table 4.

Retina Blood Vessel Segmentation

Blood vessels are an essential indicator for diagnosing whether a patient has diabetic retinopathy or not. A sample picture for blood vessel segmentation is shown in Fig. 11.

For the traditional machine learning methods, Tolias et al. [99] used FCM to segment blood vessels and achieved a detection rate of 79.41% on three images. By using ridge detection and k-NN classifier, Soares et al. [100] achieved an AUC score of 0.952 on 40 images. Ricci et al. [101] used line operators and SVM to segment blood vessels, and achieved 95.95% accuracy on the DRIVE dataset and 96.46% accuracy on the STARE dataset. Goatman et al. [102] used the threshold and watershed transform to segment blood vessels. In addition, the method can also identify newly generated blood vessels around the optic disk, and achieved 84.20% sensitivity and 85.90% specificity. Rossant et al. [103] used Top-hat transform and path opening morphological filter for unsupervised vessel segmentation, which achieved 70.31% sensitivity and 97.88% specificity.

For deep learning methods, Hu et al. [104] proposed a retinal blood vessel segmentation method for fundus images based on CNNs and fully connected conditional random fields (CRF). They designed a multi-scale CNN by combining the features of each intermediate layer to generate probability distributions. Thereafter,

Table 4 Performance measures of optic disk segmentation

Methods	Dataset	Accuracy	Sensitivity	Specificity	Overlapping error
Morphological operations [90]	–	100%	–	–	–
Watershed transform [91]	30 images	90%	–	–	–
Active contour model and Hough transform [92]	159 images	72.3%	–	–	–
Morphological operations and spline interpolation [93]	–	92%	–	–	–
DeepDisc [97]	Messidor	–	–	–	0.64%
DeepDisc [97]	ORIGA	–	–	–	0.69%
CNN [98]	DRIVE	–	87.9%	99.27%	–
Equiripple low pass FIR filter [96]	DRIVE	–	87.07%	99.39%	–
Equiripple low pass FIR filter [96]	DIARETDB0	–	74.60%	99.61%	–
Equiripple low pass FIR filter [96]	DIARETDB1	–	75.11%	99.54%	–
Markov random fields [95]	DRIVE	98.49%	–	–	–

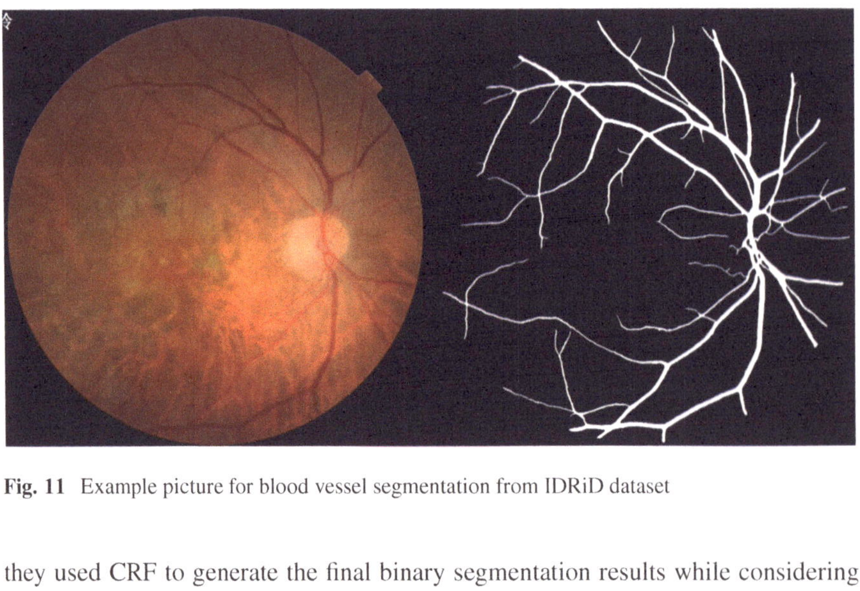

Fig. 11 Example picture for blood vessel segmentation from IDRiD dataset

they used CRF to generate the final binary segmentation results while considering the interaction between pixels. This technique resulted in a sensitivity of 77.72%

and a specificity of 97.93% on the DRIVE dataset, and a sensitivity of 0.7543 and a specificity of 89.14% on the STARE dataset. Since retinal vessels vary in length and width, they behaved as multi-scale tubular structures. Xu et al. [105] proposed a multi-scale deep-guidance neural network, which took advantage of the characteristics of the tubular structure. The model was a typical encoder-decoder structure, where skip-connection was added for feature transfer. It achieved an ROC score of 0.9793 and precision-recall curve (PR) score of 0.9104 on the DRIVE dataset.

Wu et al. [106] adopted a divide-and-conquer strategy for blood vessels with different scales. At first, the image was divided into multiple similar-sized pieces. These patches were input into the network in network (NIN) model. The NIN model divided the patch into three categories - wide blood vessels, medium blood vessels and capillaries. Then, three U-Net models were used to segment the image. Finally, the images were combined. The model achieved a sensitivity of 77.21%, a specificity of 98.30%, and an AUC score of 0.9793 on the DRIVE dataset. Oliveira et al. [107] used a model of combining stationary wavelet transform and multi-scale fully CNN to solve the multi-scale problem, which obtained 80.30% sensitivity and 98.04% specificity on the DRIVE dataset, 83.15% sensitivity and 98.58% specificity on the STARE dataset, and 77.79% sensitivity, 98.64% specificity on the CHASE_DB1 dataset. Jiang et al. [108] applied a pre-trained CNN segmentation method. An image was divided into multiple patches for processing, and the region segmentation results were combined into a complete segmentation yielding 75.40% sensitivity and 98.25% specificity on the DRIVE dataset, 83.52% sensitivity and 98.46% specificity on the STARE dataset, and 86.40% sensitivity and 97.45% specificity on the Chase_DB1 dataset. The performance measures of blood vessel segmentation are shown in Table 5.

5 Applications in Real-Life Scenarios

This section mainly introduces the application of computer-aided diagnosis of diabetic retinopathy in real-life scenarios. Abramoff et al. [64] proposed a diabetic retinopathy diagnostic system, and performed diagnostic tests with fundus images from different races and age groups in primary clinics. The system was authorized by the Federal Drug Administration (FDA) of USA. It was also the first FDA-authorized diagnostic system for diabetic retinopathy.

Ting et al. [109] evaluated the deep learning system on fundus pictures in a multi-ethnic diabetic population. The experiment obtained a sensitivity of 90.5% and a specificity of 91.6% for the detection of referable diabetic retinopathy on a dataset of 14,880 patients. It also had a sensitivity of 100% and a specificity of 91.1% for vision-threatening diabetic retinopathy. In addition, the system could also be used for diagnosis of glaucoma and age-related macular degeneration.

Table 5 Performance measures of blood vessel segmentation

Methods	Dataset	Accuracy	Sensitivity	Specificity	Detection rate	AUC	ROC
Fuzzy C-Means [99]	3 images	–	–	–	79.41%	–	–
Ridge detection and k-NN [100]	40 images	–	–	–	–	0.952	–
Line operators and SVM [101]	DRIVE	95.95%	–	–	–	–	–
Line operators and SVM [101]	STARE	96.46%	–	–	–	–	–
Thresholding and watershed transform [102]		–	84.20%	85.90%	–	–	–
Top-hat transform [103]		–	70.31%	97.88%	–	–	–
CNN and CRF [104]	DRIVE	–	77.72%	97.93%	–	–	–
CNN and CRF [104]	STARE	–	75.43%	89.14%	–	–	–
Multiscaled deeply-guidance neural network [105]	DRIVE	–	–	–	–	–	0.9793
NN [106]	DRIVE	–	77.21%	98.30%	–	0.9793	–
Combining Method [107]	DRIVE	–	80.30%	98.04%	–	–	–
Combining Method [107]	STARE	–	83.15%	98.58%	–	–	–
Combining Method [107]	CHASE_DB1	–	77.79%	98.64%	–	–	–
Pretrained CNN [108]	DRIVE	–	75.40%	98.25%	–	–	–
Pretrained CNN [108]	STARE	–	83.52%	98.46%	–	–	–
Pretrained CNN [108]	CHASE_DB1	–	86.40%	97.45%	–	–	–

Bellemo et al. [110] used deep learning models, consisting of VGGNet and ResNet, for a project on Diabetic Retinopathy Screening in Zambia. They trained the model with the patient data from the Singapore Integrated Diabetic Retinopathy Program, and validated it on patient data in Zambia. It achieved an AUC score of 0.973 and 92.25% sensitivity.

Tufail et al. [111] evaluated the possibility of applying automated diabetic retinopathy diagnosis to provide clinically valid detection and replace manual diagnosis. The project collected fundus images of 20,258 patients who participated in routine screening from June 1, 2012 to November 4, 2013. The collected images were manually graded and processed through three diagnostic models (iGradingM, Retmarker, EyeArt) for evaluation. The experimental results show that the iGradingM model classifies all images as DR or non-gradable. EyeArt and RetMarker can replace initial manual grading. It was shown that computer-aided diagnosis can reduce the cost of public health care and provide early diabetic retinopathy screening in low- and middle-income countries.

Kanagasingam et al. [112] evaluated the performance of an artificial intelligence (AI) system for diabetic retinopathy. They collected 386 images from 193 patients in Western Australia between December 1, 2016 and May 31, 2017. Both the AI model and ophthalmologists were used in the evaluation process. The results show that for 193 patients, 17 patients were diagnosed by the model to have diabetic retinopathy and needed a referral. However, of the 17 patients, only 2 patients had diabetic retinopathy, while the other 15 patients were incorrectly diagnosed. They believed that reflections and smudges on the lens interfered with the correct judgment of the model.

Bouhaimed et al. [113] evaluated the operational characteristics of a fundus image reading software - Retinalyze System. The software was used to analyze fundus images of patients in the Bro Taf DR diabetic retinopathy screening program in Wales, UK, between 2002 and 2004. A total of 400 fundus photographs from 192 eyes of 96 diabetic patients were analyzed. For high-quality images, the detection sensitivity of red and bright lesions was 93%, and the specificity was 78%. The performance measures of diabetic retinopathy classification for real-life scenarios are shown in Table 6.

6 Discussions on Future Research

6.1 Light-Weight Model

At this time, to extract more representative features, the models used for diabetic retinopathy diagnosis tend to be deeper, which imposes high computing power requirements. However, the computing resources available in many medical institutions may be insufficient. On the other hand, with the rise of the mobile internet, a large number of handheld devices and mobile computing platforms are widely used.

Table 6 Performance measures of diabetic retinopathy classification in real scenarios

Method	Dataset	Sensitivity	Specificity	AUC
Abramoff et al. [64]	Messidor-2	96.8%	87%	0.98
Ting et al. [109]	71,896 images in Singapore	90.5% (referable DR) 100% (visive threatening DR)	91.6% (referable DR) 91.1% (visive threatening DR)	0.936
Bellemo et al. [110]	4504 images in Zambia	92.25%	89.04%	0.973
Kanagasing-am et al. [112]	386 images in Australia	100%	92%	–
Bouhaimed et al. [113]		88%	52%	–

If the diagnosis model of diabetic retinopathy can be deployed on mobile platform, it can make mobile consultation a reality [114–116].

6.2 Interpretability Model

Machine learning models for medical image analysis not only require precise diagnosis, but also supply solid explanations for the conclusions. The models used in deep learning have more parameters and more complex structures than the traditional machine learning methods. So it is difficult to provide model-based interpretation. The interpretability of deep learning is often based on post-hoc interpretability, which does not explicitly explain the model itself, but proves that the network is interpretable after the training process. Examples include outputting a heat map in the model, trying to illustrate which areas of the image have an important impact on the model's judgment, or adding/removing a factor in the input to prove the importance of the factor to the model.

Kumar et al. [117] proposed the CLEAR-DR model, which used the class activation mapping (CAM) method to explain the deep learning-based diabetic retinopathy diagnosis model. The CAM shows the lesions that the model focuses on and proved that the model can make effective classification decisions in clinical applications. Liu et al. [118] used a CAM visualization method to address the interpretability issue of black-box models, demonstrating the feasibility of their proposed diagnostic method. Martins et al. [119] used Grad-Cam activation maps to demonstrate the effectiveness of their GFI-C network in glaucoma diagnosis, where the model focused on the same regions where the radiologists are interested to look at. Meng et al. [120] used Grad-CAM to display the lesion area and also designed a strategy to update these areas to make them closer to the actual lesion areas. This strategy improved the network accuracy and the interpretability of the network. Quellec et al. [121] used few-shot learning to improve the model's ability to identify rare diseases, and used the back propagation method when generating the heatmap. The generated images illustrate the diagnostic ability of the model.

6.3 Self-Supervised Learning

Supervised learning methods have greatly benefitted the diagnosis of diabetic retinopathy. However, these methods have still some deficiencies. For example, supervised learning heavily relies on a large number of labeled samples. But in real life, the number of manually labeled datasets is actually not so rich. So, if an institution or individual intends to train a model with their own dataset, a lot of time and labor costs are needed to label the data. It is practical to apply self-supervised learning in clinical diagnosis.

The Rubik's Cube+ method proposed by Zhu et al. [122] segmented 3D images and disrupted the order. Then, image enhancement operations such as masking and rotation were performed. Next, the features of the images were supplied into the network. The network was applied to classify the Cerebral Hemorrhage dataset and it achieved 78.68% accuracy. Chen et al. [123] used image context restoration for self-supervised learning, which extracted medical image features, and then tried to reconstruct the image. They trained the model by comparing the original image and the generated image, and achieved 89.39% accuracy on the classification of a standard scan plane (standard 2D ultrasound image of the fetus). Yellapragada et al. [124] used a model with self-supervised non-parametric instance discrimination and achieved 94% accuracy on a two-step classification of age-related macular degeneration.

References

1. Chan JCN, Lim LL, Wareham NJ, Shaw JE, Orchard TJ, Zhang P et al (2020) The lancet commission on diabetes: using data to transform diabetes care and patient lives. Lancet 396(10267):2019–2082
2. Cheung N, Mitchell P, Wong TY (2010) Seminar diabetic retinopathy. Lancet [Internet] 376:124–136. Available from: www.thelancet.com. [cited 2019 Dec 12]
3. Mookiah MRK, Acharya UR, Chua CK, Lim CM, Ng EYK, Laude A (2013) Computer-aided diagnosis of diabetic retinopathy: a review. Comput Biol Med 43:2136–2155
4. Peng JJ, Xiong SQ, Ding LX, Peng J, Xia XB (2019) Diabetic retinopathy: focus on NADPH oxidase and its potential as therapeutic target. Eur J Pharmacol 853:381–387
5. Wilkinson CP, Ferris FL, Klein RE, Lee PP, Agardh CD, Davis M et al (2003) Proposed international clinical diabetic retinopathy and diabetic macular edema disease severity scales. Ophthalmology 110(9):1677–1682
6. Tsiknakis N, Theodoropoulos D, Manikis G, Ktistakis E, Boutsora O, Berto A et al (2021) Deep learning for diabetic retinopathy detection and classification based on fundus images: a review. Comput Biol Med [Internet] 135:104599. Available from: https://linkinghub.elsevier.com/retrieve/pii/S0010482521003930. [cited 2021 Jul 28]
7. Fourcade A, Khonsari RH (2019) Deep learning in medical image analysis: a third eye for doctors. J Stomatol Oral Maxillofac Surg [Internet] 120(4):279–288. Available from: https://doi.org/10.1016/j.jormas.2019.06.002
8. Li T, Bo W, Hu C, Kang H, Liu H, Wang K et al (2021) Applications of deep learning in fundus images: a review. Med Image Anal 69:101971

9. Kohner EM, Dollery CT (1970) Fluorescein angiography of the fundus in diabetic retinopathy. Br Med Bull [Internet] 26(2):166–170. Available from: https://academic.oup.com/bmb/article/26/2/166/262348. [cited 2022 Jul 26]
10. Hormel TT, Hwang TS, Bailey ST, Wilson DJ, Huang D, Jia Y (2021) Artificial intelligence in OCT angiography. Prog Retin Eye Res 85:100965
11. Bélanger Nzakimuena C. Automated analysis of retinal and choroidal OCT and OCTA images in AMD. 2020;
12. Hurrah NN, Loan NA, Parah SA, Sheikh JA, Muhammad K, de Macedo ARL et al (2021) INDFORG: industrial forgery detection using automatic rotation angle detection and correction. IEEE Trans Industr Inform 17(5):3630–3639
13. Hurrah NN, Parah SA, Sheikh JA, Al-Turjman F, Muhammad K (2019) Secure data transmission framework for confidentiality in IoTs. Ad Hoc Netw 95:101989
14. Afzal I, Parah SA, Hurrah NN, Song OY (2020) Secure patient data transmission on resource constrained platform. Multimed Tools Appl [Internet]:1–26. Available from: https://link.springer.com/article/10.1007/s11042-020-09139-3. [cited 2022 Sep 19]
15. Kaur A, Rashid M, Bashir AK, Parah SA (2022) Detection of breast cancer masses in mammogram images with watershed segmentation and machine learning approach. Artif Intell Innov Healthcare Inform [Internet]:35–60. Available from: https://link.springer.com/chapter/10.1007/978-3-030-96569-3_2. [cited 2022 Sep 19]
16. Hacisoftaoglu RE, Karakaya M, Sallam AB (2020) Deep learning frameworks for diabetic retinopathy detection with smartphone-based retinal imaging systems. Pattern Recogn Lett 135:409–417
17. Bhat GM, Mustafa M, Parah SA, Ahmad J (2010) Field programmable gate array (FPGA) implementation of novel complex PN-code-generator-based data scrambler and descrambler. Maejo Int J Sci Technol 4(1):125–135
18. Bhat GM, Mustafa M, Ahmad S, Ahmad J (2009) VHDL modeling and simulation of data scrambler and descrambler for secure data communication. Indian J Sci Technol 2(10):41–44
19. Cortes C, Vapnik V, Saitta L (1995) Support-vector networks. Mach Learn 20:3. [Internet]. 1995;20(3):273–97. Available from: https://link.springer.com/article/10.1007/BF00994018. [cited 2022 Jul 22]
20. Rumelhart DE, Hinton GE, Williams RJ (1986) Learning representations by back-propagating errors. Nature 323(6088):533–536
21. Subasi A (2020) Clustering examples. In: Practical machine learning for data analysis using python, pp 465–511
22. Krizhevsky A, Sutskever I, Hinton GE (2017) ImageNet classification with deep convolutional neural networks. Commun ACM [Internet] 60(6):84–90. Available from: https://dl.acm.org/doi/10.1145/3065386. [cited 2022 Jul 22]
23. Szegedy C, Vanhoucke V, Ioffe S, Shlens J, Wojna Z (2016) Rethinking the inception architecture for computer vision. In: Proceedings of the IEEE computer society conference on computer vision and pattern recognition. 2016 Dec 9;2016-December:2818–26
24. Howard AG, Zhu M, Chen B, Kalenichenko D, Wang W, Weyand T et al (2017) MobileNets: efficient convolutional neural networks for mobile vision applications. [cited 2022 Jul 22]. Available from: http://arxiv.org/abs/1704.04861
25. Sandler M, Howard A, Zhu M, Zhmoginov A, Chen LC (2018) MobileNetV2: inverted residuals and linear bottlenecks. Proc IEEE Comput Soc Conf Comput Vis Pattern Recogn [Internet]:4510–4520. Available from: https://arxiv.org/abs/1801.04381v4. [cited 2022 Jul 30]
26. Weng W, Zhu X (2015) U-net: convolutional networks for biomedical image segmentation. IEEE Access [Internet] 9:16591–16603. Available from: https://arxiv.org/abs/1505.04597v1. [cited 2022 Jul 22]
27. GitHub - dair-ai/ml-visuals: ML Visuals contains figures and templates which you can reuse and customize to improve your scientific writing. [Internet]. [cited 2022 Oct 9]. Available from: https://github.com/dair-ai/ml-visuals

28. APTOS 2019 Blindness Detection | Kaggle [Internet]. [cited 2022 Jul 28]. Available from: https://www.kaggle.com/c/aptos2019-blindness-detection
29. Decencière E, Zhang X, Cazuguel G, Laÿ B, Cochener B, Trone C et al (2014) Feedback on a publicly distributed image database: the Messidor database. Image Analysis and Stereology 33(3):231–234
30. Diabetic Retinopathy Detection | Kaggle [Internet]. [cited 2022 Jul 28]. Available from: https://www.kaggle.com/c/diabetic-retinopathy-detection/
31. Porwal P, Pachade S, Kokare M, Deshmukh G, Son J, Bae W et al (2020) IDRiD: diabetic retinopathy – segmentation and grading challenge. Med Image Anal 59:101561
32. Introduction - Grand Challenge [Internet]. [cited 2022 Jul 28]. Available from: https://drive.grand-challenge.org/
33. Ocular Disease Recognition | Kaggle [Internet]. [cited 2022 Jul 28]. Available from: https://www.kaggle.com/datasets/andrewmvd/ocular-disease-recognition-odir5k
34. Retinal image database | Retinal Image Analysis [Internet]. [cited 2022 Jul 28]. Available from: https://blogs.kingston.ac.uk/retinal/chasedb1/
35. DIARETDB0 - Standard Diabetic Retinopathy Database [Internet]. [cited 2022 Jul 28]. Available from: https://www.it.lut.fi/project/imageret/diaretdb0/index.html
36. DIARETDB1 - STANDARD DIABETIC RETINOPATHY DATABASE [Internet]. [cited 2022 Jul 28]. Available from: https://www.it.lut.fi/project/imageret/diaretdb1/
37. Hoover A (2000) Locating blood vessels in retinal images by piecewise threshold probing of a matched filter response. IEEE Trans Med Imaging 19(3):203–210
38. Decencière E, Cazuguel G, Zhang X, Thibault G, Klein JC, Meyer F et al (2013) TeleOphta: machine learning and image processing methods for teleophthalmology. IRBM 34(2):196–203
39. Phillips R, Forrester J, Sharp P (1993) Automated detection and quantification of retinal exudates. Graefes Arch Clin Exp Ophthalmol 231:2. [Internet]. 1993 Feb [cited 2022 Jul 22];231(2):90–4. Available from: https://link.springer.com/article/10.1007/BF00920219
40. Shilpa Bv, Nagabhushan TN (2016) An ensemble approach to detect exudates in digital fundus images. In: Proceedings – 2016 2nd international conference on cognitive computing and information processing, CCIP 2016. 30 Dec 2016
41. Asha PR, Karpagavalli S (2015) Diabetic retinal exudates detection using Extreme Learning Machine. Adv Intell Syst Comput [Internet] 338:573–578. Available from: https://link.springer.com/chapter/10.1007/978-3-319-13731-5_62. [cited 2022 Jul 22]
42. Lin W, Liu H, Xu M, Zhang J (2008) Automated detection of exudates on color fundus image using region merging by k-NN graph. IFMBE Proc [Internet] 19 IFMBE:216–220. Available from: https://link.springer.com/chapter/10.1007/978-3-540-79039-6_56. [cited 2022 Jul 22]
43. Khojasteh P, Passos Júnior LA, Carvalho T, Rezende E, Aliahmad B, Papa JP et al (2019) Exudate detection in fundus images using deeply-learnable features. Comput Biol Med 104:62–69
44. Li G, Zheng S, Li X (2018) Exudate detection in fundus images via convolutional neural network. Commun Comput Inf Sci [Internet] 815:193–202. Available from: https://link.springer.com/chapter/10.1007/978-981-10-8108-8_18. [cited 2022 Jul 22]
45. Prentašić P, Lončarić S (2016) Detection of exudates in fundus photographs using deep neural networks and anatomical landmark detection fusion. Comput Methods Prog Biomed 137:281–292
46. Adem K (2018) Exudate detection for diabetic retinopathy with circular Hough transformation and convolutional neural networks. Expert Syst Appl 114:289–295
47. Chudzik P, Majumdar S, Calivá F, Al-Diri B, Hunter A (2018) Microaneurysm detection using fully convolutional neural networks. Comput Methods Prog Biomed 158:185–192
48. Yadav D, Karn AK, Giddalur A, Dhiman A, Sharma S, Muskan et al (2021) Microaneurysm detection using color locus detection method. Measurement 176:109084
49. Liao Y, Xia H, Song S, Li H (2021) Microaneurysm detection in fundus images based on a novel end-to-end convolutional neural network. Biocybern Biomed Eng 41(2):589–604

50. Murugan R, Roy P (2022) MicroNet: microaneurysm detection in retinal fundus images using convolutional neural network. Soft Comput [Internet] 26(3):1057–1066. Available from: https://link.springer.com/article/10.1007/s00500-022-06752-2. [cited 2022 Jul 22]
51. Niemeijer M, van Ginneken B, Cree MJ, Mizutani A, Quellec G, Sanchez CI et al (2010) Retinopathy online challenge: automatic detection of microaneurysms in digital color fundus photographs. IEEE Trans Med Imaging 29(1):185–195
52. Abràmoff MD, Reinhardt JM, Russell SR, Folk JC, Mahajan VB, Niemeijer M et al (2010) Automated early detection of diabetic retinopathy. Ophthalmology 117(6):1147–1154
53. Noronha K, Acharya UR, Nayak KP, Kamath S, Bhandary S (2013) Decision support system for diabetic retinopathy using discrete wavelet transform. Proc Inst Mech Eng H 227(3):251–261
54. Acharya UR, Ng EYK, Tan JH, Sree SV, Ng KH (2012) An integrated index for the identification of diabetic retinopathy stages using texture parameters. J Med Syst [Internet] 36(3):2011–2020. Available from: https://link.springer.com/article/10.1007/s10916-011-9663-8. [cited 2022 Jul 22]
55. Acharya UR, Chua CK, Ng EYK, Yu W, Chee C (2008) Application of higher order spectra for the identification of diabetes retinopathy stages. J Med Syst [Internet] 32(6):481–488. Available from: https://link.springer.com/article/10.1007/s10916-008-9154-8. [cited 2022 Jul 22]
56. Acharya UR, Lim CM, Ng EYK, Chee C, Tamura T (2009) Computer-based detection of diabetes retinopathy stages using digital fundus images. Proc Inst Mech Eng H 223(5):545–553
57. Osareh A, Mirmehdi M, Thomas B, Markham R (2002) Classification and localisation of diabetic-related eye disease. Lect Notes Comput Sci (including subseries Lecture Notes in Artificial Intelligence and Lecture Notes in Bioinformatics) [Internet] 2353:502–516. Available from: https://link.springer.com/chapter/10.1007/3-540-47979-1_34. [cited 2022 Jul 22]
58. Dupas B, Walter T, Erginay A, Ordonez R, Deb-Joardar N, Gain P et al (2010) Evaluation of automated fundus photograph analysis algorithms for detecting microaneurysms, haemorrhages and exudates, and of a computer-assisted diagnostic system for grading diabetic retinopathy. Diabetes Metab [Internet] 36(3):213–220. Available from: https://pubmed.ncbi.nlm.nih.gov/20219404/. [cited 2022 Jul 22]
59. Quellec G, Russell SR, Abramoff MD (2011) Optimal filter framework for automated, instantaneous detection of lesions in retinal images. IEEE Trans Med Imaging 30(2):523–533
60. Philip S, Fleming AD, Goatman KA, Fonseca S, Mcnamee P, Scotland GS et al (2007) The efficacy of automated "disease/no disease" grading for diabetic retinopathy in a systematic screening programme. Br J Ophthalmol [Internet] 91(11):1512–1517. Available from: https://pubmed.ncbi.nlm.nih.gov/17504851/. [cited 2022 Jul 22]
61. Gardner GG, Keating D, Williamson TH, Elliott AT (1996) Automatic detection of diabetic retinopathy using an artificial neural network: a screening tool. Br J Ophthalmol [Internet] 80(11):940–944. Available from: https://pubmed.ncbi.nlm.nih.gov/8976718/. [cited 2022 Jul 22]
62. Sinthanayothin C, Kongbunkiat V, Phoojaruenchanachai S, Singalavanija A. Automated screening system for diabetic retinopathy. 2004;915–920
63. Antal B, Hajdu A (2014) An ensemble-based system for automatic screening of diabetic retinopathy. Knowl Based Syst 60:20–27
64. Abràmoff MD, Lou Y, Erginay A, Clarida W, Amelon R, Folk JC et al (2016) Improved automated detection of diabetic retinopathy on a publicly available dataset through integration of deep learning. Invest Ophthalmol Vis Sci 57(13):5200–5206
65. Gargeya R, Leng T (2017) Automated identification of diabetic retinopathy using deep learning. Ophthalmology [Internet] 124(7):962–969. Available from: https://doi.org/10.1016/j.ophtha.2017.02.008

66. Islam SMS, Hasan MM, Abdullah S (2018) Deep learning based early detection and grading of diabetic retinopathy using retinal fundus images. Available from: http://arxiv.org/abs/1812.10595. [cited 2022 Jul 22]
67. Ghosh R, Ghosh K, Maitra S (2017) Automatic detection and classification of diabetic retinopathy stages using CNN. In: 2017 4th international conference on signal processing and integrated networks, SPIN 2017, pp 550–554
68. Xu K, Feng D, Mi H (2017) Deep convolutional neural network-based early automated detection of diabetic retinopathy using fundus image. Molecules 22(12):2054. Available from: https://www.mdpi.com/1420-3049/22/12/2054/htm. [cited 2022 Jul 22]
69. Pratt H, Coenen F, Broadbent DM, Harding SP, Zheng Y (2016) Convolutional neural networks for diabetic retinopathy. Procedia Comput Sci 90:200–205
70. Lin GM, Chen MJ, Yeh CH, Lin YY, Kuo HY, Lin MH et al (2018) Transforming retinal photographs to entropy images in deep learning to improve automated detection for diabetic retinopathy. J Ophthalmol [Internet] 2018. Available from: https://pubmed.ncbi.nlm.nih.gov/30275989/. [cited 2022 Jul 22]
71. de la Torre J, Valls A, Puig D (2020) A deep learning interpretable classifier for diabetic retinopathy disease grading. Neurocomputing 396:465–476
72. Gulshan V, Peng L, Coram M, Stumpe MC, Wu D, Narayanaswamy A et al (2016) Development and validation of a deep learning algorithm for detection of diabetic retinopathy in retinal fundus photographs. JAMA 316(22):2402–2410
73. Krause J, Gulshan V, Rahimy E, Karth P, Widner K, Corrado GS et al (2018) Grader variability and the importance of reference standards for evaluating machine learning models for diabetic retinopathy. Ophthalmology [Internet] 125(8):1264–1272. Available from: https://doi.org/10.1016/j.ophtha.2018.01.034
74. Li Z, Keel S, Liu C, He Y, Meng W, Scheetz J et al (2018) An automated grading system for detection of vision-threatening referable diabetic retinopathy on the basis of color fundus photographs. Diabetes Care 41(12):2509–2516
75. Tsighe M, Student HMT, Kant S (2019) Transfer learning based detection of diabetic retinopathy from small dataset. Available from: https://arxiv.org/abs/1905.07203v2. [cited 2022 Jul 22]
76. Wang X, Lu Y, Wang Y, Chen WB (2018) Diabetic retinopathy stage classification using convolutional neural networks. In: Proceedings - 2018 IEEE 19th international conference on information reuse and integration for data science, IRI 2018, pp 465–471
77. Rakhlin A (2018) Diabetic Retinopathy detection through integration of Deep Learning classification framework. bioRxiv [Internet] 225508. Available from: https://www.biorxiv.org/content/10.1101/225508v2. [cited 2022 Jul 22]
78. García G, Gallardo J, Mauricio A, López J, del Carpio C (2017) Detection of diabetic retinopathy based on a convolutional neural network using retinal fundus images. Lect Notes Comput Sci (including subseries Lecture Notes in Artificial Intelligence and Lecture Notes in Bioinformatics) [Internet] 10614 LNCS:635–642. Available from: https://link.springer.com/chapter/10.1007/978-3-319-68612-7_72. [cited 2022 Jul 22]
79. Zhang W, Zhong J, Yang S, Gao Z, Hu J, Chen Y et al (2019) Automated identification and grading system of diabetic retinopathy using deep neural networks. Knowl Based Syst 175:12–25
80. Zhou K, Gu Z, Liu W, Luo W, Cheng J, Gao S et al (2018) Multi-cell multi-task convolutional neural networks for diabetic retinopathy grading. Annu Int Conf IEEE Eng Med Biol Soc [Internet] 2018:2724–2727. Available from: https://pubmed.ncbi.nlm.nih.gov/30440966/. [cited 2022 Jul 22]
81. Zhao Z, Zhang K, Hao X, Tian J, Heng Chua MC, Chen L et al (2019) BiRA-net: bilinear attention net for diabetic retinopathy grading. In: Proceedings - international conference on image processing, ICIP. 1 September 2019, pp 1385–1389
82. Li X, Hu X, Yu L, Zhu L, Fu CW, Heng PA (2020) CANet: cross-disease attention network for joint diabetic retinopathy and diabetic macular edema grading. IEEE Trans Med Imaging 39(5):1483–1493

83. Zhou Y, Wang B, He X, Cui S, Shao L (2022) DR-GAN: conditional generative adversarial network for fine-grained lesion synthesis on diabetic retinopathy images. IEEE J Biomed Health Inform [Internet] 26(1):56–66. Available from: https://pubmed.ncbi.nlm.nih.gov/33332280/. [cited 2022 Jul 22]
84. Vives-Boix V, Ruiz-Fernández D (2021) Diabetic retinopathy detection through convolutional neural networks with synaptic metaplasticity. Comput Methods Prog Biomed 206:106094
85. Liu YP, Li Z, Xu C, Li J, Liang R (2019) Referable diabetic retinopathy identification from eye fundus images with weighted path for convolutional neural network. Artif Intell Med 99:101694
86. Chen W, Yang B, Li J, Wang J (2020) An approach to detecting diabetic retinopathy based on integrated shallow convolutional neural networks. IEEE Access 8:178552–178562
87. Zago GT, Andreão RV, Dorizzi B, Teatini Salles EO (2020) Diabetic retinopathy detection using red lesion localization and convolutional neural networks. Comput Biol Med 116:103537
88. Shaban M, Mahmoud AH, Shalaby A, Ghazal M, Sandhu H, El-Baz A (2020) Low-complexity computer-aided diagnosis for diabetic retinopathy. In: Diabetes and retinopathy, pp 133–149
89. Wang Y, Wang GA, Fan W, Li J (2018) A deep learning based pipeline for image grading of diabetic retinopathy. Lect Notes Comput Sci (including subseries Lecture Notes in Artificial Intelligence and Lecture Notes in Bioinformatics) [Internet] 10983 LNCS:240–248. Available from: https://link.springer.com/chapter/10.1007/978-3-030-03649-2_24. [cited 2022 Jul 22]
90. Mendels F, Heneghan C, Harper PD, Reilly RB, Thiran JP (1999) Extraction of the optic disk boundary in digital fundus images. In: Annual international conference of the IEEE engineering in medicine and biology - proceedings, vol 2, p 1139
91. Walter T, Klein JC (2001) Segmentation of color fundus images of the human retina: Detection of the optic disc and the vascular tree using morphological techniques. Lect Notes Comput Sci (including subseries Lecture Notes in Artificial Intelligence and Lecture Notes in Bioinformatics) [Internet] 2199:282–287. Available from: https://link.springer.com/chapter/10.1007/3-540-45497-7_43. [cited 2022 Jul 22]
92. Chrástek R, Wolf M, Donath K, Niemann H, Paulus D, Hothorn T et al (2005) Automated segmentation of the optic nerve head for diagnosis of glaucoma. Med Image Anal 9(4):297–314
93. Lee SS, Rajeswari M, Ramachandram D, Shaharuddin B (2007) Screening of diabetic retinopathy - Automatic segmentation of optic disc in colour fundus images. In: Proceedings of DFMA 2006 – 2nd international conference on distributed frameworks for multimedia applications, pp 37–43
94. Mahajan A, Kumar S, Bansal R (2019) A novel approach of optic disk detection for diagnosis of diabetic retinopathy. Lect Notes Netw Syst [Internet] 46:393–406. Available from: https://link.springer.com/chapter/10.1007/978-981-13-1217-5_39. [cited 2022 Jul 22]
95. Gopi VP, Anjali MS, Niwas SI (2017) PCA-based localization approach for segmentation of optic disc. Int J Comput Assist Radiol Surg [Internet] 12(12):2195–2204. Available from: https://link.springer.com/article/10.1007/s11548-017-1670-x. [cited 2022 Jul 22]
96. Bharkad S (2017) Automatic segmentation of optic disk in retinal images. Biomed Signal Process Control 31:483–498
97. Gu Z, Liu P, Zhou K, Jiang Y, Mao H, Cheng J et al (2018) DeepDisc: optic disc segmentation based on atrous convolution and spatial pyramid pooling. Lect Notes Comput Sci (including subseries Lecture Notes in Artificial Intelligence and Lecture Notes in Bioinformatics) [Internet] 11039 LNCS:253–260. Available from: https://link.springer.com/chapter/10.1007/978-3-030-00949-6_30. [cited 2022 Jul 22]
98. Tan JH, Acharya UR, Bhandary S v, Chua KC, Sivaprasad S (2017) Segmentation of optic disc, fovea and retinal vasculature using a single convolutional neural network. J Comput Sci 20:70–79
99. Tolias YA, Panas SM (1998) A fuzzy vessel tracking algorithm for retinal images based on fuzzy clustering. IEEE Trans Med Imaging 17(2):263–273

100. Soares JVB, Leandro JJG, Cesar RM, Jelinek HF, Cree MJ (2006) Retinal vessel segmentation using the 2-D Gabor wavelet and supervised classification. IEEE Trans Med Imaging 25(9):1214–1222
101. Ricci E, Perfetti R (2007) Retinal blood vessel segmentation using line operators and support vector classification. IEEE Trans Med Imaging 26(10):1357–1365
102. Goatman KA, Fleming AD, Philip S, Williams GJ, Olson JA, Sharp PF (2011) Detection of new vessels on the optic disc using retinal photographs. IEEE Trans Med Imaging 30(4):972–979
103. Rossant F, Badellino M, Chavillon A, Bloch I, Paques M (2011) A morphological approach for vessel segmentation in eye fundus images, with quantitative evaluation. J Med Imaging Health Inform 1(1):42–49
104. Hu K, Zhang Z, Niu X, Zhang Y, Cao C, Xiao F et al (2018) Retinal vessel segmentation of color fundus images using multiscale convolutional neural network with an improved cross-entropy loss function. Neurocomputing 309:179–191
105. Xu R, Jiang G, Ye X, Chen YW (2018) Retinal vessel segmentation via multiscaled deep-guidance. Lect Notes Comput Sci (including subseries Lecture Notes in Artificial Intelligence and Lecture Notes in Bioinformatics) [Internet] 11165 LNCS:158–168. Available from: https://link.springer.com/chapter/10.1007/978-3-030-00767-6_15. [cited 2022 Jul 22]
106. Wu Y, Xia Y, Zhang Y (2018) Deep classification and segmentation model for vessel extraction in retinal images. Lect Notes Comput Sci (including subseries Lecture Notes in Artificial Intelligence and Lecture Notes in Bioinformatics) [Internet] 11257 LNCS:250–258. Available from: https://link.springer.com/chapter/10.1007/978-3-030-03335-4_22. [cited 2022 Jul 22]
107. Oliveira A, Pereira S, Silva CA (2018) Retinal vessel segmentation based on fully convolutional neural networks. Expert Syst Appl 112:229–242
108. Jiang Z, Zhang H, Wang Y, Ko SB (2018) Retinal blood vessel segmentation using fully convolutional network with transfer learning. Comput Med Imaging Graph 68:1–15
109. Ting DSW, Cheung CYL, Lim G, Tan GSW, Quang ND, Gan A et al (2017) Development and validation of a deep learning system for diabetic retinopathy and related eye diseases using retinal images from multiethnic populations with diabetes. JAMA [Internet] 318(22):2211–2223. Available from: https://jamanetwork.com/journals/jama/fullarticle/2665775. [cited 2022 Jul 22]
110. Bellemo V, Lim ZW, Lim G, Nguyen QD, Xie Y, Yip MYT et al (2019) Artificial intelligence using deep learning to screen for referable and vision-threatening diabetic retinopathy in Africa: a clinical validation study. Lancet Digit Health 1(1):e35–e44
111. Tufail A, Rudisill C, Egan C, Kapetanakis V, Salas-Vega S, Owen CG et al (2017) Automated diabetic retinopathy image assessment software: diagnostic accuracy and cost-effectiveness compared with human graders. Ophthalmology 124(3):343–351
112. Kanagasingam Y, Xiao D, Vignarajan J, Preetham A, Tay-Kearney ML, Mehrotra A (2018) Evaluation of artificial intelligence-based grading of diabetic retinopathy in primary care. JAMA Netw Open [Internet] 1(5):e182665. Available from: https://pubmed.ncbi.nlm.nih.gov/30646178/. [cited 2022 Jul 22]
113. Bouhaimed M, Gibbins R, Owens D (2008) Automated detection of diabetic retinopathy: results of a screening study. https://home.liebertpub.com/dia [Internet] 10(2):142–148. Available from: https://www.liebertpub.com/doi/10.1089/dia.2007.0239. [cited 2022 Jul 22]
114. Parah SA, Sheikh JA, Ahad F, Bhat GM (2018) High capacity and secure electronic patient record (EPR) embedding in color images for IoT driven healthcare systems. In: Internet of things and big data analytics toward next-generation intelligence, pp 409–437
115. Gull S, Parah SA, Muhammad K (2020) Reversible data hiding exploiting Huffman encoding with dual images for IoMT based healthcare. Comput Commun 163:134–149
116. Kaw JA, Gull S, Parah SA (2022) SVIoT: a secure visual-IoT framework for smart healthcare. Sensors 22(5):1773. https://doi.org/10.3390/s22051773
117. Kumar D, Taylor GW, Wong A (2019) Discovery radiomics with CLEAR-DR: interpretable computer aided diagnosis of diabetic retinopathy. IEEE Access 7:25891–25896

118. Liu C, Han X, Li Z, Ha J, Peng G, Meng W et al (2019) A self-adaptive deep learning method for automated eye laterality detection based on color fundus photography. PLoS One [Internet] 14(9):e0222025. Available from: https://journals.plos.org/plosone/article?id=10.1371/journal.pone.0222025. [cited 2022 Jul 22]
119. Martins J, Cardoso JS, Soares F (2020) Offline computer-aided diagnosis for glaucoma detection using fundus images targeted at mobile devices. Comput Methods Prog Biomed 192:105341
120. Meng Q, Housden J, Matthew J, Rueckert D, Schnabel JA, Kainz B et al (2019) Weakly supervised estimation of shadow confidence maps in fetal ultrasound imaging. IEEE Trans Med Imaging 38(12):2755–2767
121. Quellec G, Lamard M, Conze PH, Massin P, Cochener B (2020) Automatic detection of rare pathologies in fundus photographs using few-shot learning. Med Image Anal 61:101660
122. Zhu J, Li Y, Hu Y, Ma K, Zhou SK, Zheng Y (2020) Rubik's cube+: a self-supervised feature learning framework for 3D medical image analysis. Med Image Anal 64:101746
123. Chen L, Bentley P, Mori K, Misawa K, Fujiwara M, Rueckert D (2019) Self-supervised learning for medical image analysis using image context restoration. Med Image Anal 58:101539
124. Yellapragada B, Hornauer S, Snyder K, Yu S, Yiu G (2022) Self-supervised feature learning and phenotyping for assessing age-related macular degeneration using retinal fundus images. Ophthalmol Retina 6(2):116–129

CNN-SVM with Data Augmentation for Robust Blur Detection of Digital Breast Tomosynthesis Images

Nur Athiqah Harron, Siti Noraini Sulaiman, Muhammad Khusairi Osman, Noor Khairiah A. Karim, and Iza Sazanita Isa

1 Introduction

Early diagnosis is key to achieve impressive survival rates for breast cancer patients. Therefore, the accuracy of proper diagnosis relies on sufficient radiologic image quality to obtain high-quality, artefact-free, and blur-free images [1]. Determining if a digital image is of adequate quality to be investigated further when taken in an uncontrolled setting is one of the most challenging components of the process. The most common causes of digital image quality reduction are images obtained with

N. A. Harron · I. S. Isa
Electrical Engineering Studies, College of Engineering, Universiti Teknologi MARA, Cawangan Pulau Pinang, Penang, Malaysia

S. N. Sulaiman (✉)
Electrical Engineering Studies, College of Engineering, Universiti Teknologi MARA, Cawangan Pulau Pinang, Penang, Malaysia

Advanced Rehabilitation Engineering in Diagnostic and Monitoring Research Group (AREDiM), Electrical Engineering Studies, College of Engineering, Universiti Teknologi MARA, Cawangan Pulau Pinang, Penang, Malaysia

Integrative Pharmacogenomics Institute (iPROMISE), Puncak Alam, Selangor, Malaysia
e-mail: sitinoraini@uitm.edu.my

M. K. Osman
Centre for Electrical Engineering Studies, Universiti Teknologi MARA, Cawangan Pulau Pinang, Permatang Pauh, Penang, Malaysia

N. K. A. Karim
Department of Biomedical Imaging, Advanced Medical and Dental Institute, Universiti Sains Malaysia, Bertam, Kepala Batas, Pulau Pinang, Malaysia

Breast Cancer Translational Research Programme (BCTRP), Advanced Medical and Dental Institute, Universiti Sains Malaysia, Kepala Batas, Penang, Malaysia

S. A. Parah et al. (eds.), *Intelligent Multimedia Signal Processing for Smart Ecosystems*, https://doi.org/10.1007/978-3-031-34873-0_6

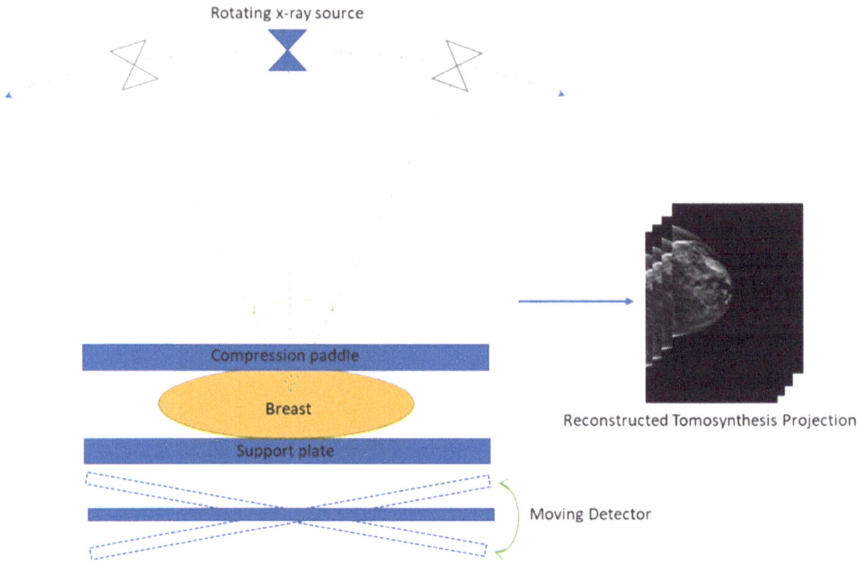

Fig. 1 Digital breast tomosynthesis system and a series of exposures result in multiple projections

a limited angular range and moving devices. Hence, automatic quality analysis of medical images has emerged as a significant area of research.

Digital breast tomosynthesis (DBT) is an imaging technique that helps characterize and diagnose abnormalities in breasts, especially in women with nonfatty, less dense breasts [2]. DBT reconstructs multiple projection pictures, enabling a visual review of thin breast sections. This technique has the ability to reveal tumors that are buried or covered by surrounding normal tissue and may be cancerous. DBT entails the capture of several projection exposures by a digital detector from a moving mammographic x-ray source [2]. Figure 1 illustrates the DBT system and the example of images projected.

DBT is known for its capability to detect breast cancer at an early stage because the advanced technology of using quasi-three-dimensional breast images and yielding comprehensive evaluations of the breast's underlying tissues [3]. Since the fundamental operation of DBT is based on an x-ray acquisition and processing technique to produce the projection images, achieving an appropriate image quality is necessary while maintaining the radiation dose as low as reasonably possible. DBT, on the other hand, exhibits out-of-plane artefacts in the z-direction due to its restricted angle sampling. In the direction of tube motion, the out-of-plane artefacts appear as blurry counterparts of in-plane objects [4]. Blurring is a well-known DBT event that occurs during image capture. It has been reported that it reduces lesion detection performance and masks tiny microcalcifications, failing to detect more minor abnormalities early until they reach more advanced stages, particularly in dense breast tissue [5]. As a result, blur detection is an important and fascinating topic in medical pictures.

Identifying functional features to discriminate between distorted and undistorted image regions is a fundamental aspect of blur detection. There are numerous methods for fixing this issue, most of which use a two-step strategy to distinguish between explicit and blurred parts. The first stage comprises manually separating image components based on a collection of empirical data in gradient format. A binary classifier is used to discriminate between deformed and clear regions [6]. As blur degrades image quality, it is distinct from other recognition problems, which allow for various degradation variables such as noise or blur. However, to identify blur efficiently, we must learn blur-related features while tolerating the significant variance supplied by varying image content. Instead of employing visual feature extraction methods such as scale-invariant feature transform (SIFT) or histograms of oriented gradients (HOG), CNN representations of features usually outperform handcrafted features.

Deep-learning CAD (AICAD) systems help reduce reading times without significantly compromising sensitivity, specificity, or recall rates, with reductions in reading times ranging from 14% to 40% [7–10]. On the other hand, the massive number of datasets with uncertain images will impact the CAD diagnostic process as calculation time is lengthened and only a minimum number of patients can be diagnosed due to exhaustion of the reader. Only acceptable data quality should be considered when optimising the CAD system. Therefore, this work aims to evaluate the capacity of a deep learning strategy to extract blur features and categorise blurred and non-blur DBT images to optimise the diagnostic process by eliminating blurry slices. A hybrid CNN-SVM model for the identification of blurry pictures is proposed in this work. On both, the public dataset and the retrospectively collected dataset, the model is tested. This approach's core idea is to train a five-layer CNN to extract resilient layer characteristics from an input picture before sending the images to SVM classifiers. Through data augmentation, the properties of DBT pictures are also analysed, and the feature representation capability of traditional deep CNN models is also researched. Comparisons with a plain CNN model and several classical deep CNN show that the proposed model outperforms other types of deep CNNs.

2 Related Works

2.1 Blur Image Detection Techniques

Most scanned images obtained have two kinds of regions: blur and sharp. Blur is divided into two types: defocus blur, also known as out-of-focus blur, which is brought on by the visual imaging system, and motion blur, which is caused by the camera's movement in relation to the scene's objects [11]. The study of [12] categorised the public domain research on blur images into three primary phases: blur detection, classification, and image restoration. Blur detection is a stage in the image restoration process known as pre-processing. Researchers have proposed many blur detection algorithms.

In addition to the previous study, [13] investigated 36 different methods or focus measure operators for calculating an image's blurriness metric. While some of these techniques are straightforward and merely use grayscale pixel statistics, others are more intricate and feature-based and evaluate the image's local binary patterns. Furthermore, [14] discusses three distinct strategies for blur detection, including the Fast Fourier Transform, the Laplacian Operator, and the Haar wavelet transform (HWT). In rapid Fourier transform, the algorithm analyses the frequencies in the image at various locations and determines whether the vision is blurry based on the frequency level. When there is a low quantity of high frequency, the image is declared fuzzy, and the programmer determines whether there will be a high or low amount of high frequency [13]. On the other hand, Haar Wavelet Transform (HWT) divides the images into NxN blocks by iteratively going over each tile in the two-dimensional HWT and grouping similar tiles into clusters with observable changes. Then, small groups of images are designated blurred.

One of the primary distinctions between Laplacian and other operators is that Laplacian operators remove inward and outward edges, whereas other operators remove edges in a defined direction. The Laplacian operator is used to determine the edges of an image. Additionally, it is a derivative operator, but since all other derivatives are first order, Sobel, Kirsch, and the Laplacian operator differ fundamentally from each other resulting in thicker edges and increased sensitivity to noise, whereas the Laplacian operator is a second order derivative mask that resulted in an increasing sharpening effect on images. The Laplacian operator is further classified into two types: Negative Laplacian operators and Positive Laplacian operators. As a result, this study examines and evaluates the Laplacian operator to detect blur in DBT images.

2.2 Deep Learning Approaches for Blur Image Detection

Along with the availability of massive data storage and evolving processing power throughout time, researchers have successfully developed pre-trained networks with various layers and depths that have been trained on huge datasets like ImageNet for public usage [15]. Due to its capability to extract rich feature representations for various images from these massive datasets, CNN is regarded as a potent deep-learning approach when trained on large datasets of various images. However, exploiting the potential of CNNs without spending time and effort on training is straightforward by using a pre-trained CNN as a feature extractor. As blur represents a deterioration of image quality, it is distinct from other issues, which allow for multiple degradation factors such as noise.

Since many computer vision tasks are dependent on learnt features, researchers in this field have turned away from heuristics-based earlier techniques and toward the Deep Learning approach. CNN is a typical discriminant model widely used to solve object detection and instant object classification problems. Most recently, [16] presented and evaluated the usage of Deep CNNs for determining whether a picture is blurred or not. However, the study discovered that the CNN approach is significantly less sensitive than the Laplacian method. CNN is a probabilistic network. However, CNN performs better in the specificity metric. It is somewhat more accurate at recognising non-blurry images. The Laplacian technique is limited in what it can accomplish, but CNN has the potential to improve. In one of the images that is considered blurred by the Laplacian approach but sharp by CNN, it can be observed that a portion where the focus disappears is too little to be relevant for the Laplacian method when compared to the remainder of the image. CNN, on the other hand, takes it into account. It may be further studied by sketching network points of interest or applying Heatman to the weight activation [17].

In 2012, Alexnet [15] won the ImageNet Large Scale Visual Recognition Competition (ILSVRC-2012). Weight sharing Rectified Linear Unit (ReLU) and dropout techniques have substantially affected future CNN model development. VGG [18] won the ILSVRC-2014 crown for detection and classification tasks in 2014, illustrating the enormous advancements that CNN-based deep learning algorithms made. By contrast, the Inception model is composed of infinite Inception structures. These Inception structures may allow for a reduction in model parameters while boosting the diversity of learned properties. He et al. [19] has developed Resnet in 2015 by cascading many residual modules to address the problem of overfitting. The result is a well-trained model with 152 levels of depth, which was awarded first place at the ILSVRC-2015. Additionally, the ensemble technique was employed as a learning paradigm to enhance the classification performance. The CNN model hasn't been applied to classify and identify blur in DBT images, as far as the authors are aware, even though it has been used to categorise objects or characters using a deep (CNN-based) representation.

2.3 *Deep Learning Approach in Digital Breast Tomosynthesis Images*

Recent years have seen much research on DBT published in radiology and the domain of artificial intelligence [20]. There are presently a number of deep-learning-based CAD systems for DBT, yet, there are still many obstacles that need to be resolved. Some of the problems that are faced in a healthcare scenario involving DBT images and their robust blur detection includes robustness to various noises [21–24], efficient realization [25, 26], authentication [27] and classification. Studies on deep learning DBT diagnostics fall into two categories: classification and localization. Due to model selection and processing resource constraints, researchers frequently construct their investigations to classify using patches or entire pictures. Typically, localization problems are divided into object detection and segmentation. Numerous research has used various forms of deep learning architectures, most notably CNN for DBT image classification, to classify positive or negative mass, malignant, benign or architectural distortion. Most research employed DBT z-stack pictures, whereas some used phantom breasts.

Cancer localisation and classification are critical issues in biomedical image analysis. Research on cancer localisation and segmentation using DBT images has been reported [28–30]. Most studies have made an effort to categorise and identify masses and microcalcifications. To discover abnormalities on DBT slices, experienced radiologists must evaluate each frame from the DBT z-Stacks and locate tumours. Given the amount of slices in each DBT scan, the process takes a long time.

Apart from immediately improving the cancer detection models, detection accuracy, the diagnosis procedure would also benefit from the consistency and quality of projection views and reconstruction images. The use of traditional image reconstruction methods on medical images is widespread. However, DBT reconstruction using deep learning is still an open area. Understanding how the objectives of such work might be transferred to the deep learning domain is critical since analysing DBT volumes differs greatly from many well researched image analysis applications. Proper study reading, for instance, calls for extremely high resolution (>1 M pixels), a broad gray-scale dynamic range and contrast, as well as information from prior research and medical sources. It also requires specialised training in image interpretation and clinical decision-making. Contrastingly, standard deep learning models, now in use for image classification call for inputs that are almost an order of magnitude lower in terms of spatial resolution and picture contrast [31].

The CNN-based methods for DBT images also have shown improvement in denoising [32, 33], deblurring [34] and in-depth contrast resolution [35] performance. However, due to the limited public DBT database, the studies only used physical or simulated breast phantom. Several studies on the deep learning approach in DBT images are tabulated in Table 1.

2.4 Features Extraction

Designing features for picture recognition has been a long-standing area of research. Powerful features have been created to enhance the creation of traditional computer vision and pattern recognition applications, such as local binary pattern (LBP), SIFT, and HOG. Feature extraction is one of the most important aspects that might influence the outcome. In this study, many methods were employed by the researchers to determine the characteristics of the images. Recent advances in deep learning have seen great success in visual recognition tasks including object and picture identification and classification [18, 36, 37]. Given that the descriptors obtained from these neural networks, which were trained on enormous natural image datasets, are rather scarce, it is common to refer to these CNNs as generic feature extractors. Even though CNN was initially trained on a dataset for visual identification, automatically extracted CNN features have proven to be quite adaptable. They are often superior to conventional handcrafted features on other graphical computing applications. These deep networks can automatically derive feature hierarchies from unprocessed pixels.

Table 1 Studies on deep learning approach in DBT images

References	Image type (Dataset)	Contributions	Method	Finding/assessment	Performance measure
Gao et al. [32]	Breast phantoms (Digital & physical) (Private)	Denoising DBT images	DNGAN (DCNN with adversarial)	improved the contrast-to-noise ratio (CNR) and detectability index (d′) of the simulated MCs in the validation phantom DBTs. *Only tested the denoiser on a small set of human subject DBTs.	CNR improved 0.097
Wu et al. [35]	Breast phantoms	Improve the contrast and in-depth resolution of DBT reconstruction images	UNet and ROI loss	improve the contrast and in-depth resolution of microcalcifications novel ROI loss function for adding attention to the region of microcalcifications *ROI loss only led to slightly worse overall RMSEs and SSIMs than L2 loss. There was no substantial visual difference in the glandular reconstruction using the two losses.	SSIM 0.0.724 ± 0.098
Sahu et al. [33]	DBT images (Private)	Denoising the DBT projections	GAN	The model contains a generator that learns the distribution of data and a discriminator, which predicts the label of input images (fake or actual). The proposed model takes patches (200×200) of DBT projection images.	The proposed GAN model has the best performance in terms of HaarPSI (0.383) compared with the MSE loss CNN (0.367), KSVD (0.295), and BM3D (0.2851) models

(continued)

Table 1 (continued)

References	Image type (Dataset)	Contributions	Method	Finding/assessment	Performance measure
Samala et al. [53]	DBT images (Private)	Classifying DBT images as mass or normal	Two stages of transfer learning (pre-trained AlexNet)	Rather than using the CNN directly as a classifier, the CNN at the second stage is used for feature extraction. Then, a feature selection method and random forest classifier are applied to the extracted features for classification	AUC of 0.90 with pruning the model and an AUC of 0.89 without pruning
Choi et al. [34]	DBT images (Private)	Method to reduce the blurring artefact in DBT images	DRCNN	Lesion detectability of DBT image is improved significantly after applying the proposed DRCNN *While the proposed network improves the visibility of the lesion, it was observed that the original delicate structures presented in the reference images were not recovered fully	–

Additionally, features retrieved by means of such networks are incredibly adaptable and frequently more efficient than classic handmade features. This serves as inspiration for the proposed CNN design, which is used to extract features. Five layers make up the architecture of this CNN, comprising a fully linked layer and five convolutional layers. The output of the second-to-last fully linked layer is used to extract features.

2.5 *Convolutional Neural Network (CNN) Architecture*

Convolutional networks have been widely used for visual object recognition in designs that take an image as input and, through alternating layers of convolution and subsampling, provide one or more output values that are the threshold to offer binary predictions about object identity. The convolution layer in the CNN model is the first layer to extract a feature from an image; inside this layer, a mathematical operation is used to determine the property of each pixel and the relationship of neighbouring pixels. The pooling layer evaluates important information by removing unnecessary data once features have been extracted. The method of subsampling, which can be either Max pooling, Avg pooling, or Sum Pooling, aims to lessen the size of the data map without losing the imported data. The feature mapping may occasionally result in overfitting when the pooling layer is applied and the important feature is mapped, necessitating the flattening of 2D arrays into 1D arrays prior to the application of a fully connected layer. Due to the fact that all networks have a suitable connection at this layer, the output of a completely linked network is often the ultimate output [38]. Figure 2 shows a typical CNN architecture.

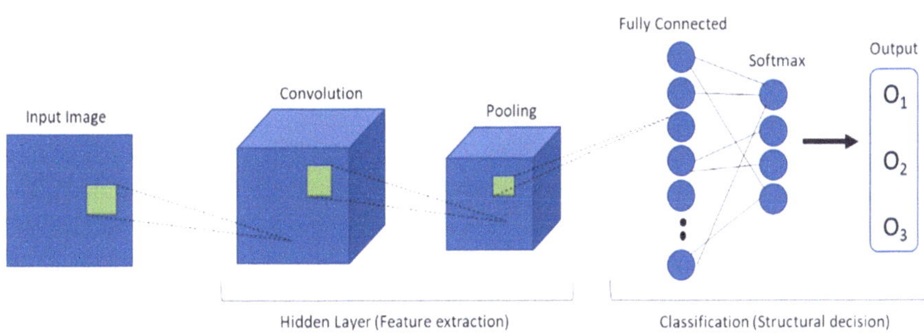

Fig. 2 General CNN architecture

2.6 Support Vector Machine (SVM)

A sophisticated classification algorithm, SVM, is one of the most popular data modelling and classification approaches. Among its benefits is its exceptional generalisation ability, which is closely tied to accurately sorting the samples outside the feature space used for training. It is a supervised machine learning that can be used to solve classification and regression issues and contains both linear and non-linear data. SVMs are applied in various applications, including face recognition, clinical diagnosis, industrial process monitoring, image processing and analysis. They are used to categorise and recognise patterns in many forms of data. The last layer of CNN's fully connected layer employs an SVM classifier to increase flexibility and successfully suit the date length to turn the kernel [39]. Additionally, a lot of focus has been placed on combining neural networks with SVM for recognition. In [40], to handle recognition tasks, a pre-built CNN representation along with a linear SVM is employed. According to the findings, it is a formidable rival to more sophisticated and finely tuned state-of-the-art algorithms on a variety of datasets.

2.7 Hybrid CNN-SVM

Several studies [41–43] have proposed the hybrid CNN-SVM approach combining the benefits of the two most effective and widely used methods for image classification, CNN and SVM. The last layer of CNN is replaced with the SVM model to create the structure of the combined CNN and SVM model. As a result, the output of the fully connected layer of CNN will be processed as input for SVM to enhance classification. Combining the benefits of CNN and SVM is primarily motivated by the CNN model's quick addition of hidden layers, which has facilitated extracting features while improving accuracy and performance. However, SVM also outperforms other algorithms in terms of efficiency and speed thanks to its distinctive characteristics for extracting features [42]. The purpose of using this function is to expedite the information extraction process based on the train and test method. The SVM model trains the feature vectors for improved classification and decision-making using the output of a fully connected layer as input. With DBT images, this combination could provide the most significant blur extraction outcomes. Table 2 summarises the previous study related to this method and its findings.

Table 2 A previous study of the CNN-SVM method

Research	Method	Application	Finding
Khairandish et al. [42]	Hybrid CNN-SVM	Threshold Segmentation Approach for Tumor Detection and Classification of MRI Brain Images	The results of earlier research are based on many models, each with varying degrees of accuracy. The overall accuracy of the hybrid CNN-SVM is obtained as 98.4959%
Wu et al. [35]	Hybrid CNN-SVM with SVM modification technique	Classification Technique for Human–Vehicle Targets in an Automotive LFMCW Radar	The suggested technique outperformed numerous state-of-the-art methods, according to experimental data, with the highest F1 score and AUC at 0.98 and 0.95, respectively
Santos and Welfer [41]	Hybrid CNN-SVM	Extracting Characteristics and Classifying Cattle Branding	The research showed better results regarding accuracy and processing time for the task of recognising cattle brandings. The accuracy percentage increased as more sample images were added to the training and feature extraction phases
Xue et al. [43]	CNN-SVM with Augmentation	Microvascular Morphological Type Recognition	The recognition rate is 92.74% on the patch level with data augmentation and classifier boosting. The synthesis results indicate that our system can assist clinical diagnosis to a certain extent

3 Materials and Methods

The proposed method evaluates work performance on two DBT image databases of public datasets and data collected retrospectively, implemented on a 2.90 GHz PC using MATLAB R2022a with (Core i5 CPU, 16 GB Ram, and 12 GB Nvidia GPU). According to the proposed methodology, the extracted feature is processed as indicated in Fig. 3 after pre-processing, taking training and testing into account. In this procedure, the CNN model extracts the features, which transfers them to a fully connected layer through a series of steps and the activation function. The CNN

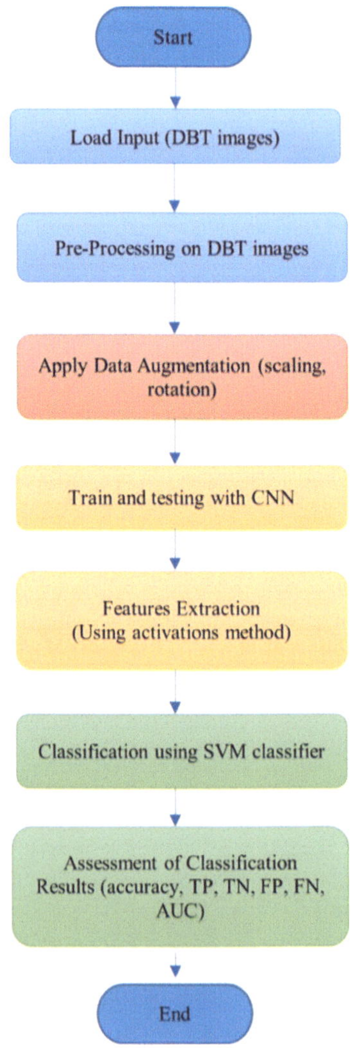

Fig. 3 Flow diagram of the proposed methodology

classifier will be trained using the extracted image features. In contrast, the SVM model processes and trains each image using the CNN output that is fully connected as an input. At the last stage, the classifier categorizes whether the input image is blurry or non-blurry. Finally, the accuracy of the suggested hybrid CNN-SVM is assessed taking into account the specified parameter.

3.1 Dataset

The dataset utilised in this study was adapted from two sources; a public dataset [30] from wiki.cancerimagingarchive.net (https://rb.gy/ygnho7), and data collected retrospectively from 2019 to 2022 at the Imaging Unit, Advanced Medical and Dental Institute (AMDI), Universiti Sains Malaysia (IPPT dataset). The committee with ethics JEPeM code USM/JEPeM/21090622 has authorised this research. In this work, 250 public dataset images were chosen at random to construct training and validation datasets. The data is split into 70% for training and 30% for testing. Then, 149 IPPT images that were not utilised for training or validating CNN models are used as test data. The basic biomedical imaging file acquisition format, DICOM, which includes meta-data essential for retaining essential image information, was used to acquire both database images [44]. However, this format cannot be used as input data for any machine learning approach that uses deep learning. The images are converted to a high quality to decrease the impact of meta-data losses during image conversion [44]. So, for this study, both images' dataset was converted to .bmp format.

3.2 Proposed Methods of CNN-SVM with Data Augmentation

This section will focus on the hybrid method that uses CNN to extract blurry features from images and SVM to detect blurry images. Figure 3 depicts the methods and phases of the proposed hybrid CNN-SVM model. The hybrid model's architecture is completed in seven steps:

1. Inputting the DBT Image

   A sample DBT image from a specified directory is imported into the image dataset to build the input imageDatastore. ImageDatastore automatically recognizes images based on folder names and saves the data into an ImageDatastore object containing the image as well as the category labels associated with each image.
2. Pre-Processing on DBT Image

   To ensure that the system can read the correct input and create a better environment for image analysis, the major functions are applied on the DBT images in this stage. The steps are as follows:

(i) Resizing of the image: Regarding the input image, the size of each image differs in height and width, but the entire dataset is in the same condition. The next stage establishes a fixed size, such as (246×189), to investigate the dataset as a whole under the same circumstances.
(ii) Balanced image per category: First, the countEachLabel function calculates each category's total number of images. Because the image object has an unequal number of images in each class, the splitEach label is used to trim the set by limiting the number of images in each category to the minimum number of images in a category.
(iii) Training and validation dataset: The dataset is divided into 70% training and 30% validation data, and the split is randomised to prevent biased results.

3. Data Augmentation

   Another way to advance recognition, data augmentation, has significantly improved prediction accuracy and system robustness. A robust pattern recognition model should provide predictions invariant to different label inputs. Regardless of the data collection and labelling problems, a simple strategy is to acquire an abundance of training samples with many variances. Instead, data augmentation techniques are used to generate new training data from the training set. Various affine transforms may be applied, depending on the characteristics and distribution of a dataset. Data augmentation using image rotation is considered to create vast volumes of data. It is important to note that data augmentation improves the efficiency of the activity at hand [45]. In addition, the problem of excessive bias (leading to underfitting) or high variance during deep neural network training is another consideration (resulting in overfitting). When the data is poorly fitted, it is difficult to generalise or fit the data points. The network must be trained for a longer time and should have a larger network of hidden layers when considerable bias is present in the training batch.

   It has a large variance when overfitting data, which means it generalizes on test sets too well. Regularisation techniques and data augmentation approaches can be used to reduce high variance [46]. In this work, the effects of augmenting data for training, such as random scaling and rotations, are compared for all training samples. Rotation and flipping are included to improve rotational invariance. Each sample is randomly augmented with rotation, flipping, and shearing ranging from 30° to 120° to produce artificial image samples.
4. Training and testing with CNN

   When developing a CNN, one of the essential characteristics is the size of the input layer. The conventional 5-layer LeNet-5 supports 28×28 pictures, whereas the 8-layer Alexnet manages 256×256 images. A deeper and broader network should learn richer hierarchical characteristics, but it will require more training samples and iterations to fine-tune. Data limitations make it challenging to train and optimise an extensive network. This study begins with three layers network by utilising standard three-stage designs (convolutional layers, pooling layers, and fully connected layers), with additional layers such as ReLU and batch

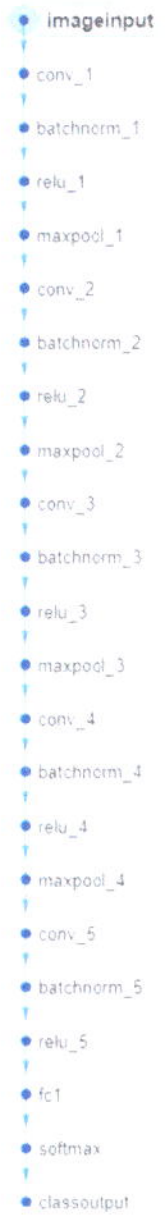

	Name	Type	Activa...	Learnables
1	imageinput	Image Input	246×189×1	-
2	conv_1	Convolution	246×189×8	Weights 3×3×1×8 Bias 1×1×8
3	batchnorm_1	Batch Normalization	246×189×8	Offset 1×1×8 Scale 1×1×8
4	relu_1	ReLU	246×189×8	-
5	maxpool_1	Max Pooling	123×94×8	-
6	conv_2	Convolution	123×94×16	Weights 3×3×8×16 Bias 1×1×16
7	batchnorm_2	Batch Normalization	123×94×16	Offset 1×1×16 Scale 1×1×16
8	relu_2	ReLU	123×94×16	-
9	maxpool_2	Max Pooling	61×47×16	-
10	conv_3	Convolution	61×47×32	Weights 3×3×16×32 Bias 1×1×32
11	batchnorm_3	Batch Normalization	61×47×32	Offset 1×1×32 Scale 1×1×32
12	relu_3	ReLU	61×47×32	-
13	maxpool_3	Max Pooling	30×23×32	-
14	conv_4	Convolution	30×23×64	Weights 3×3×32×64 Bias 1×1×64
15	batchnorm_4	Batch Normalization	30×23×64	Offset 1×1×64 Scale 1×1×64
16	relu_4	ReLU	30×23×64	-
17	maxpool_4	Max Pooling	15×11×64	-
18	conv_5	Convolution	15×11×128	Weigh... 3×3×64×1... Bias 1×1×128
19	batchnorm_5	Batch Normalization	15×11×128	Offset 1×1×128 Scale 1×1×128
20	relu_5	ReLU	15×11×128	-
21	fc1	Fully Connected	1×1×2	Weights 2×21120 Bias 2×1
22	softmax	Softmax	1×1×2	-
23	classoutput	Classification Output	1×1×2	-

Fig. 4 Proposed CNN 5-layer architecture

normalisation (BN) to address overfitting issues. Then the layers are increased to five with an incremented number of filters. The architecture of the proposed CNN is summarised in Fig. 4 for the training option setting, stochastic gradient descent with momentum (SGDM) optimiser is utilised to examine the system's performance. The minimum mini-batch size is 32, the maximum epoch is 50, and the initial learning rates are 0.01. Every 50 training iterations is the frequency of validation that is set. During training, the testing and validation data is shuffled for each epoch.

5. Feature Extraction Using CNN

Each layer of a CNN reacts to an incoming image. However, for visual feature extraction, just a few layers of CNN are required. The network's first layer gather basic visual characteristics like edges and blobs. Following the initial convolutions, image features may be extracted for training the classifier. The collection of filters learned by the CNN during training is in charge of recognizing features in the input image when a query is issued. Note how the network's first layer has developed filters for collecting blob and edge characteristics. Deeper network layers then process these "basic" data, combining them to generate higher-level image features. Because they incorporate all fundamental elements into a richer image representation, these higher-level features are more suited for recognition tasks [46]. The CNN model in the first stage made it possible to extract the features of the input image at the time of the query (see Fig. 5 for an illustration of the feature extraction).

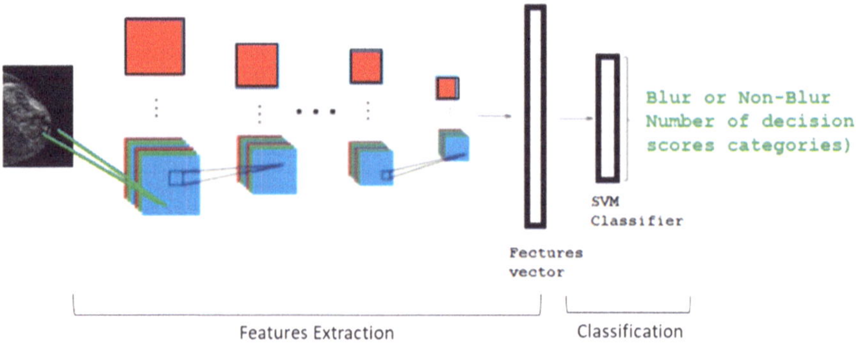

Fig. 5 Diagram of feature extraction and classification

6. Classifier Boosting with SVM

   Regarding the tool proposed in this study, as in Fig. 5, the classifier is utilised after extracting characteristics from DBT images from different sample sets. In the proposed research, the collection of images is randomly divided into two sections; one for the training stage (70%) and the other for the validation stage (30%) to eliminate polarisation in the findings. The final result is an average of the validation stage results. The training data set contains feature vectors taken from images and their associated blur grading. The training data includes feature vectors generated from pictures in CNN activation layers. The classifier then evaluates the properties of a test image and returns a probability indicating the presence or absence of blur in the image. The method is repeated to extract image characteristics from the testing dataset. The features gathered during testing are subsequently supplied to the classifier to assess the trained classifier's accuracy.
7. Evaluation Parameters

   The obtained findings are utilised to compare execution time, accuracy, ROC, and AUC. The system performance accuracy is calculated using the true positive (TP), true negative (TN), false positive (FP), and false negative (FN) values from the available confusion matrix. The ROC probability curve and AUC are calculated by comparing TPR rates against FPR rates. The objective is to simultaneously attain high precision, low recall, and high accuracy.

   (i) **Accuracy:** The proportion of all successfully recognized pixels to all the pixels in the image is used to calculate the accuracy of image classification. It looks at all of the pixels in the image.

$$\text{Accuracy} = \frac{\text{TP} + \text{TN}}{\text{TP} + \text{TN} + \text{FP} + \text{FN}} \tag{1}$$

   (ii) **Precision (TPR):** TP shows the percentage of adequately categorised pixels.

$$\text{Precision} = \frac{\text{TP}}{\text{TP} + \text{FP}} \tag{2}$$

(iii) **Recall (FPR):** The probability pixel identification that logically displays typical characteristics but is classified as abnormal is known as a TN.

$$\text{Recall} = \frac{\text{TP}}{\text{TP} + \text{FN}} \quad (3)$$

4 Experimental Results and Discussion

The proposed method with multiple architectures, including the plain CNN model, proposed hybrid CNN-SVM models, and several traditional deep CNNs is evaluated in this section. This study assessed the performance based on the system's ability to distinguish between blurry and non-blurry DBT images. The experimental outcomes were evaluated using the recognition rate attained in the confusion matrices built using the classification gained during the validation phase of both trials. The performance was also assessed based on its capacity to maintain losses while generating excellent testing and validation accuracies. The proposed method's whole processing time was looked at as well.

4.1 Plain CNN Model Assessment

The dataset was trained and evaluated on a simple CNN model to assess the effectiveness of the constructed plain CNN model and the impact of layer quantity. Figure 6 illustrates the confusion matrix for the best result obtained in the experi-

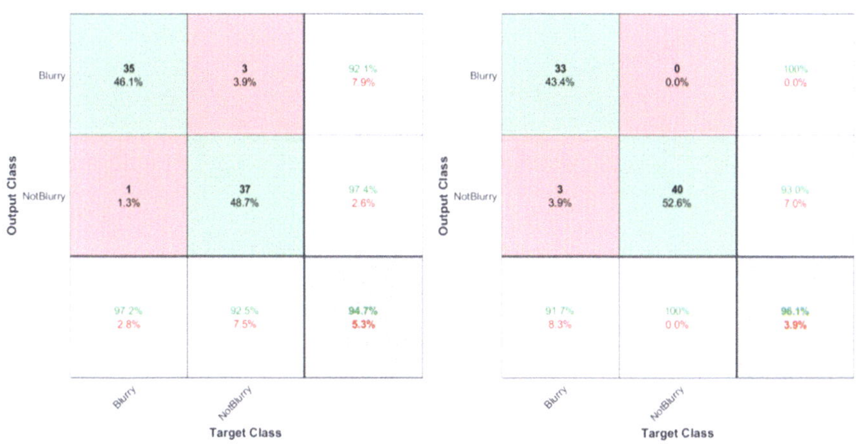

Fig. 6 Confusion Matrix of both plain CNN models (3 layers and 5 layers respectively)

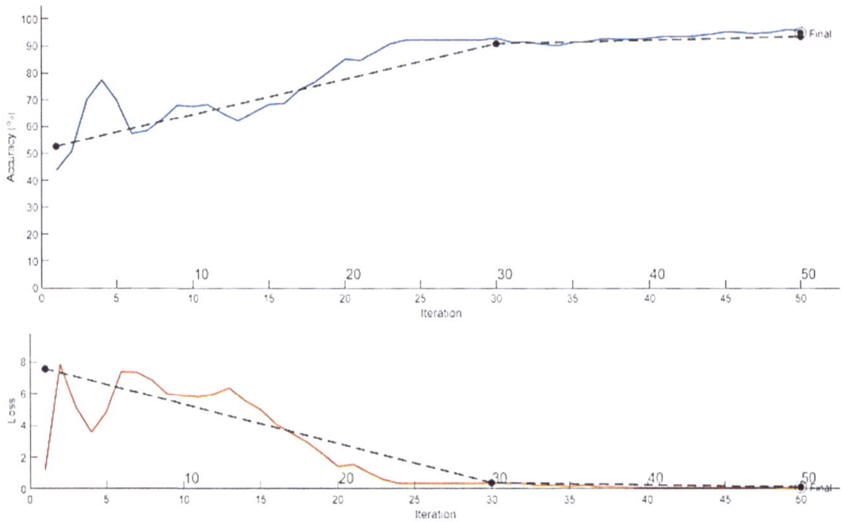

(a)

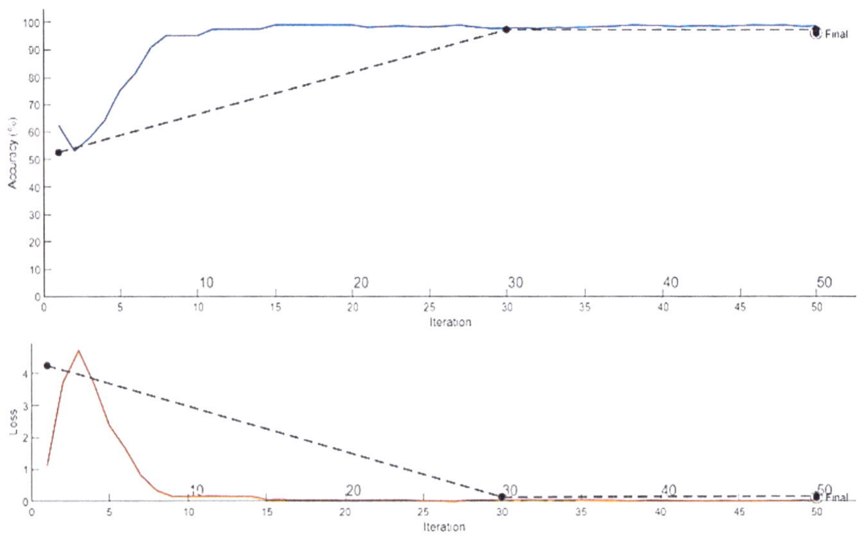

(b)

Fig. 7 Training progress for both plain CNN models (**a**) 3 layers and (**b**) 5 layers

ment for the proposed CNN models (a) 3-layer and (b) 5-layer, respectively, with an overall validation accuracy rate of up to 96.1%. Figure 7 is the training progress for both plain networks. Based on the training progress, the proposed model is neither underfitting nor overfitting and is well-trained.

Table 3 Accuracy comparison on both plain CNN (3 and 5 layers)

Dataset	Measurement (accuracy)	CNN (3-layers) (%)	CNN (5-layers) (%)
Buda [30]	Training	94.74	96.01
	Validation	94.70	96.05
IPPT	Testing	57.04	57.05

Table 4 Accuracy comparison on plain CNN and proposed hybrid CNN-SVM

Dataset	Measurement (accuracy)	CNN (3-layer) (%)	CNN (5-layer) (%)	CNN-SVM (3-layer) (%)	CNN-SVM (5-layer) (%)
Buda [30]	Training	94.74	96.01	96.35	97.22
	Validation	94.70	96.05	94.74	97.44
IPPT	Testing	57.04	57.05	58.39	65.10

Table 3 contains a summary of each performance metric that was measured. An additional layer improves the model accuracy by up to 2%. The increment of layers and filters affects performance. However, since alternative data sources are employed and are not previously trained, the result accuracy for testing is maintained at the same level.

4.2 Proposed CNN-SVM Model Assessment

As per the proposed methodology, after pre-processing, the CNN model extracts the features and then uses the activation function to move it to the fully connected layer in a series of steps. The CNN classifier is trained using the image features that were retrieved. Contrarily, the SVM model trains each picture using the fully connected output of CNN as an input to the process. Finally, the classifier is determined if the input image is blurry or not. The accuracy of the proposed hybrid CNN-SVM is assessed using the evaluation-specified parameter. The goal of this experiment was to demonstrate that the fusion of CNN and SVM slightly contributes significantly to the accuracy performance of the CNNs method.

In Table 4, a comparison of plain CNN with softmax and evolved CNN-SVM is presented. Combining CNN and SVM led to marginal improvements in training and validation accuracy of 97.22% and 97.44%, respectively (96% with plain CNN). The use of a different optimization criterion is primarily responsible for the gain. Softmax's learning algorithm, which aims to reduce prediction loss on the training set, is based on empirical risk minimization. On the other hand, SVM uses structural risk minimization concepts for the testing set to reduce the generalization error. The generalization ability of SVM is better than that of softmax due to a maximized margin [43].

The comparative examination of the proposed model demonstrates that considering the most vital points of both CNN and SVM may lead to improved models

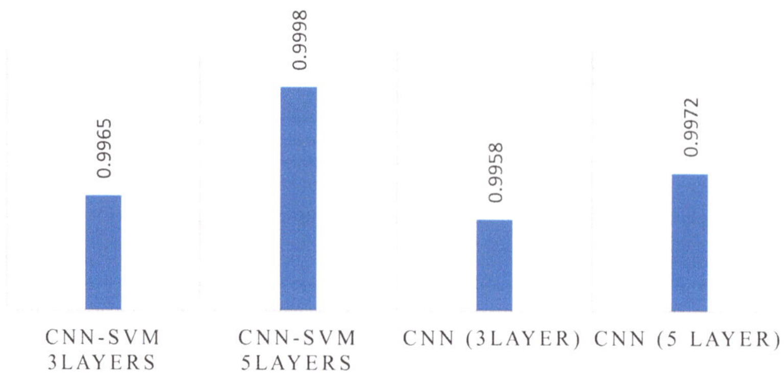

Fig. 8 Comparison of AUC for proposed CNN-SVM and CNN (3 and 5 layers, respectively)

Table 5 Comparison of execution time

	CNN (3-layer)	CNN (5-layer)	CNN-SVM (3-layer)	CNN-SVM (5-layer)
Training time	3 min 13 s	3 min 39 s	3 min 11 s	3 min 28 s

with better classification results. According to this study, the total average testing accuracy is as follows: CNN 3-layers with 57.04%, CNN 5-layers with 57.05%, and the proposed CNN-SVM model slightly increase the accuracy with 58.39% for three layers and 65.10% for the 5-layer model. The testing accuracy is a bit low compared to training and validation because testing data are from a different dataset source. The effect of varying image quality from an additional source, such as contrast and resolution, might affect the recognition process. However, simultaneously, the sensitivity degree of the proposed CNN-SVM express significant progress in the same way.

As shown in Fig. 8, the overall AUC is as follows; plain CNN 3-layers with 0.9958, plain CNN five layers with 0.9972, and the hybrid model obtained 0.9965 for three layers model and the highest 0.9998 for the 5-layers model. It is observed that the proposed hybrid CNN-SVM method has the largest AUC of ROC compared to the standard CNN without the SVM classifier. The majority class frequently tends to drive network weight updates, which often causes the class errors to decrease at the expense of rising class errors in the conventional CNN method. Additionally, Table 5 highlights the corresponding execution time. It can be seen that the proposed CNN-SVM (5 layers) time is comparable to that of the regular CNN. However, the CNN 5-layer method's running time is most significant due to random oversampling. Lastly, Table 6 represents the comparison result between the proposed method and three state-of-the-art blur detection approaches. The state-of-the-art approaches depend on threshold values to identify the blurry or non-burry image. The proposed method result outperforms the three state-of-the-art approaches.

Table 6 Comparison of the proposed method and state-of-the-art approaches

Approaches	Method	Dataset (number of data)	Result	Comparable finding
Kamona et al. (2020) [1]	Machine-learning algorithms and blur measure (BM) operators	Digital mammography (678 images)	Accuracy: >93% (434 images) >85% (244 images)	Identification of the blur magnitude at which diagnostic accuracy noticeably declines is essential.
Francis et al. (2019) [47]	PET-DEBI Laplacian with Gaussian filter	Digital image (2480 images)	Accuracy: 87.57%	Depending on a threshold value.
Szandala et al. (2020) [16]	Laplacian	Digital photo (2000 images)	Accuracy: >80%	Depending on a threshold value. The identical mask-defined Laplacian approach consistently yields the same result.
Proposed approach	CNN-SVM	DBT images (299 images)	Accuracy: **97.44%**	Outperform the state-of-the-art approaches

4.3 Comparison of Proposed Model with Classical Deep CNNs

The proposed hybrid CNN-SVM model is compared to a number of classical deep CNNs, which excel at performing on a variety of visual recognition tasks. The classical deep CNNs for comparisons include Resnet18, Resnet50, Alexnet, VGG16 and InceptionV3. A comparison was made between pre-trained models and the proposed CNN-SVM trained on the DBT image set from both datasets with the same pre-processing and setting. DBT images were supplied to trained deep CNN feature extractors for testing. Following fully connected layers, linear SVMs were used because their features are more discriminative than those of convolutional layers. The training-relevant parameters used are similar to those of the proposed model training parameters. The size of input images was modified using data augmentation to adapt to different network architectures. The experimental results are depicted in Table 7.

In summary, the proposed hybrid CNN-SVM performs better than other pre-trained deep CNNs for DBT blur detection. The accuracy of the CNN-SVM reached up 97–98% for training accuracy and 65% for testing accuracy. Meanwhile, the training and testing accuracies of the other five deep CNNs are not higher than 97% and 59% respectively. The result showed that the proposed CNN-SVM has better

Table 7 Classification result comparison of classical deep CNN model and proposed model

Architecture	Total number of layers	Accuracy		AUC	Execution time
		Training (Dataset 1)	Testing (Dataset 2)		
Resnet18 – SVM	71	0.9580	0.5338	0.9938	3 min 40 s
Resnet50 – SVM	177	0.9583	0.5980	0.9977	5 min 2 s
Alexnet – SVM	25	0.9306	0.5772	0.9861	3 min 30 s
VGG16 – SVM	41	0.9444	0.5772	0.9900	4 min 11 s
InceptionV3 – SVM	315	0.9583	0.4497	0.9946	6 min 52 s
Proposed Model CNN-SVM					
CNN-SVM (3-layer)	15	0.9597	0.5839	0.9965	3 min 11 s
CNN-SVM (5-layer)	23	0.9744	0.651	0.9998	3 min 28 s

performance in terms of training and testing accuracy, AUC, and execution time, with fewer layers compared to all deep CNNs. Thus, the suggested CNN-SVM uses less processing power and has better convergence to capture recognition capability for blur detection. Instead of using a complicated pre-trained network, the proposed CNN-SVM model showed a significant achievement with the highest accuracy and AUC. For the next experiment, only 5-layer CNN-SVM is considered.

4.4 Experiment Data Augmentation

Another technique for determining a classifier's robustness is the impact of data augmentation on classification. Thus, a robust classifier has a low variance in predictions across augmentations. For example, if an image is rotated 20°, the forecast should not alter significantly. Minh et al. [48] examined the robustness of research looking for augmentations with Reinforcement Learning by distorting images with a 50% probability and compared the accuracy of the un-enhanced data with the augmented data. This research examined data augmentation techniques such as rotation, scaling, and flipping. ImageDataAugmenter function in MATLAB is used to generate batches of augmented images. Data augmentation is dynamically performed exclusively on the training data; neither the validation set nor the test set is touched to investigate the effects of augmentation strategies on model robustness, over-fitting, and network convergence.

As demonstrated in Figs. 9 and 10, the training accuracy of the algorithm with data augmentation is not significantly different from that without data augmentation. A closer examination reveals that the curve for the augmented model is smoother and more stable than the curve for the base model. Therefore, data augmentation prevents the network from overfitting and retaining the precise characteristics of the training images. Besides, data augmentation increases and demonstrates the robustness of the proposed model. The results suggest that data augmentation affects performance.

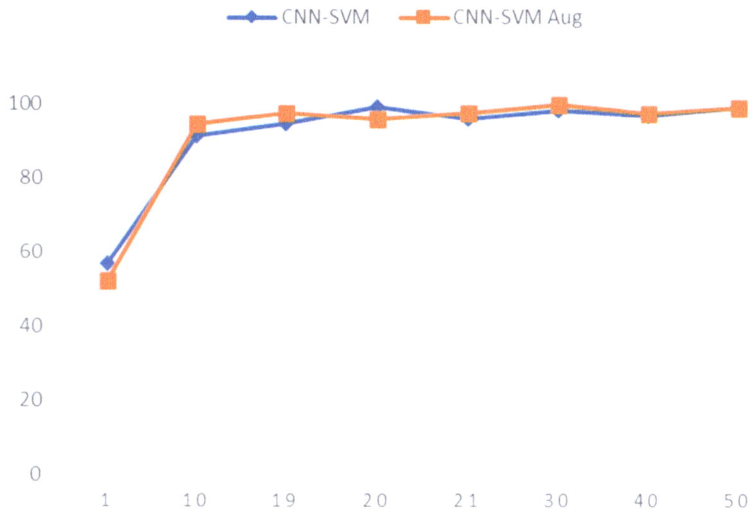

Fig. 9 Training accuracy comparison of CNN-SVM without augmentation and CNN-SVM with augmentation model

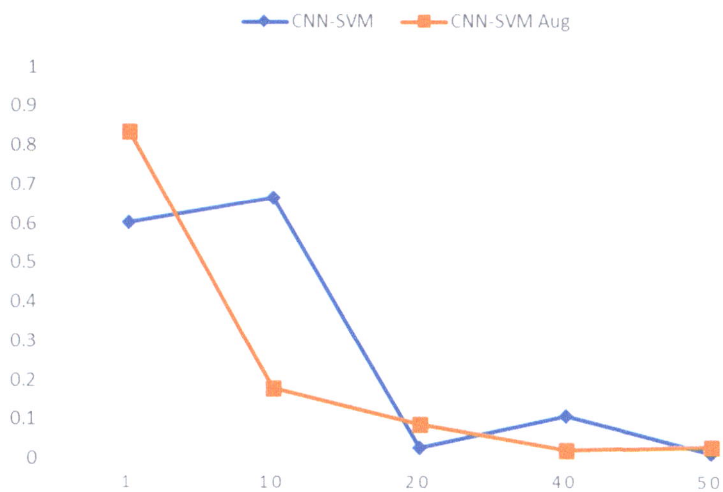

Fig. 10 Training loss comparison of CNN-SVM without augmentation and CNN-SVM with augmentation model

5 Conclusion

A hybrid CNN-SVM is proposed to improve blur detection performance using a deep learning approach. Batch normalisation is used in the network to hasten training and improve blur detection precision. The proposed model can automatically extract characteristics for blur detection in contrast to approaches that rely on customised features. According to experimental findings, the proposed method simultaneously achieves high detection rates and low false alarm rates. Data augmentation's importance is also evaluated to justify the proposed CNN-SVM robustness. Future research will concentrate on current deep learning innovations to enhance deep learning-based blur detection techniques. The DBT images may be blurred due to various causes, including an erroneous system geometry, focal point motion, and patient mobility, all of which reduce the robustness of feature extraction and recognition. Therefore, single features extracted from the CNN model cannot accurately identify them. The misclassified image may limit the cancer detection system in terms of diagnostic accuracy and efficiency. Additionally, the robustness analysis has to be further investigated using various embedding factor values and feasibility for usage in electronic healthcare system as in [49–52]. Future research will explore guided training, a method for solving these types of detection problems, which integrates numerical features (blurry vectors) and blurry features (obtained using the CNN method).

Acknowledgements This research work is financially supported by the Ministry of Higher Education Grant Scheme (FRGS) "A New Cascaded Convolutional Neural Network Model for Deblurring and Contrast Enhancement of Extremely Dense Breast Tissue in Digital Breast Tomosynthesis Images" (Ref: FRGS/1/2021/TK0/UiTM/02/19). The Advanced Control System and Computing Research Group (ACSCRG), Advanced Rehabilitation Engineering in Diagnostic and Monitoring Research Group (AREDiM), Integrative Pharmacogenomics Institute (iPROMISE), and Centre for Electrical Engineering Studies, Universiti Teknologi MARA, Cawangan Pulau Pinang are among the groups for which the authors are grateful for their support and direction during the fieldwork. Finally, the authors thank Universiti Teknologi MARA, Cawangan Pulau Pinang, Malaysia, for their immense administrative and financial support.

References

1. Kamona N, Loew M (2020) Automatic detection of simulated motion blur in mammograms. Med Phys 47(4):1786–1795. https://doi.org/10.1002/mp.14069
2. Helvie MA (2010) Digital mammography imaging: breast tomosynthesis and advanced applications. Radiol Clin North Am 48(5):917–929. https://doi.org/10.1016/j.rcl.2010.06.009
3. Fan M et al (2020) Mass detection and segmentation in digital breast tomosynthesis using 3D-mask region-based convolutional neural network: a comparative analysis. Front Mol Biosci 7:1–15. https://doi.org/10.3389/fmolb.2020.599333
4. Maldera A, De Marco P, Colombo PE, Origgi D, Torresin A (2017) Digital breast tomosynthesis: dose and image quality assessment. Phys Med 33:56–67. https://doi.org/10.1016/j.ejmp.2016.12.004
5. Hogg P, Szczepura K, Kelly J, Taylor M (2012) Blurred digital mammography images. Radiography 18(1):55–56. https://doi.org/10.1016/j.radi.2011.11.008
6. Huang R, Feng W, Fan M, Wan L, Sun J (2018) Multiscale blur detection by learning discriminative deep features. Neurocomputing 285:154–166. https://doi.org/10.1016/j.neucom.2018.01.041
7. Chae EY, Kim HH, Jeong J-W, Chae SH, Lee S, Choi YW (2019) Decrease in interpretation time for both novice and experienced readers using a concurrent computer-aided detection system for digital breast tomosynthesis. Eur Radiol 29(5):2518–2525. https://doi.org/10.1007/s00330-018-5886-0

8. Balleyguier C et al (2017) Improving digital breast tomosynthesis reading time: a pilot multi-reader, multi-case study using concurrent Computer-Aided Detection (CAD). Eur J Radiol 97:83–89. https://doi.org/10.1016/J.EJRAD.2017.10.014
9. Benedikt RA, Boatsman JE, Swann CA, Kirkpatrick AD, Toledano AY (2018) Concurrent computer-aided detection improves reading time of digital breast tomosynthesis and maintains interpretation performance in a multireader multicase study. Am J Roentgenol 210(3):685–694. https://doi.org/10.2214/AJR.17.18185
10. Kaur A, Rashid M, Bashir AK, Parah SA (2022) Detection of breast cancer masses in mammogram images with watershed segmentation and machine learning approach. In: Artificial intelligence for innovative healthcare informatics. Springer, Cham, pp 35–60. https://doi.org/10.1007/978-3-030-96569-3_2
11. Ali U, Mahmood MT (2018) Analysis of blur measure operators for single image blur segmentation. Appl Sci 8(5):807. https://doi.org/10.3390/app8050807
12. Koik BT, Ibrahim H (2013) A literature survey on blur detection algorithms for digital imaging. In: Proceedings – 1st international conference on artificial intelligence, modelling and simulation, AIMS 2013. IEEE, pp 272–277. https://doi.org/10.1109/AIMS.2013.50
13. Pertuz S, Puig D, Garcia MA (2013) Analysis of focus measure operators for shape-from-focus. Pattern Recogn 46(5):1415–1432. https://doi.org/10.1016/j.patcog.2012.11.011
14. Bansal R, Raj G, Choudhury T (2017) Blur image detection using Laplacian operator and Open-CV. In: Proceedings of the 5th international conference system modeling & advancement in research trends, SMART 2016. IEEE, pp 63–67. https://doi.org/10.1109/SYSMART.2016.7894491
15. Krizhevsky A, Sutskever I, Hinton GE (2017) ImageNet classification with deep convolutional neural networks. Commun ACM 60(6):84–90. https://doi.org/10.1145/3065386
16. Szandala T (2020) Convolutional neural network for blur images detection as an alternative for Laplacian method. In: 2020 IEEE symposium series on computational intelligence, SSCI 2020. IEEE, pp 2901–2904. https://doi.org/10.1109/SSCI47803.2020.9308594
17. Samek W, Binder A, Montavon G, Lapuschkin S, Müller KR (2017) Evaluating the visualization of what a deep neural network has learned. IEEE Trans Neural Netw Learn Syst 28(11):2660–2673. https://doi.org/10.1109/TNNLS.2016.2599820
18. Simonyan K, Zisserman A (2015) Very deep convolutional networks for large-scale image recognition. arXiv:1409.1556v6. [Online]. Available: https://arxiv.org/abs/1409.1556v6. Accessed 11 Jan 2022
19. He K, Zhang X, Ren S, Sun J (2016) Deep residual learning for image recognition. In: Proceedings of the IEEE Computer Society conference on computer vision and pattern recognition. IEEE, pp 770–778. https://doi.org/10.1109/CVPR.2016.90
20. Sujlana PS, Mahesh M, Vedantham S, Harvey SC, Mullen LA, Woods RW (2019) Digital breast tomosynthesis: image acquisition principles and artifacts. Clin Imaging 55:188–195. https://doi.org/10.1016/j.clinimag.2018.07.013
21. Parah SA, Sheikh JA, Loan NA, Ahad F, Bhat GM (2018) Utilizing neighborhood coefficient correlation: a new image watermarking technique robust to singular and hybrid attacks. Multidim Syst Sign Process 29:1095–1117
22. Parah SA, Sheikh JA, Bhat GM (2014) A secure and efficient spatial domain data hiding technique based on pixel adjustment. Am J Eng Technol Res 14(2):33
23. Kamili A, Hurrah NN, Parah SA, Bhat GM, Muhammad K (2021) DWFCAT: dual watermarking framework for industrial image authentication and tamper localization. IEEE Trans Industr Inform 17(7):5108–5117. https://doi.org/10.1109/TII.2020.3028612
24. Parah SA, Sheikh JA, Dey N, Bhat GM (2017) Realization of a new robust and secure watermarking technique using DC coefficient modification in pixel domain and chaotic encryption. J Glob Inf Manag 25(4):80–102
25. Bhat GM, Mustafa M, Parah SA, Ahmad J (2010) Field programmable gate array (FPGA) implementation of novel complex PN-code-generator-based data scrambler and descrambler. Maejo Int J Sci Technol 4(1):125–135

26. Bhat GM, Mustafa M, Ahmad S, Ahmad J (2009) VHDL modeling and simulation of data scrambler and descrambler for secure data communication. Indian J Sci Technol 2(10):41–43
27. Hurrah NN, Parah SA, Sheikh JA (2020) Embedding in medical images: an efficient scheme for authentication and tamper localization. Multimed Tools Appl 79:21441–21470
28. Zheng J et al (2021) 3D context-aware convolutional neural network for false positive reduction in clustered microcalcifications detection. IEEE J Biomed Health Inform 25(3):764–773. https://doi.org/10.1109/JBHI.2020.3003316
29. Fan J, Wu L, Wen C (2020) Sharp processing of blur image based on generative adversarial network. In: *ICARM* 2020 – *2020 5th IEEE international conference on advanced robotics and mechatronics*, vol 2. IEEE, pp 437–441. https://doi.org/10.1109/ICARM49381.2020.9195305
30. Buda M et al (2020) Detection of masses and architectural distortions in digital breast tomosynthesis: a publicly available dataset of 5,060 patients and a deep learning baseline. arXiv:2011.07995:1–14
31. Bai J, Posner R, Wang T, Yang C, Nabavi S (2021) Applying deep learning in digital breast tomosynthesis for automatic breast cancer detection: a review. Med Image Anal 71:102049. https://doi.org/10.1016/j.media.2021.102049
32. Gao M, Fessler JA, Chan H-P (2021) Deep convolutional neural network with adversarial training for denoising digital breast tomosynthesis images. IEEE Trans Med Imaging 40(7):1805–1816. https://doi.org/10.1109/tmi.2021.3066896
33. Sahu P, Huang H, Zhao W, Qin H (2019) Using virtual digital breast tomosynthesis for de-noising of low-dose projection images. In: 2019 IEEE 16th international symposium on biomedical imaging (ISBI 2019). IEEE, pp 1647–1651. https://doi.org/10.1109/ISBI.2019.8759408
34. Choi Y, Shim H, Baek J (2018) Image quality enhancement of digital breast tomosynthesis images by deblurring with deep residual convolutional neural network. In: 2018 IEEE nuclear science symposium and medical imaging conference proceedings, NSS/MIC 2018. IEEE, pp 31–33. https://doi.org/10.1109/NSSMIC.2018.8824402
35. Wu J, Li Q, Liang S, Kuang SF (2020) Convolutional neural network with Squeeze and Excitation modules for image blind deblurring. In: 2020 Information Communication Technologies Conference, ICTC 2020. IEEE, pp 338–345. https://doi.org/10.1109/ICTC49638.2020.9123259
36. Sermanet P, Eigen D, Zhang X, Mathieu M, Fergus R, LeCun Y (2014) Overfeat: integrated recognition, localization and detection using convolutional networks. arXiv:1312.6229v3:1–17
37. Szegedy C et al (2015) Going deeper with convolutions. In: Proceedings of the IEEE Computer Society conference on computer vision and pattern recognition. IEEE, pp 1–9. https://doi.org/10.1109/CVPR.2015.7298594
38. Demirhan A, Toru M, Guler I (2015) Segmentation of tumor and edema along with healthy tissues of brain using wavelets and neural networks. IEEE J Biomed Health Inform 19(4):1451–1458. https://doi.org/10.1109/JBHI.2014.2360515
39. Alkhaleefah M, Wu CC (2019) A hybrid CNN and RBF-based SVM approach for breast cancer classification in mammograms. In: Proceedings – 2018 IEEE international conference on systems, man, and cybernetics, SMC 2018. IEEE, pp 894–899. https://doi.org/10.1109/SMC.2018.00159
40. Razavian AS, Azizpour H, Sullivan J, Carlsson S (2014) CNN features off-the-shelf: an astounding baseline for recognition. In: IEEE Computer Society conference on computer vision and pattern recognition workshops. IEEE, pp 512–519. https://doi.org/10.1109/CVPRW.2014.131
41. Santos CA, Welfer D (2019) A novel hybrid SVM-CNN method for extracting characteristics and classifying cattle branding. Lat Am J Comput VI(1):9–16. [Online]. Available: https://lajc.epn.edu.ec/index.php/LAJC/article/view/157
42. Khairandish MO, Sharma M, Jain V, Chatterjee JM, Jhanjhi NZ (2021) A hybrid CNN-SVM threshold segmentation approach for tumor detection and classification of MRI brain images. IRBM 1:1–10. https://doi.org/10.1016/j.irbm.2021.06.003

43. Xue DX, Zhang R, Feng H, Wang YL (2016) CNN-SVM for microvascular morphological type recognition with data augmentation. J Med Biol Eng 36(6):755–764. https://doi.org/10.1007/s40846-016-0182-4
44. Santos M, Bastião L, Silva A, Rocha N (2016) DICOM metadata analysis for population characterization: a feasibility study. Procedia Comput Sci 100:355–361. https://doi.org/10.1016/j.procs.2016.09.169
45. Perez L, Wang J (2017) The effectiveness of data augmentation in image classification using deep learning. arXiv:1712.04621:1–8. [Online]. Available: http://arxiv.org/abs/1712.04621
46. Donahue J et al (2014) DeCAF: a deep convolutional activation feature for generic visual recognition. In: 31st International conference on machine learning, ICML 2014, vol 2, pp 988–996. https://doi.org/10.48550/arxiv.1310.1531
47. Francis LM, Sreenath N (2019) Pre-processing techniques for detection of blurred images, vol 28. Springer, Singapore
48. Minh TN, Sinn M, Lam HT, Wistuba M (2018) Automated image data preprocessing with deep reinforcement learning. arXiv:1806.05886:1–9. [Online]. Available: http://arxiv.org/abs/1806.05886
49. Ahad F, Parah SA, Sheikh JA, Bhat GM (2015) On the realization of robust watermarking system for medical images. In: 2015 Annual IEEE India conference (INDICON), New Delhi, India. IEEE, pp 1–5. https://doi.org/10.1109/INDICON.2015.7443363
50. Sarosh P, Parah SA, Bhat GM, Heidari AA, Muhammad K (2021) Secret sharing-based personal health records management for the Internet of Health Things. Sustain Cities Soc 74:103129
51. Parsa S, Parah SA, Bhat GM, Khan M (2021) A security management framework for big data in smart healthcare. Big Data Res 25:100225
52. Aljuaid H, Parah SA (2021) Secure patient data transfer using information embedding and hyperchaos. Sensors 21(1):282
53. Samala RK, Chan H-P, Hadjiiski LM, Helvie MA, Richter C, Cha K (2018) Evolutionary pruning of transfer learned deep convolutional neural network for breast cancer diagnosis in digital breast Tomosynthesis. Physiol Behav 176(5):139–148. https://doi.org/10.1088/1361-6560/aabb5b.Evolutionary

Lung Lesion Identification Using Geometrical Feature and Optical Flow Method from Computed Tomography Scan Images

Mohd Firdaus Abdullah, Siti Noraini Sulaiman, Muhammad Khusairi Osman, Noor Khairiah A. Karim, Samsul Setumin, and Adi Izhar Che Ani

1 Introduction

The second most frequent disease in the world for both genders, lung cancer, can be fatal [1]. The illness known as lung cancer starts in the lungs but has the potential to spread to other body organs. Lung lesions originate in the lungs and can spread to neighbouring lymph nodes or organs, including the brain [2]. According to the World Health Organization, 8.8 million people die of cancer annually, which is anticipated to rise to 18 million by 2030. The mortality rate attributable to lung

M. F. Abdullah · M. K. Osman (✉) · S. Setumin · A. I. C. Ani
Centre for Electrical Engineering Studies, Universiti Teknologi MARA, Cawangan Pulau Pinang, Permatang Pauh, Penang, Malaysia
e-mail: khusairi@uitm.edu.my

S. N. Sulaiman
Electrical Engineering Studies, College of Engineering, Universiti Teknologi MARA, Cawangan Pulau Pinang, Penang, Malaysia

Advanced Rehabilitation Engineering in Diagnostic and Monitoring Research Group (AREDiM), Electrical Engineering Studies, College of Engineering, Universiti Teknologi MARA, Cawangan Pulau Pinang, Penang, Malaysia

Integrative Pharmacogenomics Institute (iPROMISE), Puncak Alam, Selangor, Malaysia
e-mail: sitinoraini@uitm.edu.my

N. K. A. Karim
Department of Biomedical Imaging, Advanced Medical and Dental Institute, Universiti Sains Malaysia, Bertam, Kepala Batas, Pulau Pinang, Malaysia

Breast Cancer Translational Research Programme (BCTRP), Advanced Medical and Dental Institute, Universiti Sains Malaysia, Kepala Batas, Penang, Malaysia

S. A. Parah et al. (eds.), *Intelligent Multimedia Signal Processing for Smart Ecosystems*, https://doi.org/10.1007/978-3-031-34873-0_7

cancer is 27% of all cancer-related deaths [3]. When the cell begins to spread and grow atypically, cancer develops. A lesion may originate from the unregulated division of a cell [4]. The lesion could expand in any body area; if it does, it is called lung cancer.

After analyzing earlier studies and consulting an expert, computed tomography (CT) scan images can be used to diagnose lung cancer. Early detection can boost therapy effectiveness and patient's survival [5]. Before administering treatment to a lung cancer patient, the doctor can determine the lesion's location, shape, and size using imaging modalities. Among the imaging techniques are CT [6] and Magnetic resonance imaging (MRI) [7, 8]. However, a CT scan is favoured as it is simple and gives accurate classification and placement of foreign masses [9]. Additionally, MRI diagnostics are more expensive, time-consuming, and unavailable than CT scans [10].

Processing and analyzing clinical images are crucial, especially for lung applications. Due to the advantages of non-invasive procedures and the capacity to study complicated organs such as arteries, veins, bronchi, and bronchioles in the lung region, image processing for clinical analysis and diagnostic reasons has gained appeal among medical experts. Since the introduction of scientific computer and imaging technology, numerous attempts have been made to automate the human element of lung lesion identification [5].

The development of computer-aided diagnostic (CAD) systems for lung cancer employing various imaging modalities have generally been studied in a multitude of renowned articles [6, 11]. It is a very efficient technique for helping radiologists increase their diagnostic precision. A typical CAD system includes image filtering, segmentation, nodule recognition, and classification. Any CAD system that can help with the early identification of lung cancer must consist of the crucial step of lung segmentation from chest images. However, automating lung segmentation was challenging due to variations in volume and margin caused by changes in pulmonary inflation in combination with an elastic chest wall.

The next phase in lung cancer CAD systems is the detection of lesions. The characteristics of a lung lesion concentrate on image characteristics such as texture, geometry, and appearance [7]. The geometry of a lung nodule is depicted via geometric graphics without reference to intensity data. A lesion's volume, area, perimeter, diameter, surface area, and aspect ratio have all been applied to characterize it [7].

Numerous methods are documented in the literature, demonstrating that [8, 9], the discovered traits were based solely on a single image. The radiologist must compare the CT scan before and after to confirm cancer images. As a result, a feature that contrasts before and after images was required. The optical flow approach was chosen due to these problems because it stores the direction and speed of a moving item from one image or video frame to the next. The measuring of optical flow was unquestionably a fundamental issue in processing image sequences. Using the optical flow approach, serial CT scans of lung volumes can be compared to evaluate lesions or functional changes in the lung. Its algorithms characterize the spatial movements of each image over time to estimate the vector field using motion analysis [10]. Behbahani et al. claims that motion vectors are assigned to known values for some distinct locations at the boundary of monitored objects, which can restrict traditional optical flow techniques [12]. It was anticipated that this would improve and impact nearby places' optical flow computations.

This chapter aims to study image processing and machine learning strategies for predicting lung lesion slices. An automated method for quantifying lung lesions with an emphasis on the lung region is proposed, as data suggests that changes in lung lesions may be a cancer risk factor [8, 13]. The following are the significant contributions of this chapter:

- Most CAD-based nodule detectors require the radiologist to manually mark the lungs. In the proposed approach, CT images are automatically segmented to isolate the lungs;
- Lesions can have both regular and unusual sizes and forms. Existing algorithms detect lesions using a small number of shape templates. However, the proposed algorithm is shape-independent;
- The new method has been applied to find the possible slices for lung lesion detection based on the standard deviation from the threshold value.

2 A Brief Review of Lung Lesion Detection

Lung cancer ranks second among men and women [14]. Addiction to smoking cigarettes is the primary cause of lung cancer [12, 15, 16], followed by air pollution and radioactive gases. In addition, lung cancer is also influenced by genetic factors. The development of lung cancer might happen gradually over a short time. Different symptoms may be present depending on where the nodule is [17]. Lung cancer grows and compresses thoracic tissues, hindering lung function. From 2007 to 2011, the incidence of cancer in Malaysia was 86.9 per 100,000 among men and 99.3 per 100,000 among women [14]. Cancer patients, their families, and the communities in which they reside bear a high financial cost. Cancer has substantial emotional consequences on patients and their families and their financial expenditures. A total of 72,884 cases were examined based on the analysis done among Malaysians [14], including 64,302 solid lesions and 8582 haematological malignancies. 11.22% of all cancer cases were lung cancer, the leading cause of mortality among men [14].

This chapter aims to improve the image quality and determine the most effective approach for lesion detection to classify cancer as benign or malignant and reduce the number of fatalities. This study's findings will contribute to our understanding of the biology of lung cancer disease and provide significant information for the creation of disability prediction metrics. As a result of this study, clinicians could access to more techniques for the early detection and treatment of lung cancer.

2.1 Issues with Diagnosing Lung Cancer

Aberrant alterations cause lung cancer in the cells that line the airways. As the abnormal cells proliferate and multiply, they have the potential to create malignant lesions that invade neighbouring tissues. Lung cancer can spread to distant organs if lesion cells penetrate the circulation or lymphatic system. There are numerous challenges in identifying lung cancer, which includes:

Challenges in Symptom-Based Diagnosis

Since many of its symptoms do not manifest until the illness has progressed to an advanced stage, lung cancer is the primary cause of cancer-related mortality. Because lung disorders cannot be seen or felt, they are challenging to identify. Chronic cough, chest pain, shortness of breath, and abrupt weight loss are symptoms that indicate something is wrong. Those who smoke or have smoked can take a proactive approach to lung health by getting an annual lung cancer test [18].

Obstacles in the Screening Procedure

Screening is a highly effective method for detecting cancer in its earliest stages. While there are regular screening recommendations for breast and cervical cancers, the same cannot be said for lung cancer, for which none exist. A low-dose CT scan has been recommended for lung cancer screening because it is an effective modality with a mortality reduction benefit. Most countries and organizations have not developed guidelines for lung cancer screening due to concerns regarding the cost-effectiveness and morbidity of low dose CT screening. Considering the current situation, all organizations and institutions have recommended delaying routine lung cancer screenings for smokers at high risk for the time being.

Obstacles Posed by the Age Factor

Cancer screening in older persons needs careful consideration of the patient's overall health, life expectancy, cognitive function, disease risk, and personal preferences [19]. These variables may enhance screening-related hazards while delivering just a little benefit. Typically, the dangers of cancer screening are related to false-positive results or overdiagnosis, resulting in unneeded testing, including radiation exposure, invasive procedures, increased anxiety, and financial stress. Concerns over screening for various solid lesions in the elderly, particularly for malignancies with a more benign clinical course, such as prostate, breast, and colon, have prompted substantial debate. Nevertheless, lung cancer is typically aggressive and, regardless of age or comorbidity burden, which can result in significant morbidity and mortality if left untreated. As a result, interest in lung cancer screening has increased over the past few years, and national recommendations now advocate low-dose CT screening for select individuals.

2.2 Image Processing Technique for CT Scan of Lung Cancer

Radiologists analyze CT scan images to locate and classify lesions as malignant or benign [20]. CT scans can be analyzed using digital image processing techniques to determine lesions' size, kind, and other characteristics. However, these techniques require radiologists with extensive training who are in low supply, particularly in remote and impoverished areas. Manual examination provides a high risk of human error, necessitating CAD tools to help radiologists diagnose and eliminate erroneous reports.

The development of expert assistance systems for identifying lung cancer has made extensive and rising use of medical image processing [13]. A large amount of research is being undertaken on using CAD systems for the early detection of lung cancer [21, 22]. Due to the nature of the data utilized to identify lung cancer, primarily CT scans, automated systems are required. A CT scan of the lungs typically produces approximately 250 images per scan. Examining each patient's enormous dataset is difficult, time-consuming, and arduous. The shape and extent of lesions affecting a patient's prognosis vary from slice to slice. They sometimes involve pulmonary arteries and bronchioles. Similarly, the color in which they appear on CT scans can vary. These traits render them more difficult to identify. However, once recognized, the same qualities help researchers to determine the direction of their methodology.

Segmenting the lungs from a CT scan is critical for detecting lung lesions. Several lung segmentation methods have been presented. Some ways, like [23], uses a small number of seed pixels on the lung portion of the image to segment the lungs. Magalhães Barros Netto et al. [24] re-covers the chest using CT scans and a region-growing algorithm with four seed spots. Lungs and chest were separated similarly. Cascio et al. [25] split the lungs using region-growing. The slice's centre pixel determines the seed. Keshani et al. [26] organises chest CT images into lung walls, parenchyma, bronchioles, and lesions. CT lung removal uses active contour segmentation.

Detection of lesions on CT images is achieved by a variety of approaches described in the literature that involve thresholding based on the intensity or color of the lesion [27, 28]. The resultant image is threshold for extracting related components. In [29] the size of a lesions is utilized as the criterion for identifying probable nodular locations. Local maxima with greater values and dimensions than the typical lesions in three-dimensional space are determined for each subvolume. Messay et al. [30] apply thresholding for initial segmentation, followed by a rule-based examination of interpulmonary bronchioles' anatomical properties. Unlike previously stated approaches, it simultaneously detects and segments potential lesions. The algorithms reported in [31, 32] are thresholding-based techniques for recognizing lung lesions.

It has been established that lesions possess particular qualities, such as shape, color, and intensity, which have been utilized to construct a composite distinguishing feature [33]. The characteristics are then identified using various classification techniques to identify the lesion. Methods [34, 35] are a couple of such examples. The methodology suggested in [36] extracted lung cancer features using a rotation- and size-independent feature extraction method, ultimately generating 33 features for each class of lung cancer CT images. Tun and Khaing [10] performs feature extraction and classification of lung cancer lesions using image processing techniques. Extraction of geometric and intensity-based statistical characteristics. The physical dimensions that define an object's appearance are its, form measurements. Extraction only considers the following characteristics: area, perimeter, and eccentricity. In the method suggested in [22] histogram-based statistical textural features are extracted with changing feedforward backpropagation neural network to identify lesions from non-lesions for the detection of lesions candidates. Recent studies have also investigated deep learning approaches for lung lesions diagnosis, such as [37, 38]. Recent advancements in automatic lung lesions detection are described in [37, 39, 40] research.

2.3 Optical Flow Method for Lung Cancer Identification

CT images are often hard to compare because the patient's position and breathing rate can change between scans, and the imaging process can vary by accident. The hardest part is finding changes in time between successive CT images when the patient is scanned using volume scanning or at an angle compared to the first image. The process gets more complicated when the serial image is taken from more than one point of view. Researchers have been working on many ways to automatically find lung cancer because the number of CT lung scans has increased significantly, and they need to get better at finding it, [8, 36]. These techniques, however, do not account for lesion changes between serial images.

Ouksili and Batatia [41] proposed a method to remedy this problem using 4D CT approaches by acquiring CT images synchronized to respiration. Based on the results, this proposed method shows good potential for image reconstruction. In a recent article, Jan Ehrhardt et al. [42] developed an optical flow-based way to improve the reconstruction of 4D CT datasets. Respiratory motion decreases the repeatability of anatomic position, leading to difficulties with image acquisition, treatment planning, and radiation delivery. They may generate interpolated CT scans at the precise set breathing phase. The optical flow method suggested by Lawrence et al. [43] is a coarse approximation of motion-based techniques used to estimate the first global parametric transformation, then the local deformation. Five pairs of mismatched CT pictures of lung volumes due to patient posture, respiration, or different fields of view were utilised to test the approach. The optical flow technique compares serial CT scans of lung volumes to evaluate lesions or functional changes.

3 Methodology

The segmentation of lesions is one of the most crucial aspects of the medical field, as the outcome will significantly impact the decisions made by medical professionals. The hospital uses manual delineation and qualitative analysis to identify lung lesions using radiologists' visual evaluation of scanned clinical images, including CT images. Typically, radiologists use visual interpretations and a grading scale to evaluate the manual delineation of lesions. The operation is performed on a computer with built-in software that allows the radiologist to map the perimeter of the lesion. This process is inefficient since numerous specialists must evaluate each image (at least two radiologists). In addition, the lesion's manual delineation procedure resulted in segmentation findings that varied across two or more radiologists.

This chapter provides an adaptable, fully automated method for detecting lung abnormalities in CT scans. The proposed method consists of three stages: pre-processing, segmentation, and post-processing. The pre-processing phase includes thresholding, masking, and image enhancement. A clustering algorithm was utilized to identify lung areas and lesions during the segmentation step. The suggested technology automatically segments the lungs without user input. Segmentation determines the threshold automatically. The image's outer portion is detected and removed using this threshold, and the remaining area is processed to extract the lungs. This strategy uses morphology and associated parts. Next, lung lesions are found. Lesions, bronchus, and blood vessel interiors are isolated from parenchyma. Using statistical and shape-based lesions, a model is trained. The goal is to create an automated system to recognise left and right lung slices in CT scans with lesions based on a threshold value. Variations in image data sets, such as contrast, resolution, disease kind, data size, tissue types, and scanning procedures, render automated systems inaccurate [44]. The details of each process are explained in the subsection below.

3.1 Data Acquisition

This section discusses the dataset and methodology used to evaluate the study's outcomes. Additionally, a unique method for segmentation and lesion identification was developed. For this investigation, CT images were obtained from the Imaging Department, AMDI, USM, Kepala Batas, Pulau Pinang. To implement genuine CT images and patient data, an institution's ethical committee approved this study on August 8, 2019. This phase also included the request for ethical approval from the

Human Research Ethics Committee of USM (JEPeM) under the School of Medical Sciences, USM, IPPT, Bertam, Pulau Pinang. CT scanning is performed on humans; hence this approval is necessary for clinical research involving humans.

Participants were lung cancer patients who received imaging at USM's AMDI. Control participants' photos from AMDI's Picture Archiving and Communication System (PACS) will reveal lesions but have normal CT scans (reported by a radiologist as having no lesions or other imaging indicators of lung cancer). Before being provided to the public, data are de-identified and anonymised. Patient Contributed Image Repository collects DICOM-sized CT scans. MATLAB's image-reading function is used to extract and analyse DICOM header data.

Table 1 displays CT scans of the thorax and non-thorax regions acquired with soft tissue density. Only the part of the thorax will be examined. CT scanner slices of 125 and 1 mm were employed to obtain 1-281 and 1-882 images from a single subject. The number of slices determines distance between subsequent body scan images. If there are more slices, the space between each slice is diminished, but more image data can be collected. The size and resolution of all these images were (1255 × 1001) pixel in (DICOM format) as shown in Table 1. The constrained context in which medical pictures were created disallowed employing PNG or JPEG, affecting image quality. Digital Imaging and Communications in Medicine (DICOM) is a standard for storing and transferring digital images that integrate scanners, servers, workstations, and printers. A DICOM image may also contain patient information, the date, and other unneeded details for lung segmentation. To simplify image processing, we convert DICOM images to PNG. Patient information and tags are deleted when a DICOM image is converted to PNG format to safeguard privacy.

3.2 Lung Segmentation

Lung segmentation is seen as an essential task in lesion CAD systems because the accuracy of segmentation has a significant effect on how well the next steps of analysis work. This section shows how to use morphological image processing techniques to make a segmentation algorithm for the lungs. According to Fig. 1, the transformed PNG image consists of four components (i) a black background, (ii) a circular zone of dark grey, (iii) a brighter region, and (iv) dark grey lungs. The lungs are the target regions, hence the first two components must be eliminated.

Image pre-processing techniques, such as image enhancement and filtering procedures, increase image quality at this stage. These approaches are advantageous for detecting/identifying the features/characteristics of CT scan images. Once the image quality has been enhanced, the subsequent identification phase will be significantly more straightforward. The majority of image processing algorithms necessitate pre-processing to enhance image quality. It is a technique for performing operations on an image to obtain an improved image or extract relevant information from it. It is a signal processing in which the input is an image and the output may

Table 1 CT images of thorax and non-thorax region

	CT scan images
Lower thorax	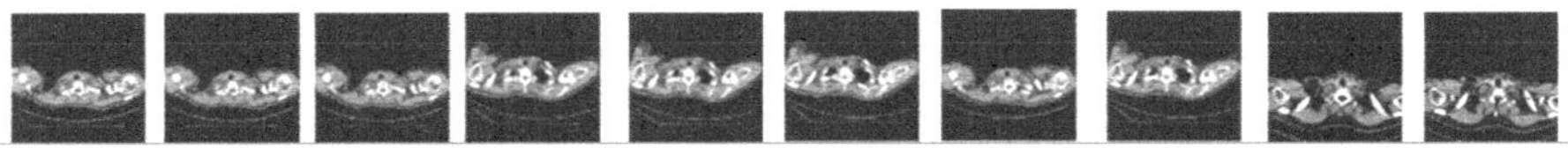
Middle thorax	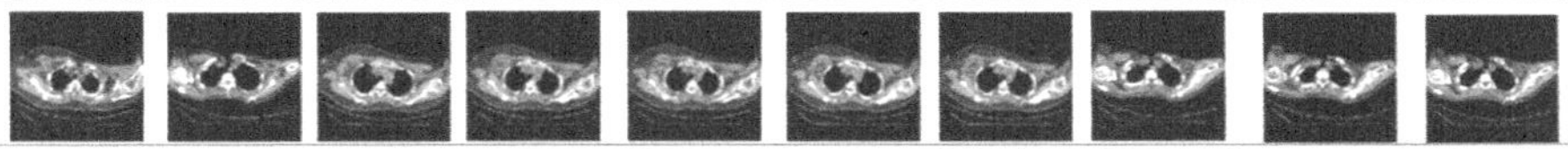
Upper thorax	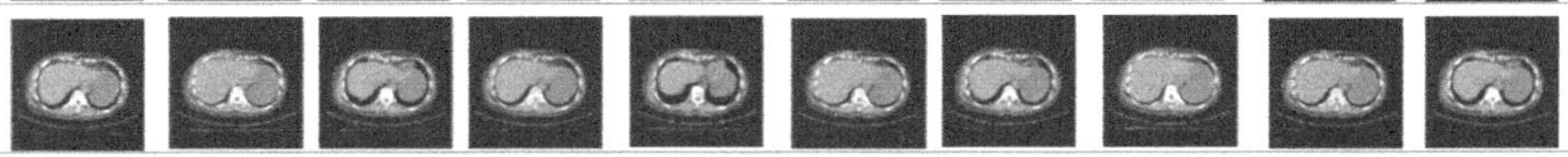
Non-thorax	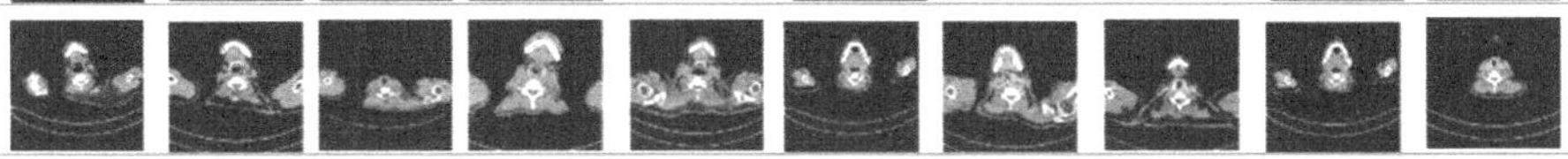

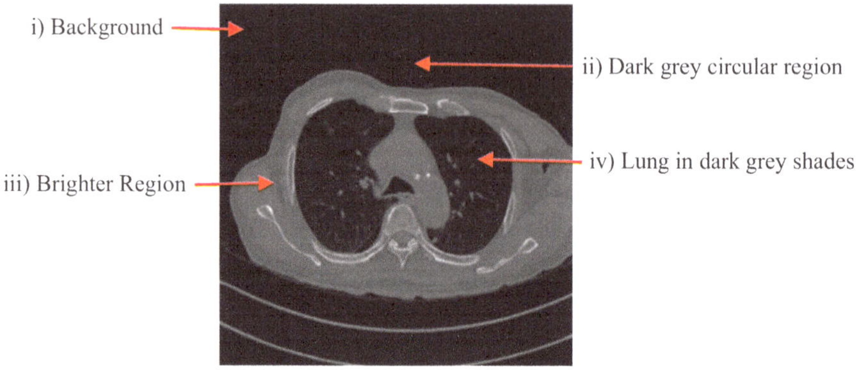

Fig. 1 A CT images of lung and proposed division into four regions

be a image or its related characteristics/features. To initiate this investigation, the photograph will be inputted initially.

Consequently, thresholding is the most fundamental method of image segmentation. The threshold value is utilised by assigning an intensity value to each pixel to differentiate lung CT scans from their backgrounds. To aid the radiologist in screening CT scan images, the thresholding approach is utilized to remove lung CT images from their backdrop. Otsu can be used to create a graph-based image threshold that automatically reduces the intensity of binary images. The background and foreground are independent media because the image has two distinct pixels. The grey scale extends from 0 to 255, with 0 representing black and 255 meaning white. Otsu can distinguish between the foreground and background because the cluster was set to match the selected color space. For example, if there were four clusters, one would be assigned a value of 255, and the others would be given a value of 0. When all the integers are set to zero, black will be produced while white will be created in a color space. As a result, the white cell image setting is 255, while the black background setting is 0. It was also increased to find out the lowest possible threshold value. Image processing can segment lung lesions, as shown in Fig. 2.

Segmentation thresholding techniques include Otsu. To create a graph-based image threshold or reduce the intensity in a binary image, Otsu can be used in conjunction with other software. It is the simplest method for segmenting an image. Pixels in the image can be divided into two groups: background and foreground. With values between 0 and 255, the grey level contains numerous shades of grey, from black at its lowest intensity to white at its highest [15]. To assist the radiologist in screening the CT scan images, the thresholding method separates the lung CT images from their background.

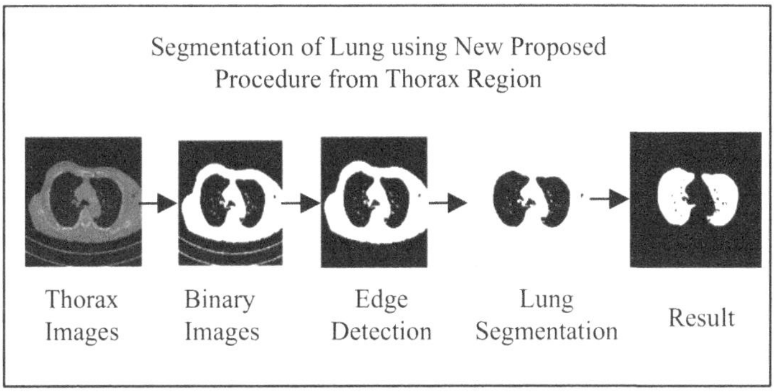

Fig. 2 Lung region identification

3.3 *Lung Lesion Segmentation*

The Otsu method [45] separates the segmented lungs' inner structural arteries, bronchi, and (if present) lesions. Lesions, bronchi, and blood vessels are separated from parenchyma. Since the parenchyma and inner structures have differing intensity levels, thresholding reveals them as bright spots. Lesions differ from other lung structures in a variety of ways. By utilizing a significant variation in shape, non-lesions feature such as blood arteries and bronchi are separated from the lesions. As depicted in Fig. 3, lesions are spherical, whereas the ship and bronchioles are cylindrical. Circular shapes in the detected collection of structures are identified using the size-invariant round/near-round shape detection algorithm [5]. It gives back the centres of the possible lesion locations. These are starting points in a region-growing algorithm to find possible lesion locations. Unlike existing strategies for lesions that use templates, the suggested approach lets us extract lesions of any shape. This makes our method template independent.

After the lung segmentation stage, the potential lesions can be discovered in this phase using the patients' image sets. The most incredible difficulty arises when the lesion indicators are too small or too close to the blood vessel. Image segmentation takes more excellent computing work to address this issue. The lung region's arteries, veins, bronchi, and bronchioles can be recognized based on their roundness to distinguish the lesion from other organs. According to past research, there are no gender differences in roundness. A roundness of 1 indicates a spherical form [46]. Evaluating these lesions in clinical practice could be enhanced by an integrated examination of CT roundness [5]. Figure 4, shows the block diagram of lung lesion and non-lesion segmentation.

The geometrical characteristics of lung lesions are essential for distinguishing benign from malignant lung lesions. For example, the greater the size of lesions, the higher the probability that it is malignant [47]. The thresholding method computes such geometric characteristics from the suspect binary region (SR). The literature [47], indicates that the growth of the malignant portion (in this case, the lesions) is

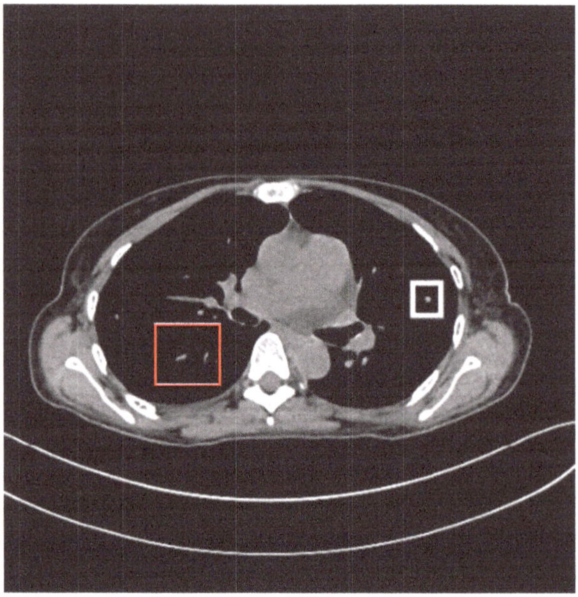

Fig. 3 The region square in white is a lesion and the region squared in red is a non-lesion

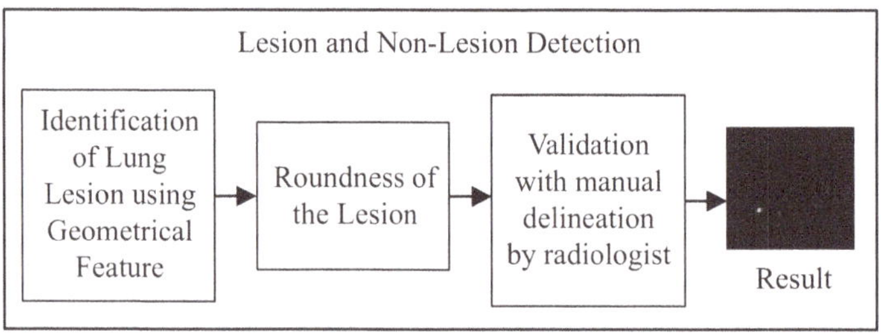

Fig. 4 Lung lesion and non-lesion segmentation

typically round; hence, the roundness of the lesions can be computed using a simple equation

$$\text{Roundness} = \frac{4\pi \times \text{area}}{\text{Perimeter}^2} \tag{1}$$

This metric value, sometimes called the roundness, circularity, or irregularity index (I), is one for circles and less than one for all other shapes. It has been assumed that the possibility that an object is a lesion increases as its circularity increases [5, 46, 47]. The roundness quantifies the irregularity of a connected component. Consequently, if a corresponding part exhibits a roundness index closer to 1, it is likely to be a lesion.

3.4 Lesions and Non-Lesions Detection Using Optical Flow Method

Figure 5 shows the whole process of this project, which was based on [48], but it used a different image segmentation method. However, the new approach has been applied to find the possible slices for lung lesion detection based on the standard deviation from the threshold value. The region in the image is recognized and removed based on the earlier work [20], and the remaining area is then further processed to isolate the lungs. The procedure begins by choosing a pair of images, classifying modes, and evaluating each mode using one of three different optical flow techniques. The effectiveness of each modality will next be assessed for each patient. The selection of the optical flow method was predicated on its presence in colocated slices of the CT scan, which was another essential factor. A lesion must

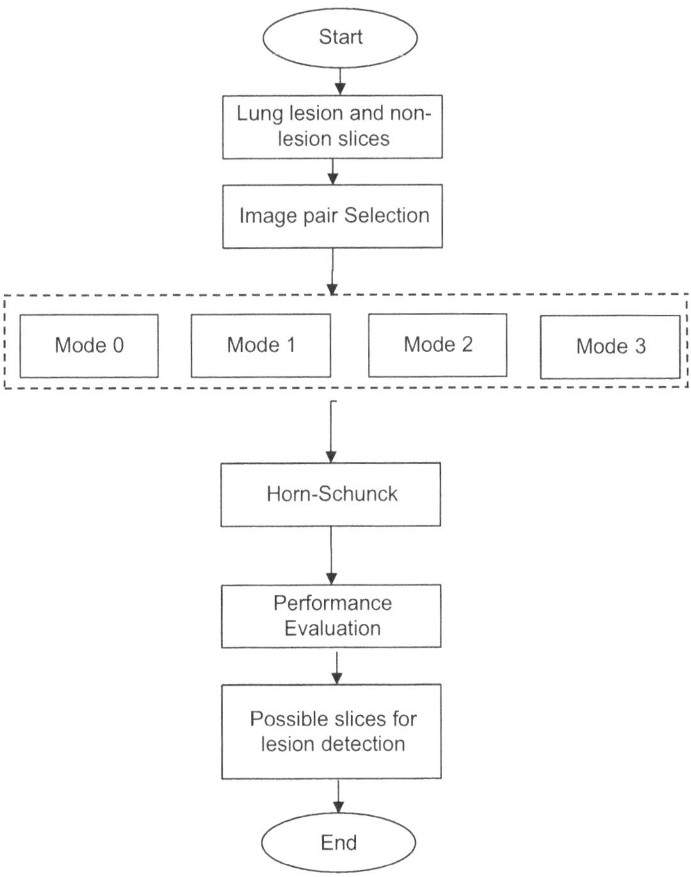

Fig. 5 Flowchart of the possible slices for lesion detection

Table 2 Combination of different modes of optical flow

Mode no.	References and current slice pair	No of pair
0	(1,2), (1,3), (1,4), (1,5), (1,6)	5
1	(1,2), (2,3), (3,4), (4,5), (5,6)	5
2	(1,3), (2,4), (3,5), (4,6)	4
3	(1,4), (2,5), (3,6)	3

be present in both the slices preceding and following its appearance on a CT scan. Conversely, the arteries and bronchi change shape and are, therefore unlikely to be discovered in the exact location in the next slice in the series if they are identified in one slice. This is a crucial feature of lesions [33].

The approach used two input images, referred to as the reference and current images, for image selection while assuming that the image backdrop did not change. Mode 0, Mode 1, Mode 2, and Mode 3 were utilized to choose the current and reference images. The optical flow-based method uses a pair of frames as data to measure optical flow. The pair of images used are not chosen at random but is related in time or space, that is why the method is used over and over in different modes. For object segmentation or classification, two frames must be used for the resulting flow vectors to include accurate data. Afterwards, various modes were tested to develop a method for selecting input frame pairings. Several metrics were used to examine the differences between backdrops that are fixed or moving and foregrounds that are moving. The combination of the various modes between the reference slice and the current slice is displayed in Table 2. Mode 0's reference slice doesn't vary, while the reference slice of the other modes does.

Horn-Schunck was selected as the optical flow method for this investigation. This method implies a steady flow over the entire image. It combines the optic flow constraint equation (OFCE) with the assumption of smoothness [49].

$$F(u, v) = \int_N (\nabla f(x, y, t)\, w(x, y))^2 + \eta \left(\|\nabla u(x, y, t)\|_2^2 + \|\nabla u(x.y.t)\|_2^2 \right) dx \tag{2}$$

The first term in the equation is the data term, which equals the OFCE squared. Smoothness limitations are the second term of the equation, which included the partial derivatives of the component's spatial and optional terms. The Euclidean norm squared is employed to express these derivatives' squared and summed form. Assume that the image sequence of a film is the grey-value function $f(x, y, t)$, where x and y are Cartesian spatial coordinates and t is the time dimension [49]. The symbol ∇ is the gradient operator.

The thresholding effect on the results is considerable. Maximum and minimum intensity data were produced during threshing by masking the lesion zone with ground truth (the consensus lesions region). This was only done to figure out how well optical flow techniques work. When a threshold length was put on the size of the optical flow to make patches, the results were the same. Because optical flow approaches are sensitive to image noise, the best way to separate lesions based on optical flow is to use thresholding. This dependence must be reduced.

This investigation used the threshold value to measure and record optical flow technique data pair images [33]. Once all the experiment's threshold values have been determined, a method is required to interpret the data easily. This is the function of descriptive statistics, which gives methods for summarising and describing sample data. Specifically, descriptive statistics assist in analyzing and comprehending the results, as well as presenting the results. Standard deviation is one of the key measurements used to summarise descriptive statistics [50]. When the standard deviation is low, the data tend to cluster close to the mean, but when it's large, the data tend to be more spread out. The unique method has been used to calculate the likely lung lesion detection slices based on the standard deviation from the threshold value.

$$\text{Standard deviation} = \sqrt{\frac{1}{(n-1)}\sum\nolimits_{i=1}^{n}(x-\overline{x})^2} \tag{3}$$

As before, n is the sample size, x is the threshold sample values, and $\overline{x}$ is the threshold mean. The novelty of this project proposed a new technique that integrates optical flow, statistical information & thresholding to determine slices containing lesions.

3.5 Manual Segmentation

The manual segmentation method used image segmentation in MATLAB software to create a ground truth image. This manual segmentation was obtained by cropping the lung image's shape without affecting or changing the format using a graph cut in the create mask tools. Hence, this makes it easier to notice the lung image's features. Furthermore, the lung's background was eliminated by marking the undesired parts to allow the manual segmentation to be more precise in their shape. Lastly, this method's performance is compared to the automated segmentation used in this research, such as recall value, accuracy, and F-score.

3.6 Performance Evaluation

This section assessed the lung lesion diagnosis approach using statistical markers. The recommended approach has four test image outcomes: true positive (TP), true negative (TN), false positive (FP), and false negative (FN). True positives (TP) indicate lesions are present and appropriately detected, while true negatives (TN) indicate no lesions. False positive (FP) is when an algorithm discovers nonexistent lesions, and false negative (FN) is when it misses a lesion. The performance includes accuracy, precision, recall, and F-score. The actual or known value informed the precise measurement. Distance between two points is calculated accurately. This relationship hasn't changed. It could be wrong but right, or vice versa. Accuracy is the ratio of lung lesion pixels successfully identified to all lung lesion pixels. To calculate recall, divide total lung lesion pixels by lung lesion and background pixels. Here's each test's formula [51].

$$\text{Accuracy} = \frac{\text{TP} + \text{TN}}{\text{TP} + \text{TN} + \text{FP} + \text{FN}} \tag{4}$$

$$\text{Recall} = \frac{\text{TP}}{\text{TP} + \text{FN}} \tag{5}$$

$$\text{F} - \text{score} = \frac{2 \times (\text{precision} \times \text{recall})}{(\text{precision} + \text{recall})} \tag{6}$$

TP: True positive
TN: True negative
FP: False-positive
FN: False-negative

4 Results and Discussion

The development of automatic lung lesions segmentation for various CT sequences is the primary goal of this study. The system developed must process the images to find signal irregularities and extract lesion features. There are three primary sections in this chapter. The pre-processing results are presented in this chapter's opening section. This chapter's second section presents the lung segmentation phases' results. The post-processing stage's overall results, including the roundness analysis for lesion segmentation, are shown in the following section of this chapter. The statistical analysis for slices with and without lesions are then performed using the optical flow approach.

4.1 Analysis of Lung Segmentation

The threshold value is used to isolate lung CT scans from the background by adjusting each pixel's intensity. Each pixel has an object or backdrop point. Results showed a 0.2706 threshold for automatically distinguishing lung CT images from

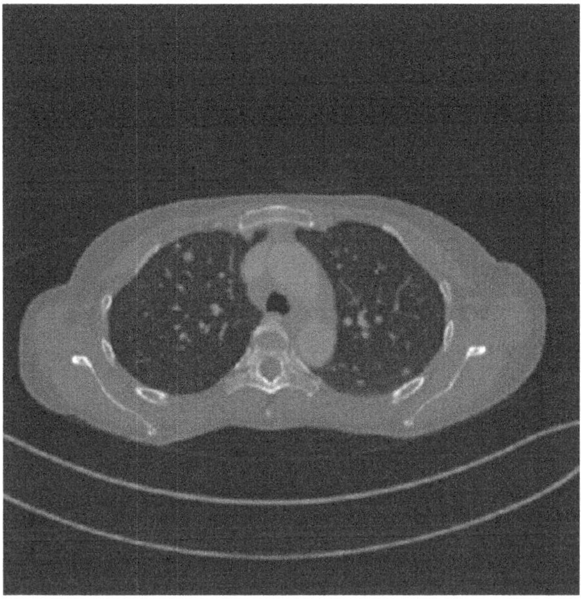

Fig. 6 The result of the threshold value in the pre-processing of a CT image

Table 3 Average performance when P = 1500

CC	Precision	Recall	Accuracy	F-score
4	99.5269	99.8908	99.5843	99.7079
8	99.5269	99.8908	99.5843	99.7079

the background. Regardless of the supplied image class, a range level of 0–1 was given. Faultless lung segmentation in image pre-processing was made possible for several reasons. The technique was significantly impacted by higher image clarity and uniformity. The pre-processed photos with the automatic threshold value of 0.2706 are shown in Fig. 6.

P was tested to determine the size of each blob's 4/8 connected neighbors (CC) and eliminate any blobs smaller than 1500 pixels, where the letter P indicates the image area used in the Matlab code. According to Table 3, the average precision, recall, and accuracy were 99.5269%, 99.8908%, and 99.5843%, respectively, while the F-score test result was 99.7079% when the CC = 4 and the P = 1500 were utilized. The average performance for CC = 8 and P = 1500 does not differ much from CC = 4 and P = 1500.

Next, 4/8 connected neighbours were tested to determine each blob's size. Blobs smaller than 2000 pixels were thrown out. The average precision, recall, and accuracy were 99.5269%, 99.8908%, and 99.5843%, respectively, based on Table 4, when the CC = 4 and the P = 2000 were employed. The F-score test result was 99.7079%. The average performance value for CC = 8 and P = 2000 is the same as

Table 4 Average performance when P = 2000

CC	Precision	Recall	Accuracy	F-score
4	99.5269	99.8908	99.5843	99.7079
8	99.5269	99.8908	99.5843	99.7079

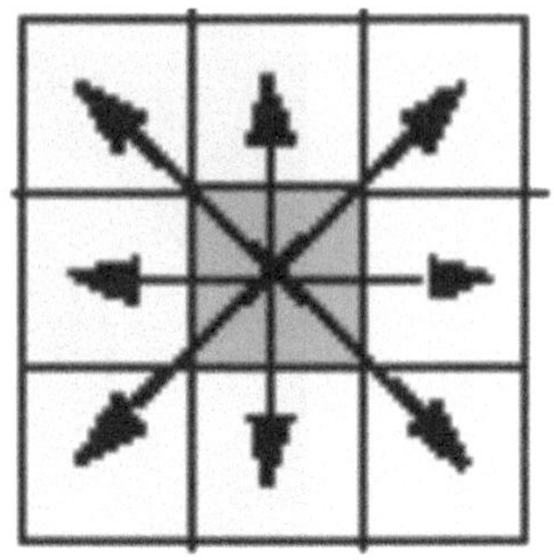

Fig. 7 Four-connected

when CC = 4 and P = 2000, with precision being 99.5269%, recall being 99.8908, accuracy being 99.5843, and F-score being 99.7079%.

The analysis results show that the image processing method is the best way to divide an image into parts that can be used to find lung cancer even when the values of CC and P change. According to analysis from the MATLAB help, which indicates that only two-dimensional connectivities images are scanned for CC = 4 and CC = 8, the results reveal average scores for all four values of CC and P with no discernible rise or decrease. Figure 7 depicts systems with four connections. If one pixel's edges touch another pixel, that pixel is said to be four-connected. Two neighbouring pixels connected in a horizontal or vertical direction form an object. In an eight-connected matrix, pixels can only be joined if their corners or advantages contact. Two neighbouring pixels are considered a single item if they relate to one another horizontally, vertically, or diagonally. The results have, therefore, fundamentally not changed. The maximum number of pixels on the items is P. Because the label matrix only influences the equation depending on the P value, the results won't change if P changes. The standard connectivity for 2-D images in this inquiry is eight-connected. For eight-connected screens, such as those in Fig. 7, the pixels are linked if a pixel's edge or corner touches another. Two neighbouring pixels connected along one of the three orthogonal axes are considered part of an object (horizontal, vertical, or diagonal).

The lung from the auto segmentation image is compared to the lung from the ground truth image. This task was essential and one of the prerequisites for performing medical image analysis. The total performance value of CC = 8 and P = 1500, which is be applied to the subsequent stage of lung cancer detection, is shown in Fig. 8. The graph in Fig. 8 demonstrates how strikingly comparable the average lung segmentation performance scores are. The average recall value, accuracy, and F-score test results for lung segmentation were 98.64%, 99.322%, and 98.975%, respectively. But among them, F-score has the most significant average. This is so that F-score can be applied when False Negative (FN) and False Positive

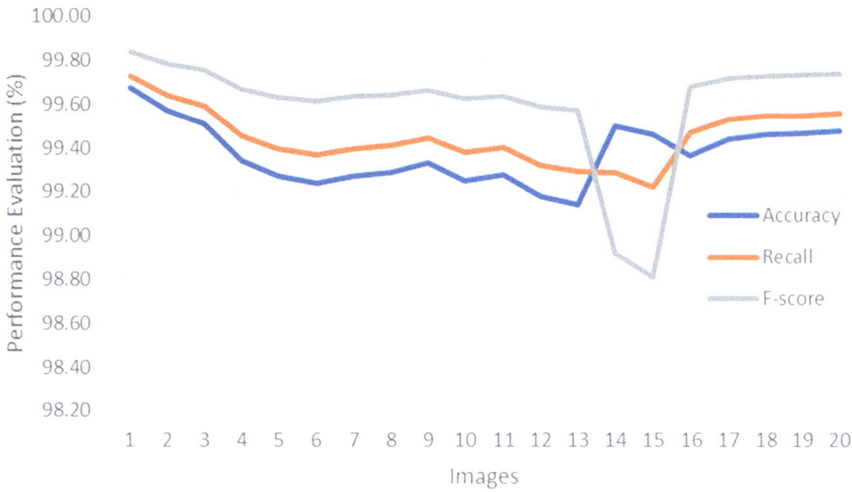

Fig. 8 Result graph when CC = 8, P = 1500

(FP) are required. A weighted average of the precision and recall numbers can also be interpreted as the F-score.

The method of segmenting an image involves dividing up the pixels into some uniform areas. This task was vital to analysing medical images and was, therefore, necessary. It is one of the fundamental methods for comprehending images. In Sect. 3.2, an investigation of lung segmentation methods based on Otsu thresholding was provided. The approach was assessed by computing the Precision, Accuracy, and F-score. The segmentation techniques were evaluated on a total of 20 lung images. The lung segmentation performance value for 20 images with P = 1500 is shown in Table 5 as CC = 8. Based on the output, the lung segmentation performance results generated an average accuracy value of 99.38%, recall at 99.45%, and F-score at 99.6%. Compared to prior works [52] the average accuracy is 99.58%, recall is 96.24%, and the F-score is 98.0%. Consequently, this study demonstrated that the novel method's lung segmentation is superior to previous methods [52].

4.2 *Analysis of Lung Lesion Segmentation*

The strategy for segmenting lung lesions based on object size effectively assisted with investigating image abnormalities for medical analysis. The complexity of the anatomy and structure of the lung made lung lesion segmentation more difficult. Therefore, the following lung lesion segmentation system approach could use the study's output. The lesions are different from other lung structures in several ways. Lesions are distinguished from other structures, such as arteries and bronchi, based

Table 5 Average performance for lung segmentation

Image	Accuracy (%)	Recall (%)	F-score (%)
1	99.67	99.73	99.84
2	99.57	99.64	99.79
3	99.51	99.60	99.76
4	99.34	99.46	99.67
5	99.27	99.40	99.63
6	99.24	99.37	99.62
7	99.27	99.40	99.63
8	99.29	99.41	99.64
9	99.33	99.45	99.67
10	99.25	99.38	99.62
11	99.28	99.41	99.64
12	99.18	99.32	99.59
13	99.14	99.29	99.57
14	99.50	99.29	98.92
15	99.46	99.22	98.81
16	99.37	99.47	99.68
17	99.44	99.53	99.72
18	99.46	99.55	99.73
19	99.47	99.55	99.73
20	99.48	99.56	99.74
Average	99.38	99.45	99.60

on their various forms. The size invariant round or nearly round form is employed to identify circular shapes in observable structures [5].

The suggested procedure allows for the extraction of lesions of any shape, making it independent of any lesions template in contrast to previous lesions template-based methodologies. According to Fig. 9a, the radiologist labels any non-lesions and considers the area inside a square a lesion. Figure 9b contrasts segmented lesions with no lesions. This led to the identification of the non-lesion, and the output will be used to complete the lesion identification process.

The segmentation of lung lesions' average performance results is shown in Table 6. The segmentation of lung lesions yielded average accuracy values of 99.92%, recall values of 42.79%, and F-score values of 56.45%. The accuracy test shows how well a diagnostic test can determine a specific condition. A recall is the percentage of the actual lung area that this method was able to separate successfully. The accuracy of an algorithm is more precisely reflected by the F-score, which is the harmonic mean of precision and recall.

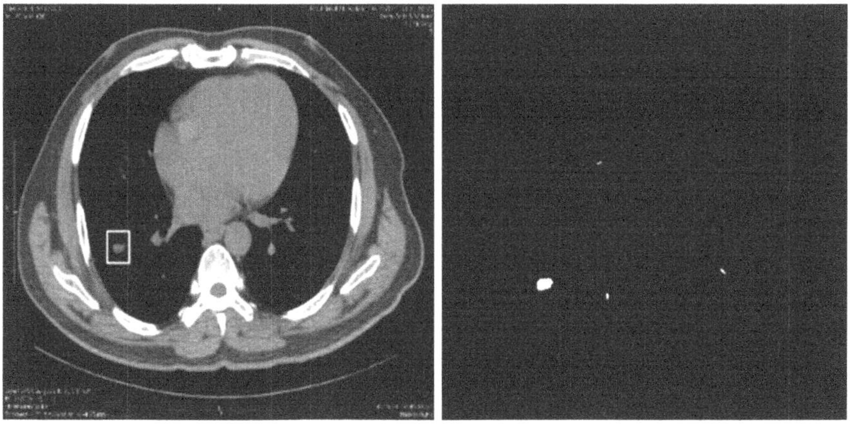

(a) (b)

Fig. 9 (**a**) The region in square is a lesion and others non lesion. (**b**) The result after lung lesion segmentation

Table 6 Average performance for lung lesion segmentation

Image	Accuracy (%)	Recall (%)	F-score (%)
1	99.97	57.74	73.21
2	99.98	83.56	91.04
3	99.92	64.88	78.70
4	99.86	54.29	70.37
5	99.96	45.03	62.10
6	99.95	40.52	57.67
7	99.87	47.05	63.99
8	99.96	45.03	62.10
9	99.90	92.93	96.33
10	99.96	32.28	48.91
11	99.95	8.27	15.27
12	99.91	20.74	34.36
13	99.91	21.80	35.81
14	99.89	16.26	27.97
15	99.93	33.75	50.46
16	99.94	42.85	60.00
17	99.94	39.26	56.38
18	99.87	18.07	30.62
19	99.87	45.60	57.50
20	99.89	45.80	56.30
Average	99.92	42.79	56.45

4.3 Analysis of Optical Flow

Figure 10a displays an example of an original lung image that a radiologist has marked. The segmented image utilized as an input for the optical flow approach and contains the lesion, vasculature, and bronchi is shown in Fig. 10c–f. The dice and Jaccard indexes are calculated and compared with the segmented and ground truth images. Table 7 presents the results of the optical flow approach.

These results should be considered in light of Table 7, which provides the dice index and Jaccard index data required to decide which mode selection will be

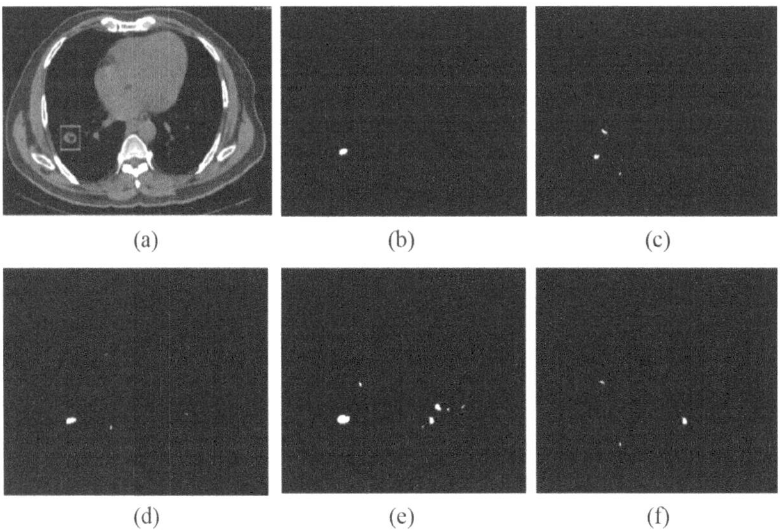

Fig. 10 (**a**) Input image for slice 145 marked by radiologist. (**b**) Ground truth image for lesion detection on slice 145. (**c**) Segmented lesion, vessels and bronchi for slice 144. (**d**) Segmented lesion, vessels and bronchi for slice 145. (**e**) Segmented lesion, vessels and bronchi for slice 146. (**f**) Segmented lesion, vessels and bronchi for slice 147

Table 7 Horn-Schunck optical flow method for lesion detection—Mode 0

Reference images	Current images	Dice index	Jaccard index	Mean	Threshold
153	154	0.85	0.73	0.0003	0.29
153	155	0.80	0.70	0.0005	0.29
153	156	0.56	0.4	0.0005	0.29
153	157	0.64	0.5	0.0005	0.28
153	158	0.56	0.39	0.0008	0.29
153	159	0.32	0.2	0.001	0.28
153	160	0.25	0.15	0.0007	0.27
153	161	0	0	0.0005	0.28
Average					0.28

Table 8 Mode comparison for Horn-Schunck optical flow method for non-lesion detection

Mode	Threshold	Stand deviation
0	0.3183	0.0246
1	0.3183	0.0237
2	0.3250	0.0309
3	0.3300	0.0330

Table 9 Mode comparison for Horn-Schunk optical flow method based on standard deviation for lesion and non-lesions detection

	Average standard deviation using Horn-Schunk	
Mode	Lesion	Non-lesions
0	0.0099	0.0246
1	0.0118	0.0237
2	0.0126	0.0309
3	0.0116	0.0330

employed for the subsequent step of lesion diagnosis. The minimum and maximum optical flow vector magnitudes for the Horn-Schunck optical flow method are shown in a table, along with the results that can be obtained. These optical flow techniques produce various u and v vectors from the same batch of images. Although the mean values are comparable for all modes, Mode 0 consistently has a higher dice index and Jaccard index for these values. The following step of the analysis are conducted in Mode 0. Horn-Schunck optical flow method mode comparison for non-lesion detection is shown in Table 8. Comparing the modes for the Horn-Schunk optical flow method based on the standard deviation for lesion and non-lesion detection as in Table 9 are in contrast. This research demonstrates a significant difference between lesion and non-lesion detection based on the standard deviation value.

When the standard deviation is close to zero, the data points are close to the mean. When it is high, the data points are above the mean. In other words, a greater value signifies that something is causing the threshold value for a response to fluctuate, which denotes erratic service performance. The range of the standard deviation for lung lesions is 0–2.0%, while the range for non-lesions is 2.1–9.2%. Based on the standard deviation from the threshold value for the subsequent stage of lung lesion detection, this result could assist the radiologist in identifying the spectrum of lesions and non-lesions. Figure 11 displays a bar chart showing the minimum and maximum standard deviation values for detecting lung lesions and non-lesions.

The quantitative assessment of the suggested strategy for lung segmentation is contrasted to ground truth images annotated by the radiologist. Table 10 displays the predictions for the potential slices with lung lesions. The CT scan images for each lesion range from 5 to 8 slices. The standard deviation range is employed based on Table 10 to determine the potential slices having lesions and non-lesion using the newly proposed method. The findings demonstrate that, compared to the photos that the radiologist had annotated, the novel proposed method could detect a variety of potential slices, including lesions and non-lesions. It is pertinent to mention that the proposed methodology needs to be tested for robustness [53–56] to various image processing attacks so that its feasibility for usage in an electronic healthcare systems [57, 58] which provides a reliable and realizable security [59, 60] could be ascertained.

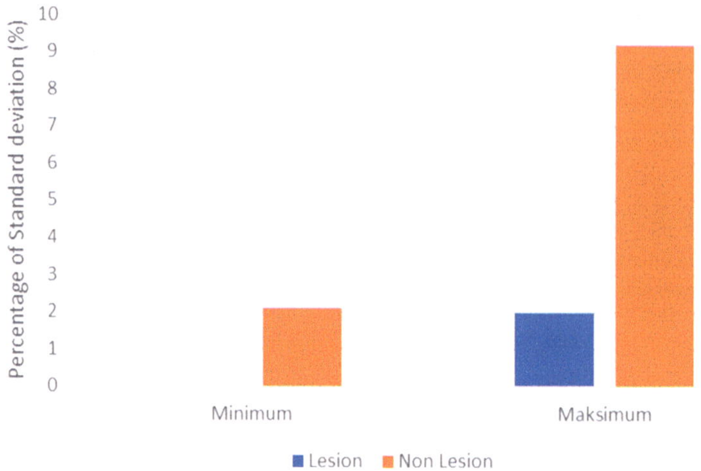

Fig. 11 Discrepancy bar chart for minimum and maximum value of standard deviation for lesion and non-lesions detection

Table 10 Predicting the possible slices containing lesion in lung

Lesion no	Standard deviation	Lesion and non-lesion detection	
		Using new proposed method	Marked by radiologist
1	1.4	Lesion	Lesion
2	0.9	Lesion	Lesion
3	1.2	Lesion	Lesion
4	0.8	Lesion	Lesion
5	1.1	Lesion	Lesion
6	1.1	Lesion	Lesion
7	0.5	Lesion	Lesion
8	0.9	Lesion	Lesion
9	2.2	Non-lesion	Non-lesion
10	2.1	Non-lesion	Non-lesion
11	2.2	Non-lesion	Non-lesion
12	2.9	Non-lesion	Non-lesion
13	3.3	Non-lesion	Non-lesion
14	2.1	Non-lesion	Non-lesion
15	2.2	Non-lesion	Non-lesion

5 Conclusion

The outcomes of the suggested process and method for lung segmentation in CT images were finally given and thoroughly reviewed. The findings were divided into three main categories: automated lung lesion diagnosis using developed software, automated image pre-processing methodology, and performance evaluation of the suggested approach against the available data. The method was created to make it easier for radiologists to segment the lung region for lung cancer detection. This work effectively assessed the performance of the image processing method for lung segmentation as a contribution to biomedical engineering, which received an average accuracy of 99.38%, recall of 99.45%, and F-score of 99.6%. Concerning the image segmentation performance standard, the results demonstrated that all performance values are above 99%, which is outstanding. The segmentation of lung lesions was done using the findings. Results were compared to the manually marked lesion segmentation by the radiologist to make sure that the proposed automated lung lesion was trustworthy. The average accuracy value for the segmentation of lung lesions was 99.92%, the average recall was 42.79%, and the average F-score was 56.45%.

This work described the significance of applying optical flow methods and statistical image analysis to medical imaging, specifically thoracic images for lesions segmentation. The flow vector in optical flow methods was calculated from two input image frames. Four distinct modes of operation were defined to select the best pair of images from a set of input images. The most effective mode was discovered to be mode 0. Based on three optical flow approaches, Horn-Schunk was the most efficient way of locating lesions. The impact of noise on optical flow computations was the fundamental difficulty in employing optical flow for segmentation. As a result, the requirement for thresholding restricted the algorithm's generalizability. The newly presented approach established that the range for lung lesions is 0–2.0% and the range for non-lesions is 2.1–9.20% by computing the standard deviation based on the threshold value. Based on the findings of this study, a standard deviation from the threshold value could be employed for the next step of lung lesion identification, helping the radiologists distinguish between lesions and non-lesions. This could be seen as a positive development in the development of a smart, quick, and accurate method for identifying lung cancer.

Acknowledgement This research work was financially supported by the Ministry of Higher Education Grant Scheme (FRGS) "A new Motion-based Optical flow Cancer Detection method for Lung Nodule Segmentation on Lung CT-Scan Images" (Ref: FRGS/1/2021/TK0/UITM/02/58) and ethics from Universiti Sains Malaysia (USM/JEPeM/19040231). The authors would like to express their gratitude to the Machine Learning Research Group (MLRG) members, Integrative Pharmacogenomics Institute (iPROMISE), and Centre for Electrical Engineering Studies, Universiti Teknologi MARA, Cawangan Pulau Pinang for their assistance and guidance during the fieldwork. Finally, the authors thank Universiti Teknologi MARA, Cawangan Pulau Pinang, for their immense administrative and financial support.

References

1. Chalasani P, Rajesh S (2020) Lung CT image classification using deep neural networks for lung cancer detection. Int J Eng Adv Technol 9(3):3998–4002. https://doi.org/10.35940/ijeat.c6409.029320

2. Bhatia S, Sinha Y, Goel L (2019) Lung cancer detection: a deep learning approach. Adv Intell Syst Comput 817:699–705. https://doi.org/10.1007/978-981-13-1595-4_55
3. Travis WD et al (2015) The 2015 World Health Organization classification of lung tumors: impact of genetic, clinical and radiologic advances since the 2004 classification. J Thorac Oncol 10(9):1243–1260. https://doi.org/10.1097/JTO.0000000000000630
4. Vas M, Dessai A (2018) Lung cancer detection system using lung CT image processing. In: 2017 International conference on computing, communication, control and automation, ICCUBEA 2017. IEEE, pp 1–5. https://doi.org/10.1109/ICCUBEA.2017.8463851
5. Heidinger BH et al (2017) Lung adenocarcinoma manifesting as pure ground-glass nodules: correlating CT size, volume, density, and roundness with histopathologic invasion and size. J Thorac Oncol 12(8):1288–1298. https://doi.org/10.1016/j.jtho.2017.05.017
6. Shaziya H, Shyamala K, Zaheer R (2018) Automatic lung segmentation on thoracic CT scans using U-Net convolutional network. In: Proceedings of the 2018 IEEE international conference on communication and signal processing, ICCSP 2018, pp 643–647. https://doi.org/10.1109/ICCSP.2018.8524484
7. Shewaye TN, Mekonnen AA (2016) Benign-malignant lung nodule classification with geometric and appearance histogram features. arXiv:1605.08350
8. Alom MZ, Hasan M, Yakopcic C, Taha TM, Asari VK (2018) Recurrent residual convolutional neural network based on U-Net (R2U-Net) for medical image segmentation. arXiv:1802.06955
9. Namitha S, Preethi R, Vinitha M, Kalaichelvi T, Hemalatha S, Subedha V (2017) Proposal on detection of lung cancer on reduced images using foggy K-means clustering algorithm. Int J Sci Res Comput Sci Eng Inf Technol 2(2):296–301
10. Tun KMM, Khaing AS (2014) Feature extraction and classification of lung cancer nodule using image processing techniques. Int J Eng Res Technol 3(3):2204–2210
11. Rani KV, Joseph Jawhar S (2018) Emerging trends in lung cancer detection scheme: a review. Int J Res Anal Rev 5(3):530–542
12. Makaju S, Prasad PWC, Alsadoon A, Singh AK, Elchouemi A (2018) Lung cancer detection using CT scan images. Procedia Comput Sci 125:107–114. https://doi.org/10.1016/j.procs.2017.12.016
13. Riti YF, Nugroho HA, Wibirama S, Windarta B, Choridah L (2016) Feature extraction for lesion margin characteristic classification from CT Scan lungs image. In: Proceedings – 2016 1st international conference on information technology, information systems and electrical engineering, ICITISEE 2016. IEEE. https://doi.org/10.1109/ICITISEE.2016.7803047
14. Asmayani Khalib AAM (2018) Malaysian study on cancer survival (MySCan), vol 4. National Cancer Institute, Health Education Division
15. Roy TS, Sirohi N, Patle A (2015) Classification of lung image and nodule detection using fuzzy inference system. In: International conference on computing, communication & automation, ICCCA 2015. IEEE, pp 1204–1207. https://doi.org/10.1109/CCAA.2015.7148560
16. Ignatious S, Joseph R (2015) Computer aided lung cancer detection system. In: Global conference on communication technologies, GCCT 2015. IEEE, pp 555–558. https://doi.org/10.1109/GCCT.2015.7342723
17. Moreno S, Bonfante M, Zurek E, Juan HS (2019) Study of medical image processing techniques applied to lung cancer. In: 14th Iberian conference on information systems and technologies, CISTI 2019. IEEE, pp 1–6. https://doi.org/10.23919/CISTI.2019.8760888
18. Pratap GP, Chauhan RP (2017) Detection of lung cancer cells using image processing techniques. In: 1st IEEE international conference on power electronics, intelligent control and energy systems, ICPEICES 2016. IEEE, pp 1–6. https://doi.org/10.1109/ICPEICES.2016.7853347
19. Kotwal AA, Walter LC (2020) Cancer screening among older adults: a geriatrician's perspective on breast, cervical, colon, prostate, and lung cancer screening. Curr Oncol Rep 22(11):108. https://doi.org/10.1007/S11912-020-00968-X
20. Bhalerao RY, Jani HP, Gaitonde RK, Raut V (2019) A novel approach for detection of lung cancer using digital image processing and convolution neural networks. In: 2019 5th International conference on advanced computing & communication systems, ICACCS 2019. IEEE, pp 577–583. https://doi.org/10.1109/ICACCS.2019.8728348

21. Dhalia Sweetlin J, Nehemiah HK, Kannan A (2018) Computer aided diagnosis of pulmonary hamartoma from CT scan images using ant colony optimization based feature selection. Alex Eng J 57(3):1557–1567. https://doi.org/10.1016/j.aej.2017.04.014
22. Vijayan A (2017) Classification of lung tumor from CT images using computer aided diagnosis scheme. Int J Sci Technol Eng 3(9):589–593
23. Abdullah MF et al (2019) A comparative study of image segmentation technique applied for lung cancer detection. In: Proceedings – 9th IEEE international conference on control system, computing and engineering, ICCSCE 2019. IEEE, pp 72–77. https://doi.org/10.1109/ICCSCE47578.2019.9068574
24. Netto SMB, Silva AC, Nunes RA, Gattass M (2012) Automatic segmentation of lung nodules with growing neural gas and support vector machine. Comput Biol Med 42(11):1110–1121. https://doi.org/10.1016/j.compbiomed.2012.09.003
25. Cascio D, Magro R, Fauci F, Iacomi M, Raso G (2012) Automatic detection of lung nodules in CT datasets based on stable 3D mass-spring models. Comput Biol Med 42(11):1098–1109. https://doi.org/10.1016/j.compbiomed.2012.09.002
26. Keshani M, Azimifar Z, Tajeripour F, Boostani R (2013) Lung nodule segmentation and recognition using SVM classifier and active contour modeling: a complete intelligent system. Comput Biol Med 43(4):287–300. https://doi.org/10.1016/j.compbiomed.2012.12.004
27. Sindhu Priya S, Ramamurthy B (2018) Lung cancer detection using image processing techniques. Res J Pharm Technol 11(5):2045–2049. https://doi.org/10.5958/0974-360X.2018.00379.7
28. Rendon-Gonzalez E, Ponomaryov V (2016) Automatic lung nodule segmentation and classification in CT images based on SVM. In: 9th International Kharkiv symposium on physics and engineering of microwaves, millimeter and submillimeter waves, MSMW 2016. IEEE, pp 1–4. https://doi.org/10.1109/MSMW.2016.7537995
29. Pu J, Zheng B, Leader JK, Wang XH, Gur D (2008) An automated CT based lung nodule detection scheme using geometric analysis of signed distance field. Med Phys 35(8):3453–3461. https://doi.org/10.1118/1.2948349
30. Messay T, Hardie RC, Rogers SK (2010) A new computationally efficient CAD system for pulmonary nodule detection in CT imagery. Med Image Anal 14(3):390–406. https://doi.org/10.1016/j.media.2010.02.004
31. Chunran Y, Yuanvuan W, Yi G (2019) Automatic detection and segmentation of lung nodule on CT images. In: Proceedings – 2018 11th international congress on image and signal processing, biomedical engineering and informatics, CISP-BMEI 2018. IEEE, pp 2–6. https://doi.org/10.1109/CISP-BMEI.2018.8633101
32. Alam J, Alam S, Hossan A (2018) Multi-stage lung cancer detection and prediction using multi-class SVM classifie. In: International conference on computer, communication, chemical, material and electronic engineering, IC4ME2 2018. IEEE, pp 1–4. https://doi.org/10.1109/IC4ME2.2018.8465593
33. Khehrah N, Farid MS, Bilal S, Khan MH (2020) Lung nodule detection in CT images using statistical and shape-based features. J Imaging 6(2):6. https://doi.org/10.3390/jimaging6020006
34. Pandian R, Ravi Kumar DNS, Kumar RR (2020) Development of algorithm for identification of maligant growth in cancer using artificial neural network. Int J Electr Comput Eng 10(6):5709–5713. https://doi.org/10.11591/ijece.v10i6.pp5709-5713
35. Farahani FV, Ahmadi A, Zarandi MHF (2018) Hybrid intelligent approach for diagnosis of the lung nodule from CT images using spatial kernelized fuzzy C-means and ensemble learning. Math Comput Simul 149:48–68. https://doi.org/10.1016/j.matcom.2018.02.001
36. Miah MBA, Yousuf MA (2015) Detection of lung cancer from CT image using image processing and neural network. In: 2nd International conference on electrical engineering and information communication technology, iCEEiCT 2015. IEEE, pp 21–23. https://doi.org/10.1109/ICEEICT.2015.7307530
37. Alakwaa W, Nassef M, Badr A (2017) Lung cancer detection and classification with 3D convolutional neural network (3D-CNN). Int J Adv Comput Sci Appl 8(8):409–417. https://doi.org/10.14569/ijacsa.2017.080853

38. Pattnaik A, Kanodia S, Chowdhury R, Mohanty S (2019) Predicting tuberculosis related lung deformities from CT scan images using 3D CNN. CEUR Workshop Proc 2380:9–12
39. Jakimovski G, Davcev D (2019) Using double convolution neural network for lung cancer stage detection. Appl Sci 9(3):427. https://doi.org/10.3390/app9030427
40. Afzal I, Parah SA, Hurrah NN, Song OY (2020) Secure patient data transmission on resource constrained platform. Multimed Tools Appl:1–26. https://doi.org/10.1007/s11042-020- 09139-3
41. Ouksili Z, Batatia H (2010) 4D CT image reconstruction based on interpolated optical flow fields. In: Proceedings – 2010 IEEE international conference on image processing, ICIP. IEEE, pp 633–636. https://doi.org/10.1109/ICIP.2010.5650059
42. Ehrhardt J et al (2007) An optical flow based method for improved reconstruction of 4D CT data sets acquired during free breathing. Med Phys 34(2):711–721. https://doi.org/10.1118/1.2431245
43. Dougherty L, Asmuth JC, Gefter WB (2003) Alignment of CT lung volumes with an optical flow method. Acad Radiol 10(3):249–254. https://doi.org/10.1016/S1076-6332(03)80098-3
44. Noor NM et al (2015) Automatic lung segmentation using control feedback system: morphology and texture paradigm. J Med Syst 39(3):22. https://doi.org/10.1007/s10916-015-0214-6
45. Chithra AS, Renjen Roy RU (2018) Otsu's adaptive thresholding based segmentation for detection of lung nodules in CT image. In: Proceedings of the 2nd international conference on trends in electronics and informatics, ICOEI 2018. IEEE, pp 1303–1307. https://doi.org/10.1109/ICOEI.2018.8553694
46. Ligthart ST et al (2013) Circulating tumor cells count and morphological features in breast, colorectal and prostate cancer. PLoS One 8(6):1–11. https://doi.org/10.1371/journal.pone.0067148
47. Patil SA, Udupi VR, Kane CD, Wasif AI, Desai JV, Jadhav AN (2009) Geometrical and texture features estimation of lung cancer and TB images using chest X-ray database. In: Proceedings of the 2nd international conference on biomedical and pharmaceutical engineering, ICBPE 2009. IEEE, pp 1–7. https://doi.org/10.1109/ICBPE.2009.5384113
48. Suji RJ, Bhadouria SS, Dhar J, Godfrey WW (2020) Optical flow methods for lung nodule segmentation on LIDC-IDRI images. J Digit Imaging 33(5):1306–1324. https://doi.org/10.1007/s10278-020-00346-w
49. Raudies F (2013) Optic flow. Scholarpedia 8(7):1–30. https://doi.org/10.4249/scholarpedia.30724
50. King AP, Eckersley RJ (2019) Descriptive statistics I: univariate statistics. In: Statistics for biomedical engineers and scientists. Academic Press, pp 1–21. https://doi.org/10.1016/b978-0-08-102939-8.00010-4
51. Baratloo A, Hosseini M, Negida A, El Ashal G (2015) Part 1: Simple definition and calculation of accuracy, sensitivity and specificity. Emergency 3:48–49
52. Marzuki NNSM, Isa IS, Karim NKA, Shuaib IL, Soh ZHC, Sulaiman SN (2020) Demarcation of lung lobes in CT scan image for lung cancer detection using watershed segmentation. In: Proceedings of the 2020 12th international conference on computer and automation engineering, ICCAE 2020. ACM, pp 70–74
53. Parah S, Sheikh J, Loan N, Bhat GM (2018) Utilizing neighborhood coefficient correlation: a new image watermarking technique robust to singular and hybrid attacks. Multidim Syst Sign Process 29:1095–1117
54. Kamili A et al (2021) DWFCAT: dual watermarking framework for industrial image authentication and tamper localization. IEEE Trans Industr Inform 17(7):5108–5117. https://doi.org/10.1109/TII.2020.3028612
55. Dey N et al (2017) Realization of a new robust and secure watermarking technique using DC coefficient modification in pixel domain and chaotic encryption. J Glob Inf Manag 25(4):80–102
56. Ahad F et al (2015) On the realization of robust watermarking system for medical images. In: 2015 Annual IEEE India conference (INDICON), New Delhi, India. IEEE, pp 1–5. https://doi.org/10.1109/INDICON.2015.7443363

57. Sarosh P, Parah S, Bhat GM, Khan M (2021) A security management framework for big data in smart healthcare. Big Data Res 25:100225
58. Hurrah NN et al (2020) Embedding in medical images: an efficient scheme for authentication and tamper localization. Multimed Tools Appl 79:21441–21470
59. Parah SA et al (2014) A secure and efficient spatial domain data hiding technique based on pixel adjustment. Am J Eng Technol Res 14(2):33
60. Parah SA et al (2012) On the realization of a secure, high capacity data embedding technique using joint top-down and down-top embedding approach. Comput Sci Eng 49:10141–10146

Performance Improvement with Optimization Algorithm in Isolating Left Ventricle and Non-Left Ventricle Cardiac

Dayang Suhaida Awang Damit, Siti Noraini Sulaiman, Muhammad Khusairi Osman, Noor Khairiah A. Karim, and Belinda Chong Chiew Meng

1 Introduction

The Cardiac MR image detects the overall sections of the heart that often include the outmost slices of the Left Ventricle (LV) segment [1], known as the non-LV cardiac segment. However, not all image slices will be taken into account when a radiologist evaluates the myocardial scar on the LV. Only the LV cardiac segment slices will be considered, as they contain useful information, while the rest will be neglected. This approach incorporates a significant amount of time and must be performed by a

D. S. A. Damit · B. C. C. Meng
Electrical Engineering Studies, College of Engineering, Universiti Teknologi MARA, Cawangan Pulau Pinang, Penang, Malaysia

S. N. Sulaiman (✉)
Electrical Engineering Studies, College of Engineering, Universiti Teknologi MARA, Cawangan Pulau Pinang, Penang, Malaysia

Advanced Rehabilitation Engineering in Diagnostic and Monitoring Research Group (AREDiM), Electrical Engineering Studies, College of Engineering, Universiti Teknologi MARA, Cawangan Pulau Pinang, Penang, Malaysia

Integrative Pharmacogenomics Institute (iPROMISE), Puncak Alam, Selangor, Malaysia
e-mail: sitinoraini@uitm.edu.my

M. K. Osman
Centre for Electrical Engineering Studies, Universiti Teknologi MARA, Cawangan Pulau Pinang, Permatang Pauh, Penang, Malaysia

N. K. A. Karim
Department of Biomedical Imaging, Advanced Medical and Dental Institute, Universiti Sains Malaysia, Bertam, Kepala Batas, Pulau Pinang, Malaysia

Breast Cancer Translational Research Programme (BCTRP), Advanced Medical and Dental Institute, Universiti Sains Malaysia, Kepala Batas, Penang, Malaysia

S. A. Parah et al. (eds.), *Intelligent Multimedia Signal Processing for Smart Ecosystems*, https://doi.org/10.1007/978-3-031-34873-0_8

skilled radiologist or pathologist with years of experience. The procedure frequently requires a second opinion or even assistance, which is where artificial intelligence is needed.

Deep learning systems replicate the workings of the human brain in expressing complicated data and aid in decision-making. Convolutional Neural Networks (CNN) are among the most popular deep neural networks and offer significant advantages [2]. In terms of computational efficiency, CNN is better than other approaches. It has been employed as an image classification algorithm, resulting in a remarkable impact on computer vision performance. Although the data that enters and the decisions that come out are both visible, its architecture and algorithms are described as a "black box" in which what occurs in between is difficult to predict. Convolutional, activation layer, pooling, and fully connected layers are three common types of CNN layers. Each layer has a distinct function [3]. Many prominent and reliable CNN designs, such as AlexNet [4], GoogleNet [5], InceptionV3 [6], and ResNet50 [7], were established as a result of the rapid advancements in GPU cards and the growing number of datasets. The transfer learning technique provides several advantages and enables the knowledge gained from previous model training to be applied to new tasks. It extends the information acquired from earlier models and saves time by starting from the end point of the most recent training rather than from scratch [8]. Transfer learning is also beneficial, especially when the training dataset is limited [9].

Studies conducted in a variety of deep learning applications [10–13] have led to the conclusion that some of the best ways to improve the performance of CNN include increasing the depth of the model architecture, expanding the dataset with data augmentation or transfer learning, increasing hyper-parameter tuning and increasing the amount of training time. Finding the optimum training criterion optimization is one of the most challenging tasks in deep learning. Optimizers are algorithms used to modify the weights and learning rate of the neural network during training in order to reduce the loss function or increase production efficiency.

Despite advancements in CNN architectures, previous studies comparing the performance of various optimizers for various deep learning networks in medical data remain limited. Ho et al. [1] used transfer learning to assess the performance of nine popular CNN architectures for the task of automatic classification on cardiac short-axis slices. Classifiers were built to take a cine cardiac MR image as input and classify it into three categories. This study used gradient descent with Nesterov Momentum at three different learning rates. Yaqub et al. [14] compared ten different gradient descent-based optimizers for CNN to benchmark brain tumor segmentation for improvement. The experiments were carried out using the Brain MRI BraTS2015 dataset. Adam optimizer has been discovered to have the highest accuracy of 99.2% in improving CNN classification and segmentation abilities. Wibowo et al. [15] investigated six algorithms, namely Gradient Descent, Momentum, AdaGrad, AdaDelta, RMSProp, and Adam, to determine the best optimization algorithm and neural network tuning parameter for cancer classification using microRNA data. Adam and RMSProp obtained the highest accuracy, with Adam achieving 98.536% and RMSProp achieving 98.54762%.

With an aim to enable this study aims to improve the reliability of the network classifier by comparing the three most widely used first-degree optimizers, namely Stochastic Gradient Descent with Momentum (SGDM), RMSProp, and Adam optimizers, to select the best network for classifying left ventricle and non-left ventricle segments in late gadolinium enhancement (LGE) cardiac MRI data.

The subsequent section in this chapter provides a concise overview of these three gradient-based optimizers, followed by the methodology section, which describes the architecture of the deep convolution network from scratch model and pre-trained network models used in this experiment, including dataset preparation. The findings of the network performance are presented in the results and discussion section. The final section concludes the research work.

2 Network Optimizer

The selection of an optimization algorithm is a crucial stage in deep learning because it significantly impacts the model's performance [16]. It generally defines how the neural network learns. The model has correctly learned the dataset if the loss function is zero, which is determined by comparing the actual input label present in the training set to the predicted label. The learning rate is the model's step to reach the global minimum. The gradient descent method involves taking small steps iteratively and using the error of the predictions to update the model until it reaches the properly learnt weights [17].

The parameters in stochastic gradient descent (SGD) are updated at each training sample instead of the entire data set. It randomly samples the training samples at each epoch before training. This method is more memory-efficient and considerably faster than the previous version of gradient descent when applied to a large training dataset [18]. However, because the parameters are regularly updated, it makes exceedingly noisy steps in the direction of the solution, which, as a result, causes the behavior of convergence to become extremely unstable. A mini-batch or small subset of training examples with no overlap is used to overcome this issue [12]. By combining the advantages of SGD, it also achieves a steady convergence. Momentum has been introduced to address the oscillation problem, which typically causes a long time to reach convergence when updating the weights by employing mini-batch SGD [19]. It helps SGD move faster in the right direction toward the optimum.

Deep learning algorithms typically deal with non-convex problems, whereas pioneer optimization has a fixed learning rate [20]. Typically, the learning rates need to be changed, and the model needs to be trained again if the learning rate does not produce good results [21]. Training the models in deep learning typically requires a lot of time. Therefore, the concept of adaptive optimization techniques that select the learning rate according to the situation is highly appreciated. Instead of setting a learning rate to initialize the learning rate parameters' values, these adaptive optimization methods will continuously update learning rates as the model is being trained [22]. Two adaptive optimizers used in this study are RMSProp and Adam optimizer. RMSProp and Adam optimizers differ only in their method of calculating gradients. RMSProp gradients accumulate into an exponentially weighted average and preserve only current gradient knowledge. In comparison, Adam upgrades its operation by considering the smooth gradient variant and providing a bias correction mechanism [23].

3 Methodology

Figure 1 presents the flowchart of the overall methodology in this research. Firstly, the cardiac LGE-MRI STACOM2012 dataset was used to train the networks with the SGDM optimizer for CNN model from scratch called CNN3 and the three pre-trained CNN, namely AlexNet, GoogleNet and SqueezeNet. Data acquisition and the architecture for all networks are shown in the next section. The network performance and training time for this dataset were observed. These networks were then compared against the RMSProp and Adam network optimizers to investigate whether training and network performance had been improved. Finally, all the aforementioned CNN networks were tested using a different dataset of five patients' cardiac LGE-MRI images obtained from the Advanced Medical and Dental Institute (AMDI) using the transfer learning method in order to assess the network's reliability.

The CNN3 and three pre-trained models were trained using 35 epochs with fivefold cross-validation, which involved the computation of hyper-parameters. Training and validation were performed on a personal computer with an Intel® Core i7™ RAM 8GB memory, and a single GPU (NVIDIA GeforceMX130). The operating system used was Windows® 10, Professional Edition.

3.1 Data Acquisition

The public dataset from the Statistical Atlases and Computational Models (STACOM2012) challenge website [24] was utilized to train all four network classifiers, which consists of 398 short-axis images of cardiac LGE-MRI of humans provided with an expert observers' consensus ground truth.

In this study, the cardiac LGE-MRI images are classified into three categories: first, the images which contain the LV Ventricle Segment (LVS); secondly, the Non-LV Apical (NLVA), which is the Non-LV cardiac segment that refers to the images which do not contain the LV region located below the apical, and lastly, the Non-LV Basal (NLVB), which is also the Non-LV cardiac segment that refers to the images which do not contain LV region located at the basal. The images labeled as LV will be taken as images of interest, while NLVA and NLVB will be neglected in clinical practice. Figure 2 shows an example of cardiac MRI short-axis slices with three labeling classes.

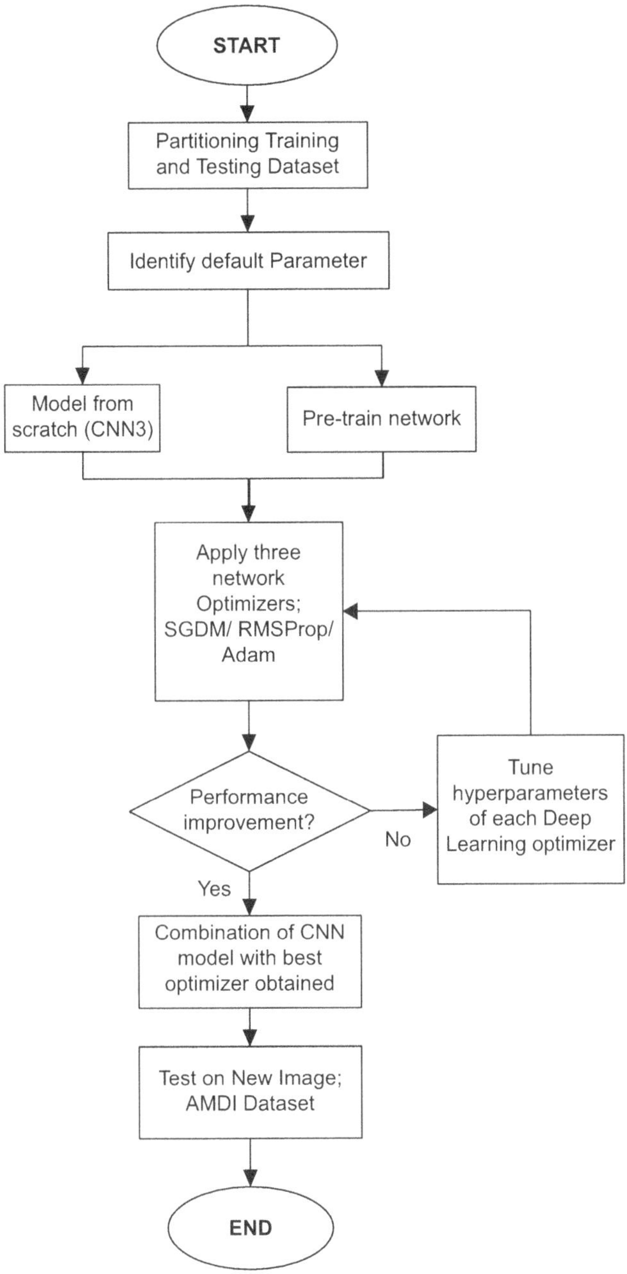

Fig. 1 Flowchart of the overall methodology

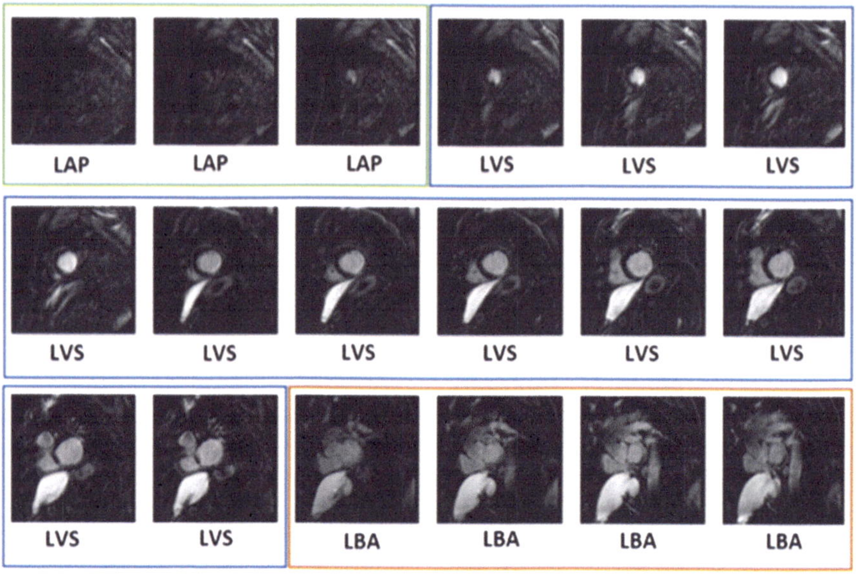

Fig. 2 Cardiac MRI slices with three labelling classes

Table 1 Number of images in each class (a) before augmentation and (b) after data augmentation

(a)				
Original dataset	LV	NLVA	NLVB	Total
Image count	255	62	81	398
Percentage	64%	16%	20%	100%
(b)				
After data augmentation	LV	NLVA	NLVB	Total
Image count	255	310	324	889
Percentage	29%	35%	36%	100%

All classification network models were trained on 80% of the images and validated on the remaining 20% of the images after the data augmentation process. Table 1 displays the total number of images in each class before and after data augmentation. The images in NLVA and NLVB classes were oversampled to correspond with images in LV class using a simple augmentation approach that applied random rotations and mirroring to overcome the class imbalance and overfitting issue [25].

Each class was prepared equally so that the CNN perceives a balanced dataset with 255 images during the fivefold cross-validation process as in Table 2. The total image containing image slices of 765 was randomly divided into five distinct subsets for training and validation set to comply with fivefold convention. One subset was used for validation for each fold, and the remaining four were used for training. Therefore, to assess the performance of a single parameter choice, a total of five models were trained and validated. Based on Table 2, the arrangement for fivefold datasets is illustrated in Fig. 3.

Table 2 Number of images in each class for training and validation

Training and validation data	LV	NLVA	NLVB	Total
Image count	255	255	255	765
Percentage	33.30%	33.30%	33.30%	100%

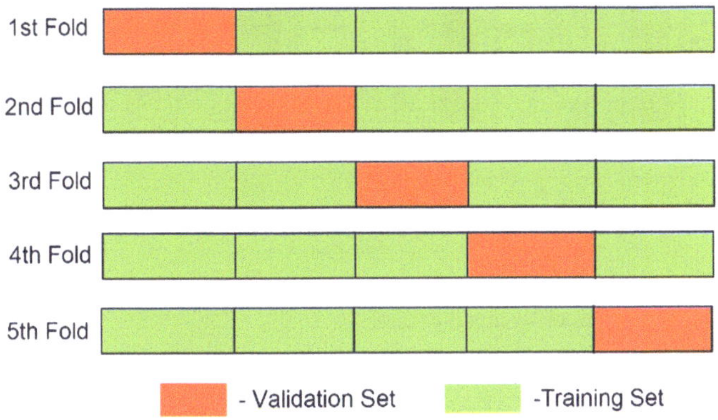

Fig. 3 Fivefold cross-validation datasets arrangement

Table 3 The total number of AMDI images for each class

Label	LV	NLVA	NLVB	Total
Image count	60	26	30	116

All CNN networks were tested on the second dataset obtained from the AMDI database at University Sains Malaysia. This dataset involves five patients, with each patient providing 15–45 slices of LGE cardiac MR images. The total number of 116 images and each class from this dataset are tabulated in Table 3.

3.2 *Model from Scratch Architecture*

CNN is well-known for its great accuracy. Thus, attempting to train the model from scratch is worth the effort. In a previous study [26], six deep CNN models with various convolution layer depths were trained to classify the LV Segment (LVS) and Non-Left Ventricle Segment (Non-LVS) and the results were compared. From these six CNN models, the authors have found that the network from scratch achieved the highest score with three convolution layer architecture. Thus, the highest score model in previous study which is the three-convolution layer network was adapted in this current study to be compared with the well-established pre-trained models by using deep learning optimization algorithms for further improvement.

The three-convolution network named CNN3 consists of different types of layers, including an input layer, convolutional layers, rectified linear unit (ReLU) activation layer, pooling layers, fully connected layers, and an output layer, as shown in Fig. 4.

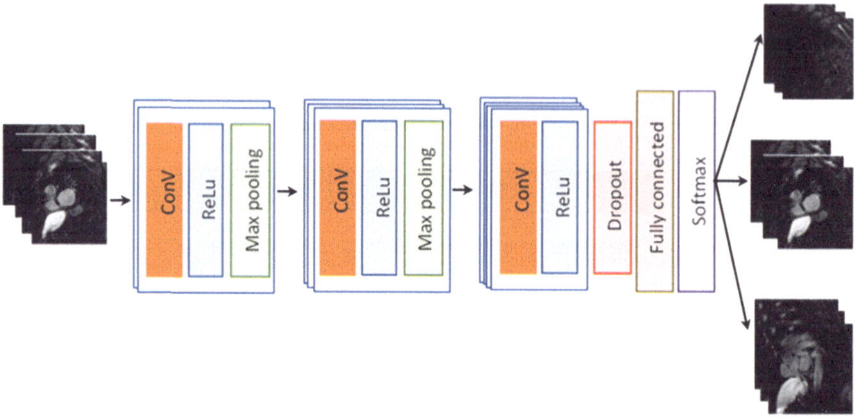

Fig. 4 CNN3 model network layer architecture

CNN architecture involves feature extraction and feature classification. Convolution and pooling layers are responsible for feature extraction. The fully connected layer and the activation function, which is put directly after the fully connected layer, plays a role in the classification process. A 3D volume is produced as the result of pooling and convolution process. The output from this process is "flattened" by the fully connected input layer into a single vector, 1D volume. The activation function receives the output from the last hidden layer and computes the probability distribution across the final set of all classes. In this study, the probability of each class to which the input belongs to is determined using a softmax activation function, where the probability, f, of each class is defined as in Eq. (1).

$$f\left(y_i\right) = \frac{e^{y_i}}{\sum_k e^{y_k}} \tag{1}$$

After passing an input image through the model for testing, each feature class of the trained network's neural output of y is determined. The probability value f is generated by combining the ratio of the exponential of a specific class with the sum of exponentials for each class, with the commensurate neural output. The final class prediction score will be determined by the class with the highest value of f.

The network structure architecture for the deep CNN3 model is depicted in Table 4. The hyper-parameters of a convolutional layer are the variables that determine the network structure, such as kernel size, number of kernels, stride, and padding, and the factors that govern how the network is trained such as learning rate, batch size, and number of epochs. These parameters must be carefully selected to deliver highly accurate and fast output. The parameters that were set up to specify the training option for the CNN3 model in the MATLAB are shown in Table 5.

Table 4 Network structure architecture for CNN3 model

CNN3 layer	Kernel size	No. of kernel	Stride	Pad
Conv1	3 × 3	8	1	"Same"
Relu	–	–	–	–
Pool1	2	8	2	0
Conv2	3 × 3	16	1	"Same"
Relu	–	–	–	–
Pool2	2	16	2	0
Conv3	3 × 3	32	1	"Same"
Relu	–	–	–	–
Pool3	2	32	2	0
fc	–	–	–	–

Table 5 Training parameter setup

CNN parameter	Specifications
Optimizer	SGDM/RMSProp/Adam
Learning rate	0.0001
MaxEpochs	35
Mini batch size	16

3.3 Transfer Learning with Pre-trained Networks

Pre-trained networks are popular and very efficient option for small image dataset [27]. Adopting a transfer learning method greatly reduces the computational cost of building and training the CNN [28]. Although there are many pre-trained models, this chapter opted to concentrate on AlexNet, GoogleNet, and SqueezeNet since they are three well-known CNN designs and because training other network models, which have deeper layers and more complex architectures, would necessitate more computational capacity. A brief overview of these three pre-trained models is provided below.

To compare these pre-trained models to CNN3, which is the model from scratch, all pre-trained networks are fixed with the same learning rate of 0.0001, mini batch size of 16, and an optimizer similar to the CNN3 model. The last fully connected layer was fine-tuned to accommodate the new classification problem which is three image categories. The number of epochs was fixed at 35 to ensure consistency in all network results. Figure 5 shows the transfer learning process.

AlexNet by Alex Krizhevsky et al. [4] was the first prominent model to incorporate the convolutional layers that define a CNN. AlexNet was the winner of 2012 ImageNet Large Scale Visual Recognition Challenge (ILSVRC) competition, and the successful development of this network was a major breakthrough in CNN development. AlexNet came out with novel approaches to cut down on the amount of overfitting in their model through data augmentation and dropout [29]. Dropout allows neurons to become less reliant on neighboring neurons. This reduces complex co-adaptations of neurons and allows each neuron to learn more valuable features. The architecture of AlexNet contains five convolutional layers and three

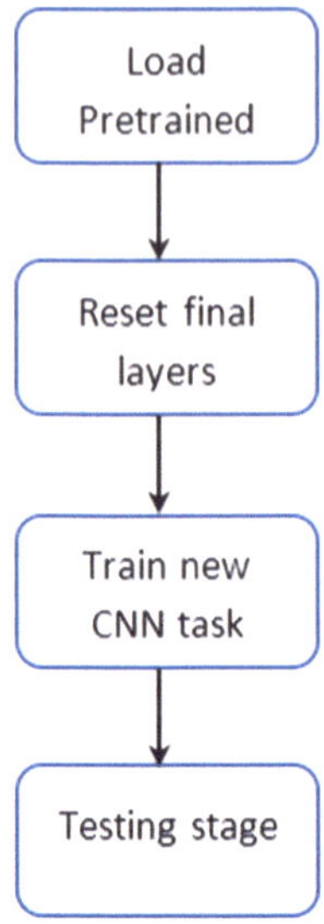

Fig. 5 Transfer learning process

fully connected layers as shown in Fig. 6. It also introduced ReLU as activation function instead of sigmoid function. In this chapter, the AlexNet pre-trained model was imported from the Matlab Deep Learning Toolbox Model. The input size of the network is fixed at 227 × 227 × 3, where 3 is the number of colored channels.

GoogleNet was the winner of the ImageNet Large-Scale Visual Recognition Challenge 2014 (ILSVRC14), which featured image classification and object detection [5]. GoogleNet introduced nine inception modules as part of its architecture, each of which has 1 × 1, 3 × 3, and 5 × 5 filters acting as multilevel feature extractors as illustrated in Fig. 7. Its architecture consists of 22 layers deep CNN or 27 layers including pooling layers as shown in Fig. 8. At its inception, all the feature maps given by different convolution channels are stacked together to produce the final feature map. The input layer of GoogleNet requires input images of size 227 × 227 × 3. For LV and non-LV segment image classification, all layers are frozen except on the output layer, where the number of predicted classes is reduced from 1000 to 3.

A SqueezeNet [30] "squeezes" the training networks and is usually used in image classification. The architecture of SqueezeNet introduced nine fire modules with pooling layers, shown in Fig. 9. The fire module contains a squeeze layer and an expand layer, as illustrated in Fig. 10. SqueezeNet was designed to create a smaller neural network with fewer parameters by using 1 × 1 filters as a bottleneck layer to reduce the depth and computation of the subsequent 3 × 3 filters. In this chapter, the SqueezeNet pre-trained model is also imported from the Matlab Deep Learning Toolbox Model, with the input size of the network fixed at 227 × 227 × 3. The number of classes predicted is fine-tuned from 1000 to 3 classes in the output layer.

Fig. 6 AlexNet model network layer architecture [4]

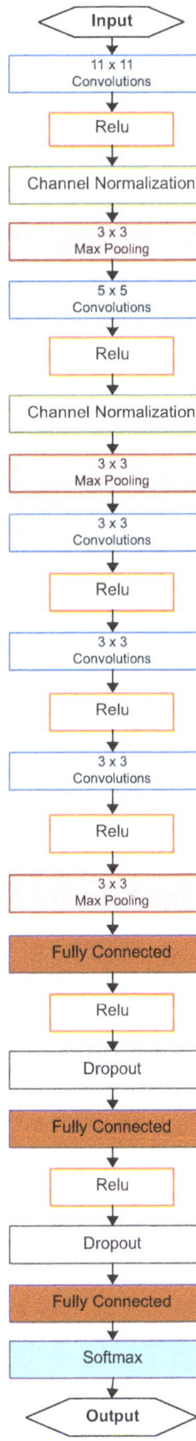

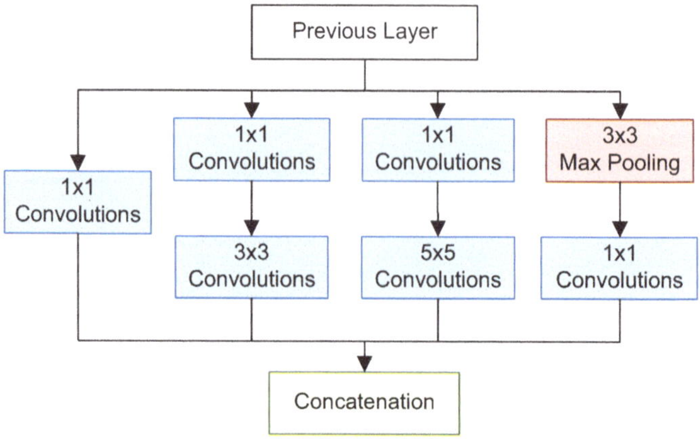

Fig. 7 GoogleNet inception modules

3.4 Performance Evaluation

Multi-class classification metrics are employed in this chapter since both datasets involved more than two classes. The training and validation dataset from the STACOM2012 is a balanced dataset, which implies that all data in each class has the same number. The AMDI testing dataset, on the other hand, is uneven. The details of these datasets have been discussed in the previous sections. Therefore, the performance of each network model will be evaluated by the accuracy percentage since it is the best evaluation for a balanced dataset [31], and macro F1-score evaluation will be added to the unbalanced dataset result since the metric better defines and quantifies imbalanced performance.

Accuracy indicates how much the model is correctly predicting the entire set of data. Generally, accuracy is mathematically expressed as follows [32]:

$$\text{Accuracy} = \frac{\text{TP} + \text{TN}}{\text{TP} + \text{FN} + \text{TN} + \text{FP}} \tag{2}$$

F1-score involves all classes by requiring a multi-class precision and recall measure to be included into the harmonic mean. Macro-Precision and Macro-Recall must first be computed in order to obtain the macro F1-score. It is mathematically expressed as follows [31]:

$$\text{Macro F1} - \text{Score} = 2\frac{MacroAverage\ \text{Precision} \times MacroAverage\ \text{Recall}}{MacroAverage\ \text{Precision} + MacroAverage\ \text{Recall}} \tag{3}$$

$$Precision_k = \frac{TP}{(TP_k + FP_k)} \tag{4}$$

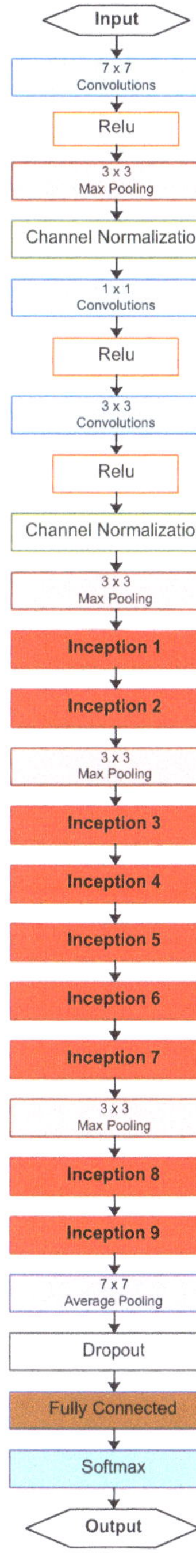

Fig. 8 GoogleNet architecture

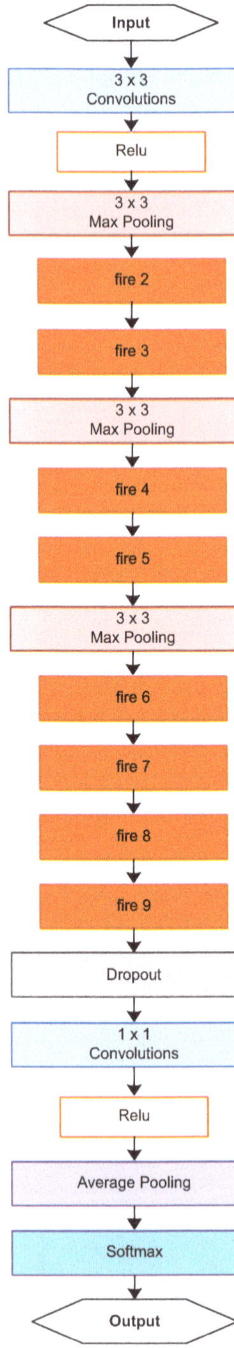

Fig. 9 SqueezeNet architecture [30]

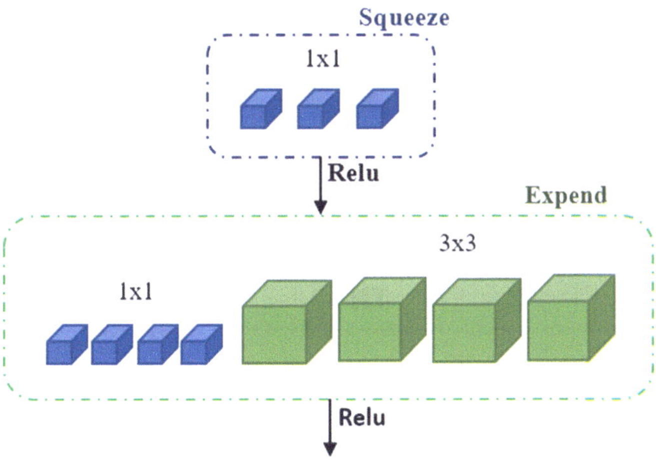

Fig. 10 Organization of convolution filters in SqueezeNet fire modules [30]

$$Recall_k = \frac{TP_k}{TP_k + FN_k} \tag{5}$$

$$MacroAverage\ \text{Precision} = \frac{\sum_{k=1}^{K} Precision_k}{K} \tag{6}$$

$$MacroAverage\ \text{Recall} = \frac{\sum_{k=1}^{K} Recall_k}{K} \tag{7}$$

Where k represents the number of each class.

The performance of all classification network models in this chapter was evaluated from the output component of the confusion matrix. Figure 11 portrays the confusion matrix for multiclass, along with a description of how the component is described [33]:

True Positive (TP): outcome of correctly identified prediction for each class.
True Negative (TN): outcome of correctly rejected prediction for each class
False Positive (FP): outcome of incorrectly identified prediction for each class.
False Negative (FN): outcome of incorrectly rejected prediction for each class.

		PREDICTED CLASS		
	CLASSES	X	Y	Z
ACTUAL CLASS	X	TN	FP	TN
	Y	FN	TP	FN
	Z	TN	FP	TN

Fig. 11 Example of multiclass confusion matrix in identifying class y for three classes

4 Results

This section presents the results produced for each network model by different optimizers.

The results tabulated in Table 6 indicate the performance in terms of validation accuracy and training time for CNN3 and three pre-trained models after training with cardiac LGE-MRI STACOM2012 dataset using the network optimizers; SGDM, RMSProp, and Adam. When Adam and RMSprop are used instead of SGDM, there is a minor change in model accuracy performance. The accuracy of CNN3 models trained with RMSprop and Adam optimizers increased from 92.3% to 93% and 93.2%, respectively, compared to SGDM. The same increment pattern was observed in GoogleNet using RMSProp and Adam. GoogleNet outperformed both the RMSprop and Adam optimizers, giving this model the highest validation accuracy in the same optimizer categories of 97.25% and 97.54%, respectively. On the other hand, when SqueezeNet was trained with the RMSprop optimizer, the validation accuracy dropped approximately to 96.67% from 96.81%. However, this model's accuracy reached 97.1% while utilizing Adam. In both RMSprop and Adam, AlexNet exhibited the same pattern of accuracy performance as SqueezeNet.

The training time in all models is comparatively less with SGDM except in the SqueezeNet model. CNN3 trained the quickest, with a total duration of around 21 minutes, and SqueezeNet trained the second-fastest with 1 hour 10 minutes 10 seconds, slightly different from AlexNet. GoogleNet took the longest training time, requiring approximately 2 hours 30 minutes to almost 2 hours 41 minutes. The comparison of results with previous work are shown in Table 7.

Next, each model was tested with an untrained AMDI dataset to determine its reliability. Table 8 presents the test accuracy, accuracy per class, macro F1-score, and overall test time for 116 images. Since the AMDI dataset provided an imbalanced dataset for each class, this chapter attempts to investigate further using the macro F1-score performance findings, as analysis from previous studies discovered bias in the accuracy performed by classes that provided fewer data

Table 6 Performance of each network model trained and validated with STACOM2012 dataset

Optimizer	Validation accuracy	No parameter (million)	Training time (H:M.S)
CNN3		0.081299	
SGDM	92.3		0:21:10
RMSProp	93.0		0:21:70
Adam	93.2		0:21:45
AlexNet		40.099203	
SGDM	94.49		1:15:35
RMSProp	93.91		1:31:35
Adam	94.64		1:21:35
GoogleNet		5.976627	
SGDM	96.52		2:30:05
RMSProp	97.25		2:40:30
Adam	97.54		2:37:05
SqueezeNet		0.724035	
SGDM	96.81		1:07:20
RMSProp	96.67		1:05:02
Adam	97.10		1:10:10

Table 7 The comparison of CNN3 result with related work

Author	Optimizer	Dataset	Accuracy result (%)
Ho et al. [1]	Nesterov momentum	Cardiac MRI (private)	86
Our (CNN3)	Adam	Cardiac MRI STACOM2012	93.2

[31]. The F1-score incorporates involvement from all classes and reaches its highest possible value of 1, and its lowest possible score of 0. Nevertheless, the difference between these evaluation metrics in this result was only in the number of values. Similar observations in terms of outcome patterns were observed in both performance measures.

The performance of all network models deteriorates when evaluated with a new and untrained AMDI dataset. GoogleNet, which had previously been trained with RMSProp, outperformed the other networks with a macro F1-score of 0.908 and accuracy per class higher than 84%. SqueezeNet was next in line with the highest macro F1-score of 0.8717 and accuracy per class higher than 75%. It has been noticed that the CNN3 model and AlexNet have comparable performances. CNN3's F1-score performance was observed to be marginally better than AlexNet.

It can be observed that all models performed effectively in classifying NLVB classes because the performance variance for each class is greater than 73%. The majority of LV class accuracy values fall between 70% and 98%, with the exception of CNN3 and AlexNet models which utilized Adam. A significant variation was observed in one of the outer slice classes, NLVA, with the values ranging from 15.38% to 84.62%.

The final column tabulated the testing time spent to predict the classes for a total of 116 images from the AMDI dataset. There is only a slight difference in

Table 8 Performance of each network model tested with AMDI dataset

Optimizer	Accuracy per class % LV (60)	NLVA (26)	NLVB (30)	Test accuracy	Macro F1-score	Test time per 116 images (seconds)
CNN3						
SGDM	75.00	30.77	93.33	69.83	0.7022	3.27
RMSProp	71.67	38.46	93.33	69.83	0.7120	4.15
Adam	58.33	15.38	93.33	57.76	0.5837	2.50
AlexNet						
SGDM	78.33	26.92	90.00	69.82	0.6967	2.22
RMSProp	76.67	23.08	73.33	63.79	0.6335	4.17
Adam	48.33	42.31	100.00	60.34	0.6845	2.25
GoogleNet						
SGDM	83.33	30.77	86.67	72.41	0.7252	3.22
RMSProp	95.00	84.62	90.00	91.38	0.9080	3.31
Adam	98.33	65.38	80.00	86.20	0.8472	3.35
SqueezeNet						
SGDM	90.00	50.00	76.67	77.59	0.7527	2.01
RMSProp	88.33	76.92	93.33	87.07	0.8717	1.9308
Adam	93.33	46.15	86.67	81.03	0.8029	2.0051

the amount of time spent predicting an image between the various optimizers used on the GoogleNet and SqueezeNet models. With only 0.017 seconds spent on each image, SqueezeNet used less predicted time than other models.

5 Discussion and Applicability to Smart Healthcare

All models performed excellently, exceeding validation accuracy by over 90% when trained and tested on the cardiac LGE-MRI STACOM2012 dataset. Each model was trained using three different network optimizers, and the results showed a notable increase in each model's accuracy depending on the optimizer. The Adam optimizer had the highest validation accuracy performance compared to the other optimizers within each model with a slight improvement of 0.3–1.1% over SGDM. The best outcome was achieved by GoogleNet, which had a validation accuracy of 97.54%, with only a 0.44% difference from SqueezeNet.

The amount of time required for training in a deep neural network is determined by various factors, some of which include the complexity of the network model [29], the number of network parameters, and the number of layers in the network. Therefore, it is not surprising that CNN3 provides the fastest training time among all models since this model has fewer layers and only 0.081 million parameters.

Even though GoogleNet provides the highest performance among all models, it also provides the longest training time due to the depth of its network architecture. It has been noticed that the RMSProp training time was slightly higher than most models that employed it.

Prior to implementation, these networks were tested on a cardiac MRI dataset from AMDI. There has been a noticeable decline in terms of accuracy and overall macro F1-score performances. Since none of the network models were trained with the AMDI's dataset, it makes sense that the results indicate a significant decline in performance for all networks due to different image resolutions and the generalization of the networks on the ability to adapt correctly to a new dataset. Theoretically, the network optimizer should only be involved in the network's training process and not the testing process. However, significant accuracy and macro F1-score performance differences are observed when testing the previously trained network with different network optimizers.

Overall, all networks' performance remained satisfactory, with a micro F1 score ranging from 0.57 to 0.9.

When analyzing the AMDI images, the CNN3 model which previously used SGDM and RMSProp optimizers outperforms the AlexNet model in terms of F1 score performance. This highlights the potential of the CNN3 model as a network classification for LGE cardiac MR images without having to deal with a complex network.

Proper diagnosis of heart disease at right time could save lots of lives. The automated approach of classifying LV and Non-LV segments in cardiac MR images by utilizing a deep CNN developed in this chapter needs to be further analyzed for time complexity, and resource analysis before it could be integrated with various digital health technologies like e-health, IoMT and smart healthcare [34–38].

6 Conclusion

In summary, this chapter intends to increase the robustness of the network classifier by comparing three widely used optimizers (SGDM, RMSprop, and Adam) to identify the best network for classifying left ventricle segment and non-left ventricle segment in LGE cardiac MR images automatically. A comparison between a baseline model, namely CNN3 and three well-known pre-trained models was made using these three different network optimizers. After training, all network models excelled by exceeding 90% accuracy in classifying LV and non-LV images in the STACOM2012 dataset. Adam was found to be the best network optimizer, with an accuracy improvement of 0.3–1.1% over SGDM.

The GoogleNet model outperformed other models with an accuracy performance of 97.54% and a macro F1-score of 0.9080 when tested with the STACOM2012 and AMDI datasets, respectively. Since the previously trained networks with RMSprop and SGDM performed better when tested with the untrained AMDI dataset, this chapter also suggests additional research into RMSprop and SGDM, such as retraining the AMDI dataset. Further research into the architecture and parameter tuning of the CNN3 model is recommended, as this uncomplicated model shows high potential in classifying the left ventricle and non-left ventricle segments in cardiac MR images.

Acknowledgements This research work ethics approval is obtained from Universiti Sains Malaysia (USM/JEPeM/21090623). The authors would like to express their gratitude to members of the Advanced Control System and Computing Research Group (ACSCRG), Advanced Rehabilitation Engineering in Diagnostic and Monitoring Research Group (AREDiM), Integrative Pharmacogenomics Institute (iPROMISE) and Centre for Electrical Engineering Studies, Universiti Teknologi MARA, Cawangan Pulau Pinang for their assistance and guidance during the fieldwork. Finally, the authors are grateful to Universiti Teknologi MARA, Cawangan Pulau Pinang for their immense administrative and financial support.

References

1. Ho N, Kim YC (2021) Evaluation of transfer learning in deep convolutional neural network models for cardiac short axis slice classification. Sci Rep 11(1):1–11. https://doi.org/10.1038/s41598-021-81525-9
2. Dhar S, Shamir L (2021) Evaluation of the benchmark datasets for testing the efficacy of deep convolutional neural networks. Vis Inform 5(3):92–101. https://doi.org/10.1016/j.visinf.2021.10.001
3. Ghosh A, Sufian A, Sultana F, Chakrabarti A, De D (2020) Fundamental concepts of convolutional neural network. Intell Syst Ref Libr 172:519–567. https://doi.org/10.1007/978-3-030-32644-9_36
4. Krizhevsky A, Sutskever I, Hinton GE (2017) ImageNet classification with deep convolutional neural networks. Commun ACM 60(6):84–90. https://doi.org/10.1145/3065386
5. Szegedy C et al (2015) Going deeper with convolutions. In: 2015 IEEE conference on computer vision and pattern recognition (CVPR). IEEE, pp 1–9. https://doi.org/10.1109/CVPR.2015.7298594
6. Szegedy C, Vanhoucke V, Ioffe S, Shlens J, Wojna Z (2016) Rethinking the inception architecture for computer vision. In: Proceedings of the IEEE Computer Society conference on computer vision and pattern recognition. IEEE, pp 2818–2826. https://doi.org/10.1109/CVPR.2016.308
7. He K, Zhang X, Ren S, Sun J (2016) Deep residual learning for image recognition. In: Proceedings of the IEEE Computer Society conference on computer vision and pattern recognition. IEEE, pp 770–778. https://doi.org/10.1109/CVPR.2016.90
8. Oztel I, Yolcu G, Oz C (2019) Performance comparison of transfer learning and training from scratch approaches for deep facial expression recognition. In: Proceedings of the 4th international conference on computer science and engineering, UBMK 2019, pp 290–295. https://doi.org/10.1109/UBMK.2019.8907203
9. Tan C, Sun F, Kong T, Zhang W, Yang C, Liu C (2018) A survey on deep transfer learning. In: Kůrková V, Manolopoulos Y, Hammer B, Iliadis L, Maglogiannis I (eds) Artificial neural networks and machine learning – ICANN 2018. ICANN 2018. Lecture Notes in Computer Science, vol 11141. Springer, Cham. https://doi.org/10.1007/978-3-030-01424-7_27
10. Parah SA, Rashid M, Vijaykumar V (2022) Artificial intelligence for innovative healthcare informatics. Springer. ISBN: 978-3-030-96568-6
11. Alzubaidi L et al (2020) Towards a better understanding of transfer learning for medical imaging: a case study. Appl Sci 10(13):4523. https://doi.org/10.3390/app10134523

12. Alzubaidi L, Al-Shamma O, Fadhel MA, Farhan L, Zhang J, Duan Y (2020) Optimizing the performance of breast cancer classification by employing the same domain transfer learning from hybrid deep convolutional neural network model. Electronics 9(3):445. https://doi.org/10.3390/electronics9030445
13. Nazeer S et al (2022) Automatic classification of COVID-19 infected patients using convolution neural network models. In: Parah SA, Rashid M, Varadarajan V (eds) Artificial intelligence for innovative healthcare informatics. Springer, Cham, pp 119–131. https://doi.org/10.1007/978-3-030-96569-3_6
14. Yaqub M et al (2020) State-of-the-art CNN optimizer for brain tumor segmentation in magnetic resonance images. Brain Sci 10(7):427. https://doi.org/10.3390/brainsci10070427
15. Wibowo A, Wiryawan PW, Nuqoyati NI (2019) Optimization of neural network for cancer microRNA biomarkers classification. J Phys Conf Ser 1217(1):012124. https://doi.org/10.1088/1742-6596/1217/1/012124
16. Yu ZL, Sun G, Lv J (2022) A fractional-order momentum optimization approach of deep neural networks. Neural Comput Appl 34(9):7091–7111. https://doi.org/10.1007/s00521-021-06765-2
17. Jaworska T (2021) Image segment classification using CNN. In: Uncertainty and imprecision in decision making and decision support: new challenges, solutions and perspectives, Advances in intelligent systems and computing, vol 1081. Springer, pp 409–425. https://doi.org/10.1007/978-3-030-47024-1_38
18. Bottou L (2010) Large-scale machine learning with stochastic gradient descent. In: Lechevallier Y, Saporta G (eds) Proceedings of COMPSTAT'2010. Physica-Verlag HD, pp 177–186
19. Jin R, He X (2020) Convergence of momentum-based stochastic gradient descent. In: IEEE 16th international conference on control and automation, ICCA, 2020. IEEE, pp 779–784. https://doi.org/10.1109/ICCA51439.2020.9264458
20. Becherer N, Pecarina J, Nykl S, Hopkinson K (2019) Improving optimization of convolutional neural networks through parameter fine-tuning. Neural Comput Appl 31(8):3469–3479. https://doi.org/10.1007/S00521-017-3285-0
21. Kumar A, Sarkar S, Pradhan C (2020) Malaria disease detection using CNN technique with SGD, RMSprop and ADAM optimizers. In: Deep learning techniques for biomedical and health informatics. Studies in big data. Springer, pp 211–230. https://doi.org/10.1007/978-3-030-33966-1_11
22. Zhang Z (2018) Improved Adam optimizer for deep neural networks. In: 2018 IEEE/ACM 26th international symposium on quality of service (IWQoS). IEEE, pp 1–2. https://doi.org/10.1109/IWQoS.2018.8624183
23. Hikmat Haji S, Mohsin Abdulazeez A (2021) Comparison of optimization techniques based on gradient descent algorithm: a review. PalArch's J Archaeol Egypt/Egyptol 18(4):2715–2743
24. Cardiac Atlas Project. Ventricular infarct segmentation. https://www.cardiacatlas.org/challenges/ventricular-infarct-segmentation/. Accessed 19 Aug 2022
25. Shorten C, Khoshgoftaar TM (2019) A survey on image data augmentation for deep learning. J Big Data 6(1):60. https://doi.org/10.1186/s40537-019-0197-0
26. Suhaida D, Damit A, Sulaiman SN, Osman MK, Khairiah N (2022) Classification of left ventricle and non-left ventricle segment for cardiac assessment using deep convolutional neural network. J Electr Electron Syst Res 21:31–38
27. Sahu P, Chug A, Singh AP, Singh D, Singh RP (2020) Implementation of CNNs for crop diseases classification: a comparison of pre-trained model and training from scratch. Int J Comput Sci Netw Secur 20(10):206–215
28. Weiss K, Khoshgoftaar TM, Wang DD (2016) A survey of transfer learning. J Big Data 3(1):9. https://doi.org/10.1186/s40537-016-0043-6
29. Alzubaidi L et al (2021) Review of deep learning: concepts, CNN architectures, challenges, applications, future directions. J Big Data 8(1):53. https://doi.org/10.1186/s40537-021-00444-8
30. Iandola FN, Han S, Moskewicz MW, Ashraf K, Dally WJ, Keutzer K (2017) SqueezeNet. arXiv:1602.07360 4:370. [Online]. Available: https://github.com/DeepScale/SqueezeNet

31. Grandini M, Bagli E, Visani G (2020) Metrics for multi-class classification: an overview. arXiv:2008.05756:1–17. [Online]. Available: http://arxiv.org/abs/2008.05756
32. Hossin M, Sulaiman M (2015) A review on evaluation metrics for data. Int J Data Min Knowl Manag Process 5(2):1–11
33. Markoulidakis I, Rallis I, Georgoulas I, Kopsiaftis G, Doulamis A, Doulamis N (2021) Multiclass confusion matrix reduction method and its application on net promoter score classification problem. Technologies 9(4):81. https://doi.org/10.3390/technologies9040081
34. Kamili A et al (2021) DWFCAT: dual watermarking framework for industrial image authentication and tamper localization. IEEE Trans Ind Inform 17(7):5108–5117. https://doi.org/10.1109/TII.2020.3028612
35. Sarosh P et al (2021) Secret sharing-based personal health records management for the Internet of Health Things. Sustain Cities Soc 74:103129
36. Parsa S, Parah SA, Bhat GM, Khan M (2021) A security management framework for big data in smart healthcare. Big Data Res 25:100225
37. Bhat GM et al (2010) Field programmable gate array (FPGA) implementation of novel complex PN-code-generator-based data scrambler and descrambler. Maejo Int J Sci Technol 4(1):125–135
38. Parah SA, Sheikh JA, Bhat GM (2014) A secure and efficient spatial domain data hiding technique based on pixel adjustment. Am J Eng Technol Res 14(2):33

Dual-Feature CNN-SVM Method for Breast Mass Tissue Abnormality Classification on Digital Mammography Images Adapted to Breast Density

Noor Fadzilah Razali, Iza Sazanita Isa, Siti Noraini Sulaiman, Noor Khairiah A. Karim, and Muhammad Khusairi Osman

1 Introduction

Breast cancer is the world's second-largest cause of cancer death, and the number of women diagnosed with the disease is increasing every year. According to the International Agency for Research on Cancer, its diagnosis among women has surpassed lung cancer, with an estimated 2.3 million new cases annually [1]. Local cases of breast cancer are among the three most common cancers in the population and have increased rapidly from 32% to 34.1% within 2 years [2]. Clinical screening examinations can identify suspicious breast lesions early and lower the risk of breast

N. F. Razali · I. S. Isa (✉)
Electrical Engineering Studies, College of Engineering, Universiti Teknologi MARA, Cawangan Pulau Pinang, Penang, Malaysia
e-mail: izasazanita@uitm.edu.my

S. N. Sulaiman
Electrical Engineering Studies, College of Engineering, Universiti Teknologi MARA, Cawangan Pulau Pinang, Penang, Malaysia

Advanced Rehabilitation Engineering in Diagnostic and Monitoring Research Group (AREDiM), Electrical Engineering Studies, College of Engineering, Universiti Teknologi MARA, Cawangan Pulau Pinang, Penang, Malaysia

Integrative Pharmacogenomics Institute (iPROMISE), Puncak Alam, Selangor, Malaysia

N. K. A. Karim
Department of Biomedical Imaging, Advanced Medical and Dental Institute, Universiti Sains Malaysia, Bertam, Kepala Batas, Pulau Pinang, Malaysia

Breast Cancer Translational Research Programme (BCTRP), Advanced Medical and Dental Institute, Universiti Sains Malaysia, Kepala Batas, Penang, Malaysia

M. K. Osman
Centre for Electrical Engineering Studies, Universiti Teknologi MARA, Cawangan Pulau Pinang, Permatang Pauh, Penang, Malaysia

S. A. Parah et al. (eds.), *Intelligent Multimedia Signal Processing for Smart Ecosystems*, https://doi.org/10.1007/978-3-031-34873-0_9

cancer in women [3, 4]. Mammography is a principal modality that uses digital image acquisition technology termed full-filed digital mammography (FFDM), along with its traditionally acquired image through screen-filmed mammography (SFM), by exposing the patient's breasts to minimal doses of X-ray radiation [5]. The breasts are flattened on a platform as the radiation is emitted to detect any presence of tissue abnormality. Since its introduction, clinical screening exams have helped women reduce their chance of developing breast cancer by detecting suspicious breast lesions sooner [6]. FFDM screening exams are the most widely used imaging modality for this purpose, with their accessibility in most medical facilities, low radiation exposure, wide breast evaluation area, excellent specificity, and high sensitivity are some advantages of mammography [7]. Even with the development of more advanced bioimaging modalities like three-dimensional digital breast tomosynthesis (DBT), ultrasound (US), magnetic resonance imaging (MRI), and computed tomography (CT) scan, the mammogram remains the best breast cancer screening method due to its low implementation cost [8].

Early diagnosis of breast cancer lesions saves lives and increases the chance of early treatment. Generally, a breast radiologist's mammography interpretation comprises information such as suspected abnormalities, breast density, Breast Imaging Reporting and Data System (BI-RADS) ratings, and patient follow-up notes. This approach requires radiologists to manually assess mammograms for abnormalities and prescribe additional imaging modalities if a lesion is not visible due to dense breasts, such as using the US and MRI [9]. It was reported that the rate of having a false positive result from misdiagnosis after an initial mammography scan ranges from 7% to 12% [10]. However, this method is prone to inter-rater false-positive and false-negative results due to factors such as image complexity caused by distributed dense tissue in the breast, especially to accurately classify whether the overlapped mass region contains the mass or the fibroglandular tissue surrounding it.

The medical world has recognized breast tissue density as a significant risk factor for breast cancer progression. It is one of the risks radiologists analyze, primarily as it is often associated with a higher likelihood of developing malignant lesions, which is also one of the most repeated results in dense breast mammography diagnostic [11]. Moreover, unreported cancer cases on regular mammography screening are related to dense breast tissue, with an increased cancerous risk of 1.6 to 2.2-fold [9]. It has been reported that 75% of patients have an enhanced risk due to undetected lesions on mammograms since adjacent fibroglandular breast tissues mostly resemble and have the same intensity level as mass lesions. Figure 1 is a sample of dense breast mammograms with dense, fatty, fibroglandular, and breast lesion region depicted in different color mappings. Undiagnosed abnormalities in dense regions may impair raters' assessments, resulting in high inter-observer variability (kappa scores = 0.43) [12].

Although inter-rater observers for the same patient may reduce the false positive rate, it is time intensive and frequently leads to additional screening. Because of this, automated breast diagnostic screening procedures are necessary to conduct a more in-depth analysis of the breast abnormalities that improve the laborious and

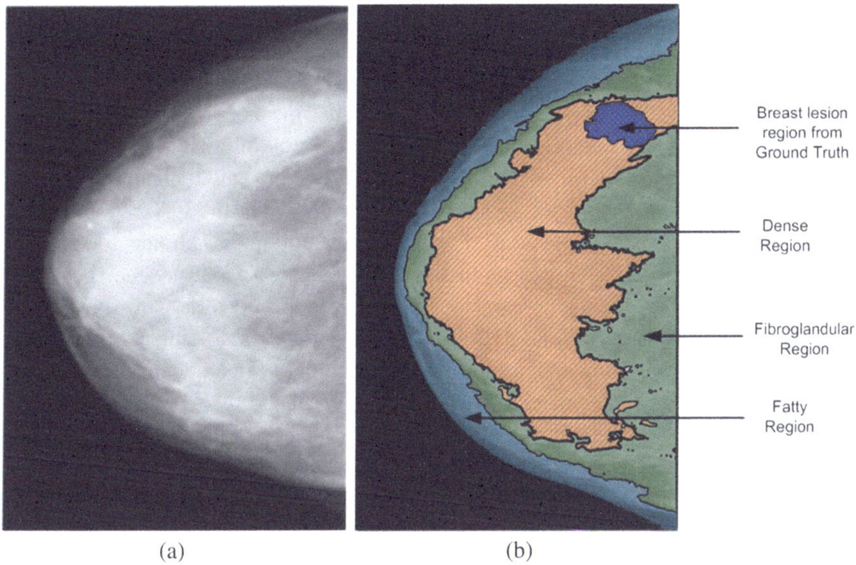

Fig. 1 Mammogram images: (**a**) Original. (**b**) Labelled breast tissue of dense region, fatty region, fibroglandular region, and mapped lesion area on the dense region from ground-truth

time-consuming process and detect malignant breast lesions, particularly in dense breast tissues.

In the past decade, various image processing methods via computer-based tools have been implemented in research to process FFDM images, eventually helping radiologists to make better clinical diagnoses. However, images taken have limitations in terms of quality, such as the presence of noise, artifacts, and inhomogeneity of intensity obtained during the acquisition of the images leading to possible misdiagnoses of patients. Therefore, techniques that combine image processing and artificial intelligence (AI) have been employed as essential elements before further analyzing mammogram images. When used to aid radiologists in clinical practice, the AI methods are implemented in a compact system known as the Computer Aided Detection (CAD) system, which is mainly used to ease the operation of assessing individual mammogram images through user-based of semi or fully automated computer-based tool. For such systems integrity authentication and security of data is of paramount importance [13–15]. An increasing application of AI in the field has led to more studies looking into how it may be used as an assistive tool for radiologists to do laborious and repetitive jobs. Various studies have focused on improving the quality of mammogram images through developing algorithms, especially for lesion detection and cancer classification [16–18], as well as estimating the repeatability and reproducibility of cancer [19]. Additional efforts in creating various CAD phases should include creating individualized breast assessments for different patients, such as automatically learning the prevalence of breast cancer gene mutations and the level of breast density that affects mass formation.

Although traditional CAD system utilizes machine learning (ML) as their primary operating algorithm, several issues arise, such as the radiologist's need to ignore the numerous false positive results that these systems generated, associated with an estimated 20% increase in the amount of time needed to interpret each study [20]. Research has demonstrated that enquiring a radiologist to make a second diagnosis for falsely CAD-detected lesions per study can skew their interpretation, making them less likely to spot an actual malignancy when the CAD system does not signal any prospective lesions, leading to its unpopular choice to be used among radiologists [21]. However, in the past decade, a CAD system has been developed incorporating deep learning as the main framework, substituting the more traditional ML-based system. For instance, a study by McKinney et al. (2020) utilized mammography images in their proposed model to detect cancer, indicating that the human rater's false detection rate has been reduced by 5.7% [22].

Nonetheless, the fact that breast density is a significant risk factor for the presence of cancer as determined by risk assessment studies [16, 20], combined with a less developed density classification model developed using limited manufacturer's images during training, emphasizes the importance of pursuing this research. Less research was done that adapts breast density consideration in the development phase [8, 14, 21]. Even more, limited studies have been found that classify the features of multiple tissues within a mammogram image that a radiologist can use to determine whether the suspected lesion is as it is or just a fatty or fibroglandular region. Therefore, methods are needed to fully automate tissue classification based on an intelligence approach and image processing methods by combining features that are deemed necessary by the radiologist during manual mammogram analysis through handcrafted (HC) features, as well as automated features captured from the input image by the AI algorithm that cannot be determined through naked eyes. At the same time, the development of this CAD system is significant for including dense tissue with its cancer risk association clinical profiling that is vital in mammogram image reporting and can be used in the future.

2 Related Studies

The use of the CAD system in determining breast tissue abnormalities has been implemented since its early introduction. However, its use is still in the precautionary stage due to the lack of sensitivity of the system to detect masses, especially in dense mammograms [7, 22], where radiologists have difficulty determining whether the suspected tissue is a type of mass lesion, or whether it is tissue derived from normal fibroglandular tissue that forms dense breast. With that, constructing a suitable CAD system to find relevant features to recognize these types of tissue is vital to produce a robust CAD.

A deep learning-based algorithm such as Convolutional Neural Network (CNN) is intended to facilitate the development stage of the CAD system, where a high number of mammogram images is required to extract essential features for tissue classification [26]. However, most studies focus on separately classifying the tissue based on its same category, for example, mass classification (into benign, malignant) or dense tissue classification (dense and non-dense breast) [23–25]. In a study by Mohamed A. A. et al. (2018), classification of heterogeneous and scattered dense breast images using transfer learning of AlexNet CNN on 14,000 typical mammogram images, such applications to classify dense and non-dense mammograms have been successfully applied by past researchers using the deep learning approach [27]. Both the scratch and transfer learning frameworks are trained, and the results demonstrated that only a scratch-based CNN with a negative margin accuracy of 0.0025% trained on a balanced dataset could be effectively adapted using transfer learning. Another strategy by Saffari N. et al. (2020) that compares thresholding methods to autoencoder networks of conditional generative adversarial networks (cGAN) to segment the region and produce binary masks for the images was proposed in [28], where Res-Net101 was used as the base feature extractor to classify the breasts into four breast density levels.

Meanwhile, Sadad T. et al. (2021) utilized the internet of things (IoT) in their development of classifying dense and non-dense breasts through DenseNet transfer learning trained on 322 images for three tissue classes (fatty, fibroglandular, dense) by providing real-time access for the end user to upload and receive output with 90.4% accuracy in classification [29]. Some researchers also comprehend the need to adapt breast density to determine mass abnormality, signifying the risk of malignancy imposed on dense breasts. In their study, Huang M. and Lin T.-Y. (2021) classify the four-density breast type according to whether the mammogram contains benign or malignant mass through augmentation procedures [24]. The result shows higher performance on fatty tissue breast to detect a mass abnormality, with 99.7% accuracy. This study is also supported by Suh Y. J. et al. (2020), showing the highest accuracy of 96.7% on the fatty breast by utilizing CNN models trained separately on different density levels to class malignancy after the concatenation of both craniocaudal (CC) and mediolateral oblique (MLO) from the single breast before training [11].

However, a problem that always arises in any CNN application especially in computer vision bioimaging is the need for datasets annotated by expert radiologists [30]. With that, the use of an ML framework remains relevant where the developers can select features that are important to be associated with the type of tissue in the breast using a lower number of the training dataset and indirectly produce a more robust system against the abnormality of the tissue in the breast during the diagnosis process [31]. Even though limited, modifications have been made in past studies to incorporate both deep learning and ML to benefit from both frameworks to solve breast mammogram-related CAD development [32, 33]. For example, eliminating the classification layer in CNN may make it possible to make superior use of automated feature extraction in conjunction with an ML classifier. By first automatically extracting features using CNN architecture, then classifying them using ML-based algorithms like Support Vector Machine (SVM) [32] and extreme machine learning (ELM) classifier [30, 31], it is possible to categorize dense regions and breast mass lesions in mammograms. For instance, Ragab D. et al. (2019) suggested fusing many features from various CNN architectures to tackle the breast lesion classification problem by combining multiple features from different CNN architectures to address the breast lesion classification problem [32]. The approach trained four pre-trained networks and then used the extracted features to develop SVM classifiers with various kernel types. Then, features were combined based on how well they performed across the first four pre-trained networks. The fused features are finally classified using the SVM classifier with a second comparison using Principal Component Analysis (PCA) feature reduction, showing improved accuracy from 94.9% to 97.9% by incorporating feature fusion based on ranking with a linear kernel SVM classifier. Ultimately, PCA decreased computation times while noticeably boosting system accuracy. The use of ML classifiers increased accuracy was also validated by [18, 34]. Both authors classified combined data from CNN architecture and handcrafted features using SVM and ELM algorithms for comparison.

In contrast to SVM (92.93%) and CNN classifier alone (89.67%), Mahmoud H. A. H. et al. (2022) in their study demonstrated higher accuracy (97.43%) when using the ELM classifier [34]. The results corroborated a prior study by Wang Z. et al. (2019) that employed a similar framework and found that ELM had a 3% greater accuracy than SVM [18]. Additionally, using segmentation in mammograms to directly predict lesion location through semantic segmentation [35] and ML-based watershed segmentation [36], as well as using an optimizing method, such as that used by Sha Z. et al. (2020), using a grasshopper optimization algorithm (GOA) with CNN architecture [37]. The GOA is a meta-heuristic function that seeks out the best CNN tuning parameters. However, it is computationally expensive to identify the best parameters since each GOA swarm agent needs to be trained 1000 times on the breast cancer dataset. The malignancy of the feature is then determined using an SVM classifier using features based on geometric, textural, and statistical data for training.

In this study, the combination of CNN and ML architecture is proposed fusing both automated, and HC features to improve the overall classification accuracy. Although the use of an ML classifier disrupts the continuity of an otherwise end-to-end classification process and requires specific human input to get the best predictor, the use of an ML classifier is expected to increase the end performance, with the addition that the image features are well-extracted by the CNN backbone itself. This is highly related to the fact that the ML algorithm is more widely attunable, with multiple optimization techniques for improvement in between that can be introduced in the framework and input feature data extracted explicitly for determining specific image feature vectors. Meanwhile, compared to CNN's classifier, which only relies on probability weightage as the final predictor for classification, all information in the input image is included, whether relevant or not, as feature training.

In summary, the expected contributions of this study are:

1. Additional handcrafted features as extra defining elements to differentiate tissue types within mammogram images based on density level.
2. Fewer augmentation properties set on dual-feature model to avoid dependency on a high number of images during CNN training.
3. Classification of four tissue types within a mammogram image that radiologists can use to determine the nature of the suspected tissue in clinical diagnosis.

3 Methodology

This work uses an established dataset of mass abnormality from publicly available INbreast [38] for system training, widely used in breast mammogram image processing and neural network algorithms [14, 34–36]. The ground-truth provided with this dataset is used as labeled data during system training, including its density level and manual lesion segmentation, for testing the trained system's performance. All the model's performance is evaluated using suitable statistical performance matrices. The work is divided into two main stages: Image Preparation and System Training. In the end, three models are trained and compared for tissue classification based on the type of input feature used for training. The main objective is achieved through this work's flowchart, as illustrated in Fig. 2.

3.1 Image Preparation

Image pre-processing steps are taken to prepare the mammogram image properly before training. The original mammogram images in Digital Imaging and Communications in Medicine (DICOM) format are converted to portable network graphics

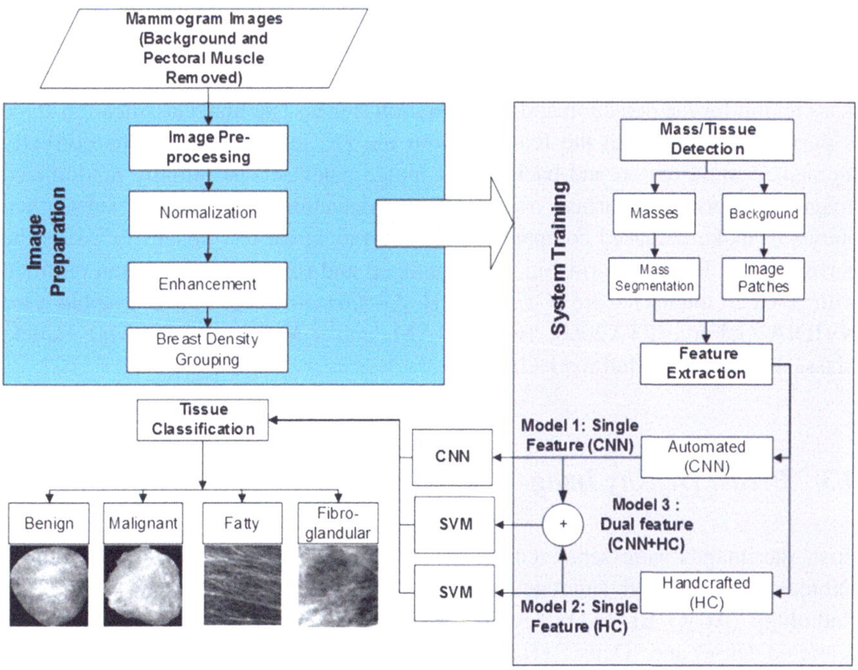

Fig. 2 Overall methodology of the proposed method

(*.png*) format to process the images. The images are normalized to ensure the pixel intensity is uniformly assigned between the 0–1 range. Next, to remove the manual annotations within the images while only collecting the breast area, morphological operations, including image binarization through object extraction, are done, and the remaining background pixel is converted to a zero-pixel value. The breast skin area is removed on a threshold of 0.01 from the binarization to lessen the unwanted pixel information within the image. The dataset is then randomly divided into 8:1:1 from all 112 images from mass cases for training, validation, and testing for all system stages. For developing a deep-learning model that does not overfit as the result of low training images, the training input images are augmented with horizontal and vertical flipping, random degrees of ration of 30–300°, and image scaling up to 1.3 factorizations. For the model with HC features, the images were pre-augmented to get the HC features of the images, but no additional augmentation settings were made during the CNN parameter setup. Finally, for tissue classification of benign, malignant, fatty, and fibroglandular classes, the training is validated using five-fold cross-validation to ensure all the datasets groups are trained and tested. The final average performance is used as the overall performance.

3.2 Experimental Settings

This work performs all pre-processing, detection, segmentation, and classification steps using the INbreast dataset and its accompanying ground-truth data. A deep learning-based experimental setup is created to accurately prepare the extracted mass region for the detection and segmentation stages. The final classification stage is performed using both the features from the HC and CNN from the correctly segmented mass region and background image patches. The initially randomized image numbering is carried over from the detection phase to the subsequent phases to make unbiased comparisons between identical test images to assess the performance. These experiments are visualized and run on a workstation outfitted with a CPU Intel(R) Core™ i7-10870H 2.3 GHz, a single GPU graphic card NVIDIA GeForce RTX2060, 16 GB RAM, and a MATLAB (R2022a) Natick, Massachusetts: The Mathworks Inc.

3.3 Breast Density Image Enhancement Consideration

First, the images were separated based on their density level as assigned by the INbreast accompanied annotated dataset according to its American College of Radiology (ACR) BI-RADS level to "Non-dense" (1: almost entirely fatty, 2: scattered dense) and "Dense" (3: heterogeneously dense, 4: dense).

In this work, image enhancement is made based on the type of classified breast density to reduce low, dense image representability during training. Image enhancement based on contrast adjustment on specific Region Of Interest (ROI) is proven to increase detection ability [40]. To do this, image adjustment based on the images' histogram is applied to the image, with a lower clip limit cap set higher on the dense images than on non-dense images. This is to reduce the effect of oversaturation on the homogenous pixel mapping on the dense region that contains overlapping mass due to these pixels falling on the same gray level range. The finalized image is then used as the input for the mass detection step.

3.4 System Training

This subsection discusses the process of getting the mass location through a mass detection algorithm; and density tissue types that are classed into two classes: fatty and fibroglandular (combining dense and fibroglandular tissue) using random image patch sizes on the remaining background image where these cropped image patches are then utilized on the next section.

Figure 3 shows an example of an enhanced mammography image (a) with its respective example of an image (b) of patches illustrated through extraction per image for training. The red region indicates the mass region obtained from the bounding-box resulting from the mass detection algorithm. In contrast, the yellow and green areas are image patches with random sizes extracted for fatty and

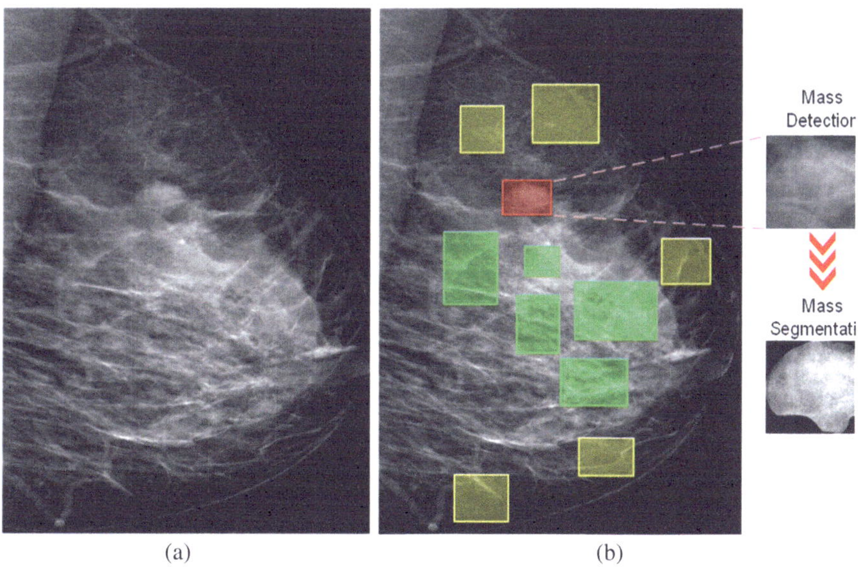

Fig. 3 Sample of enhanced mammogram image (**a**) and its patch extraction process (**b**) for mass (red region) using YOLOv3 and patch-based extraction for fatty (yellow region) and fibroglandular (green region). The detected mass region (red) will be used for mass detection and segmentation

fibroglandular tissue. As for the region containing mass (bounding boxes), other mass detection and segmentation method is done to obtain the final segmented mass as in Fig. 3b-right. Details of the steps taken are discussed in the following subsection.

3.5 Breast Mass Detection

Breast mass localization is vital in this work to detect the location of the mass within an image of the mammogram. A CNN-based algorithm of You-Only-Look-Once version 3 (YOLOv3) proposed by [41] is adopted to perform the mass detection task. Although newer versions of YOLO, including YOLOv4 and YOLOv5 that is proven to have higher detection accuracy [42], previous studies have shown that it also requires more capable hardware processing for stable training as well as operates in lower frame-per-second (fps) to be applicable in real-time application [14]. Moreover, YOLOv3 has been proven to achieve satisfactory mass detection performance in past mammogram studies [9, 33, 34, 39] that leads to a high mass malignancy classification of 96.5% [46], and able to receive simple modification in its architecture and pre-processing method to increase its final performance, such as by introducing cycleGAN [13] in the past, making it easier to be implemented in standard computer processing settings. Meanwhile, compared to other detection algorithms such as faster region-based CNN, YOLOv3 was shown to have a lower false positive rate (FPR) and better overall accuracy, making it a faster object detection algorithm to maintain its overall accuracy [42–45].

The ground-truth provided with the INbreast dataset is used to get the bounding box of the mass in training images. A typical bounding box contains the location of the mass on the image using a coordinate system of $[x,y,w,l]$, where x and y are the coordinates of the top left-most object location, while w and h are the width and height of the particular object. During training, the algorithm searches for the optimal $[x,y]$ coordinates of the object, with the cells holding the midpoint responsible for determining the target object's presence. The training mages are passed through a series of convolutional layers and an output layer that have the information of the mass location showing the probability of the detected mass is true through confidence score (CS), where the highest probability value for the final detected output for class prediction based on this prediction value on a threshold of >0.5 is chosen as the final location of the detected mass. In this stage, the final bounding box predicting the mass location is later used for mass segmentation. In contrast, the remaining background images are used to extract patch images of fatty and fibroglandular, each with five randomized sizes for tissue classification.

3.6 Mass Segmentation & Fatty/Fibroglandular Tissue Image Patch Extraction

The resulting bounding box indicating the mass location within the mammogram is extracted from the previous step and subjected to a CNN-based semantic segmentation framework compared to a five-layer U-Net segmentation to obtain only the pixel representing the mass tissue. Since the dataset has more malignant mass data than benign class, the algorithm is trained based on the percentage of overall weightage representation of each class. It is updated at the final pixel classification layer during the CNN feature extraction method. Meanwhile, an additional class of tissue representing fatty and fibroglandular tissue that affect the ability of both manual and automated CAD to identify mass is included for final classification by differentiating the mass and the background that could belong to both fatty (non-dense) and fibroglandular (dense) tissue. After the mass detection bounding box has been extracted, the remaining area of the mammogram image is used as an additional input for the algorithm's training. Five image patches per image with random sizes are extracted from both dense and non-dense breast images, where these additional datasets are trained with the segmented benign and malignant from the previous step to finally output the result of four-class classification (mass: Benign, Malignant; background: Fatty, Fibroglandular).

3.7 Feature Extraction: Automated Convolutional Neural Network Features

The four segmented image classes trained in this work are subjected to a CNN image feature extraction via ResNet50 [47] as a base network through transfer learning. ResNet50 is a CNN network introduced as an upgrade to its predecessor ResNet18 with additional training layers that utilize skip connections in its architecture to help model convergence and improve the overall vanishing gradient problem, resulting in minimal network losses reduces the overfitting effect [47]. As the networks re-use training weights from their original ImageNet dataset through transfer learning, this further reduces the impact of overfitting resulting from low training images as seen in previous studies with adequate training time and losses [14, 23, 26] compared to having to train a network from scratch. The CNN's fully connected layer (FCL) output layer is activated and stored as feature vectors for all four image classes to obtain the trained feature vector image features.

3.8 Feature Extraction: Handcrafted (HC) Features

Aside from automated features from CNN, manually extracted HC radiomics features representing the four classes are also extracted to be used in training later for tissue classification. Two additional HC features are extracted based on the knowledge of radiologists during visual mass delineation, which is the mass texture, and the regional features representing the mass are included.

3.9 Textural Feature: Gray Level Co-occurrence (GLCM)

The GLCM are textural features that utilize second-order statistics for features that can be exploited to estimate the degree of connection between pairs of pixels i and j due to varying pixel orientation and displacement when quantifying the texture features of the mass abnormality. The GLCM can emphasize aspects of the spatial distribution of grey levels in the texture image. In the earlier stage, the enhancement based on the breast density approach is used to boost the textural refinement of the dense and mass region. Since segmented benign and malignant regions do not alter in illuminance when exposed to light, textural analysis is particularly useful for distinguishing traits between two adjacent pixels [48]. The texture properties extracted from the produced matrix are four statistical feature descriptors defined as contrast, correlation, energy, and homogeneity by Eqs. (1)–(4). P reflects the total number of occurrences of pixel values i and j concerning the number of grey levels, where the standard deviation and central moments are obtained in the form of variance and means skewness.

$$Contrast = \sum\nolimits_{i,j=0}^{levels-1} P_{i,j}(i-j)^2 \tag{1}$$

$$Correlation = \sum_{i,j=0}^{levels-1} P_{i,j}\left[\frac{(i-\mu_i)\left(j-\mu_j\right)}{\sqrt{\left(\sigma_i^2\right)\left(\sigma_j^2\right)}}\right] \tag{2}$$

$$Energy = \sqrt{\sum\nolimits_{i,j=0}^{levels-1} P_{i,j}^2} \tag{3}$$

$$Homogeneity = \sum\nolimits_{i,j=0}^{levels-1} P_{i,j}\,|i-j| \tag{4}$$

3.10 Regional Feature: Solidity

An additional feature of Solidity is included as a HC feature in this work, which is the ratio of Area and Convex Area of the input image, given by the Eq. (5).

$$Solidity = \frac{A,\ Area}{A_c,\ Convex\ Area} \tag{5}$$

where A is the area of all pixels within the image, whereas A_c, Convex Area, is the area in which the interconnected pixels with $n \times n$ connectivity between neighboring pixels are considered as an object within the resulting convex hull image. The resulting value represents how closely packed the region is connected, which is related to the pathological formation of the tissue representing the four classes [49–51].

3.11 Regional Feature: Mean Intensity

Based on the assumption that both fatty and malignant cells formed more densely than benign or fibroglandular tissue, the image's mean intensity is considered one of the HC features to be extracted. Mean Intensity is given as in Eq. (6).

$$Mean\ Intensity = \frac{\sum_{n=0}^{n} (R + G + B)}{n} \tag{6}$$

where R, G, and B represent the number of pixels that belongs to each channel Red, Green and Blue on a three-channel image with the total number of pixels, n.

3.12 Tissue Classification

In this section, tissue classification into four breast tissue types in the mammogram for classification of benign, malignant, fatty, and fibroglandular classes. Based on Fig. 2, for single feature classification, both CNN and HC features are trained and classified independently and classified using CNN (Model 1) and SVM (Model 2) classifiers, respectively. Dual feature training involves automatic features from the CNN architecture and HC features collected from the previous section into one training and classification framework (Model 3).

3.13 Dual Feature: Model 3: HC and CNN Features for Classification Using Support Vector Machine Classifier

This experiment utilizes a non-linear supervised ML algorithm of SVM with dual feature input coming automatically and directly from the images (CNN feature vector) and manually extracted (handcrafted-HC) features to be trained together for

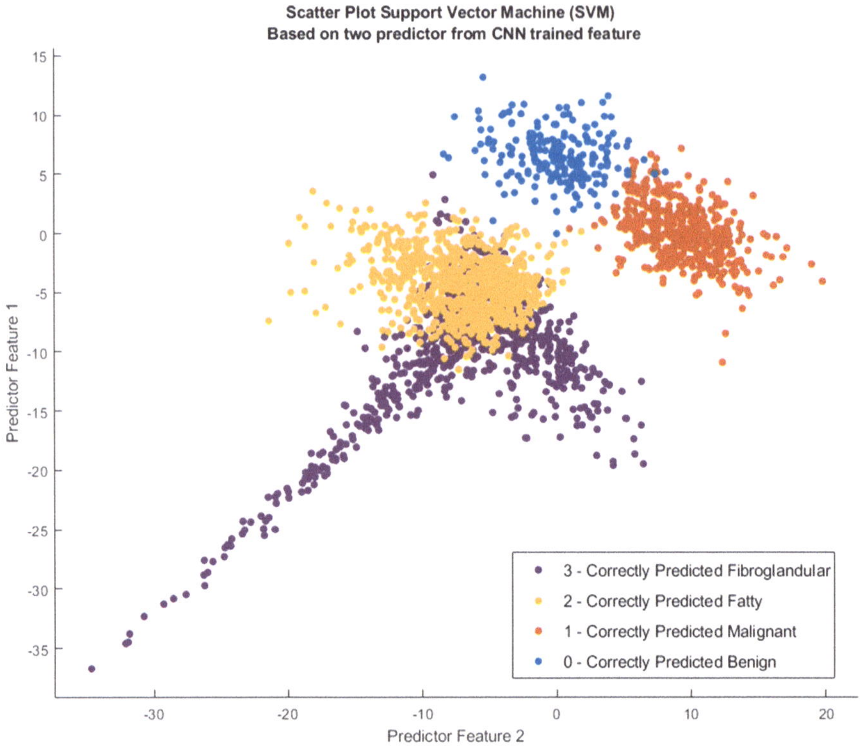

Fig. 4 Feature vector scatter plot mapping from CNN automated feature extraction of four classes depicted on different color schemes used for SVM classification

four tissue type classifications. SVM is a type of supervised ML that is primarily used for both classification and regression problems. The algorithm intuitively determines the hyperplane that separates the feature representing different classes through parallel margin planes that are easily linearly separable for the intended classes. The marginalized parallel planes will pass through at least one of the nearest features representing a class, termed the support vector features. When a generalized SVM model is created, the algorithm finds the most probable hyperplane having the maximum marginal distance between the separating hyperplanes. In this chapter, a non-linear hyperplane is used as the features between the four classes are mixed in a two-dimensional plane; the SVM kernel is used to convert the low dimension into a high-dimensional feature to enable easier separation determination of the hyperplane. Figure 4 shows an example of a two-dimensional scatterplot of the feature vector resulting from CNN feature extraction. More straightforward hyperplane determination is expected between the grouping of scattered features, as shown on malignant (orange) and benign (blue), compared to the feature vector for fatty (yellow) and fibroglandular (purple) that clearly overlapped with each other. Adding more than two feature vector predictors is expected to provide more

Table 1 Setting parameter for non-linear SVM training and classification

Training parameters	Setting
Optimizer	Bayesian
Learner	Non-linear
Coding	Onevsall
Kernel scale	3.4
Kernel function	Gaussian
K-fold	5

information for data optimization to minimize the cost function during feature learning and obtain a better-performed model.

The algorithm works by solving a convex optimization problem that maximizes the margin γ, where:

$$\max_{w,b} \gamma\ (w, b) \tag{7}$$

in such that the linear separating hyperplane Eq. (8):

$$\forall i\ y_i \left(w^T x_i + b\right) \geq 0 \tag{8}$$

where (w,b) is a set of points of the optimal hyperplane in the linear line of x and y. The SVM's simplicity may become problematic if the hyperplane cannot distinguish the points in more extensive training features. The solution to this issue entails determining the separating hyperplane in the higher dimensional space, projecting back to the original space, and expanding the training data for imbalanced class situations. Since the trained CNN network layers had already learned the input data, the computation of the margin of the hyperplane is optimized in this study. The input parameters for SVM are summarized in Table 1.

3.14 Single Feature: Model 1: CNN Feature Classification on CNN Classifier and Model 2: HC Feature Classification on ML-SVM Classifier

In this work, comparison and analysis are made between training using single feature classification using direct automated features from the training images through automated or only using HC features classified independently using both Model 1 of CNN classifier and Model 2 of SVM, respectively. Dual-feature training by concatenating both CNN and HC features for training and classification in a single Model 3 framework, as discussed previously.

For Model 1 CNN single feature classification, the classification layer of a typical CNN network consists of two layers: the activation function (*softmax*), and the classification layer is placed directly after the FCL layer. The probability, f, of each class y of the input belongs to is computed using a *softmax* function, where f of each class is defined according to Eq. (9)

$$f(y_i) = \frac{e^{y_i}}{\sum_k e^{y_k}} \tag{9}$$

A higher f value in a range of [0 1] represents the class that the image feature belongs. Meanwhile, for Model 2, using single feature classification with the SVM algorithm, as discussed in Model 3 for dual feature classification, is used for comparison. The same parameter settings in Table 1 are applied for a non-bias performance comparison.

3.15 System Evaluation

The performance of experiments done in this work is based on the calculation of True Positive (TP), True Negative (TN), False Positive (FP) and False Negative (FN), defined as:

TP – the outcome of positively predicting True class
TN – the outcome of negatively predicting True class
FP – the outcome of positively predicting False class
FN – the outcome of negatively predicting False class

These metrics calculate accuracy based on Eq. (10) for mass detection, segmentation, and classification.

$$Accuracy = \frac{TP + TN}{TP + FN + TN + FP} \tag{10}$$

Intersection over Union (IoU) index from Eq. (11) is used to evaluate the YOLOv3 mass detection by measuring the ability of the algorithm to get the ratio of overlapped detected region A to its ground-truth B. From this, the mean Average Precision (mAP) is used to measure the ability of the algorithm to detect an object based on its precision and recall rate within the image using Eq. (12), where N is the number of classes, and i is the number of observations for each interpolation points.

$$IoU = \frac{Area\ of\ Overlap}{Area\ of\ Union} = \frac{|A \bigcap B|}{|A \bigcup B|} \tag{11}$$

$$mAP = \frac{\frac{1}{11}\sum_{R_i} P(R_i)}{N} \tag{12}$$

4 Result and Discussion

The results are discussed in this section based on the stages of experimental techniques described in the methodology section. Training on the overall proposed model are 9 h and 15 min, with testing per image for 1.07 s.

4.1 Breast Mass Tissue Detection and Segmentation Using Enhanced Image

In this work, the enhanced image input is compared with the original input for mass detection of the YOLOv3 framework. Mass detection performance shows the mAP value as depicted in the Recall-Precision Graph in Fig. 5, showing a mAP score of 0.8125 for the enhanced images, compared to mAP = 0.6842 using original images on a bounding box location of threshold >0.5.

The increase in performance lies directly on the effect of enhancement of the image that increases the difference between the overall edge pixel intensity of the mass tissue group and its surrounding background by replacing the fatty tissue or non-dense region within the image with lower-level intensity while increasing

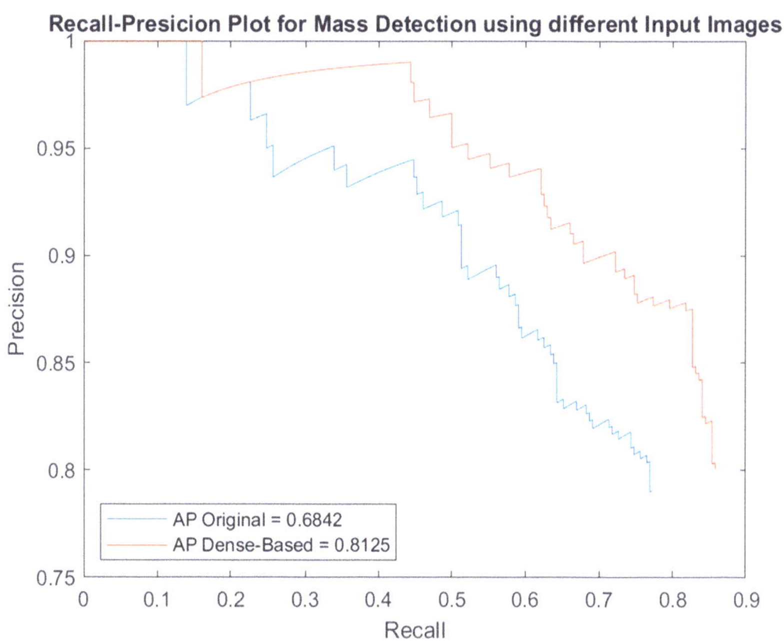

Fig. 5 Recall-precision plot for mass detection using original and enhanced images. AP is the mean Average Precision for each system

the difference of mean intensity level for each class of mass, fibroglandular and fatty tissue. This pixel replacement can be observed in an example of image analysis in Fig. 6, with red, green, and yellow regions extracted to represent mass, fibroglandular and fatty tissue, respectively.

Based on the pixel mean values, the difference between each group shows that the enhanced image has a higher difference between each class, with enhanced image: mass-fibroglandular-fatty of 49.18 and 79.65, while original: mass-fibroglandular-fatty of 30 and 39.25. The higher pixel intensity difference between each class contributes to more spatial features to be trained representing class during YOLOv3 mass detection.

Figure 7 shows mammogram samples results from the dense breast (top) and non-dense (bottom) from YOLOv3 mass detection on the original images (left) and enhanced images (right) in determining accurate bounding box location based on the highest confidence score (CS) obtained. Higher CS scores are observed in the enhanced images, proving that using enhanced images based on the mammogram's dense level can improve the overall mass detection result.

After the bounding boxes have been determined from the previous step, the mass locations are extracted and subjected to semantic and U-net segmentation to obtain the final region containing only the mass tissue.

Figure 8 shows samples of ground-truth of the mass region (white) using CNN semantic segmentation and U-net segmentation. The green and magenta color map indicates the false positive results for mass and background classes. The segmentation result comparison can be seen in Table 2, showing the comparison of evaluation metrics for both segmentation techniques.

From the result, semantic segmentation achieves higher overall performance compared to U-net, with a mean accuracy of 0.9441 compared to 0.9011 on U-net using the same training parameters. The result shows that utilizing transfer learning from the training weightage of pre-trained CNNs can benefit the segmentation result, even though past research shows good segmentation results using U-net architecture [41, 42]. From the segmented image result in Fig. 8, the detected region (white) is similar between the methods. However, semantic segmentation can capture finer details features predominantly on the background image to provide better accuracy on the overall image compared to U-net architecture. This might be caused by more complex CNN training of semantic segmentation that uses pre-trained weightage through transfer learning, compared to U-net, which is a scratch-based CNN technique that requires extensive hyperparameter tuning and optimization. Nevertheless, a higher number of preset U-net layers could increase its performance given more training data and a more capable computing environment.

4.2 *Breast Mass Tissue Detection Classification*

This section compares the classification of both two classes (benign vs. malignant and fatty vs. fibroglandular) and four classes (benign vs. malignant vs. fatty vs.

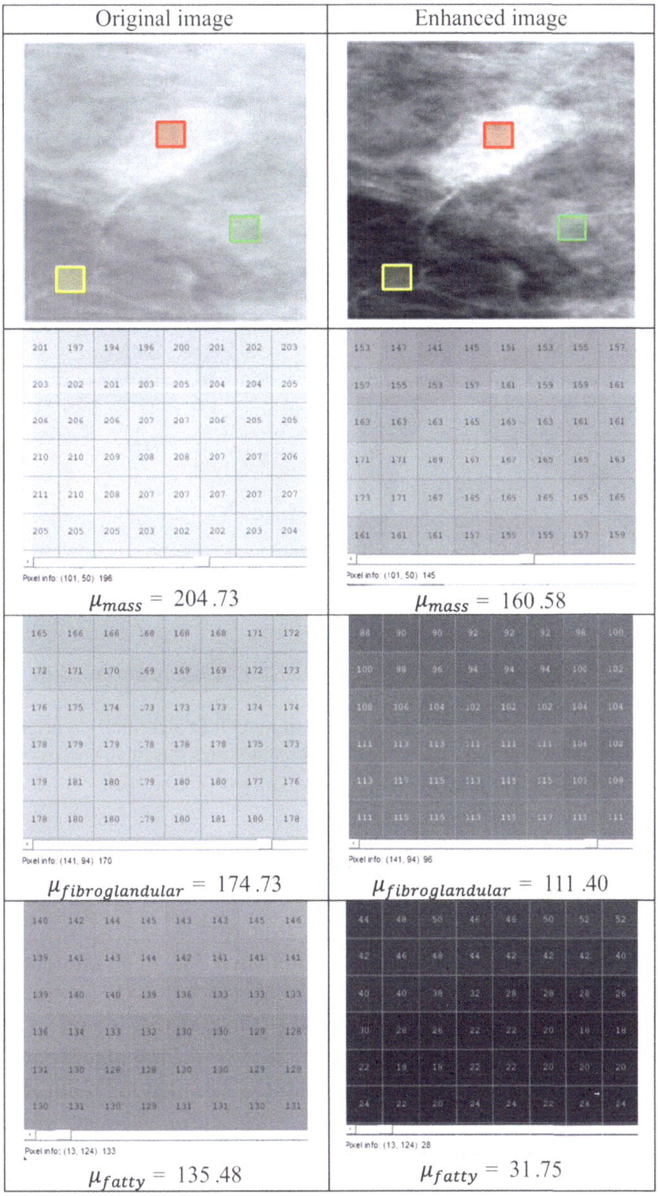

Fig. 6 Sample image analysis on the enhanced image before mass detection. Columns 1 & 2 are samples from original and enhanced images on the same pixel location. Rows 1, 2, and 3 are detailed cropped pixel regions within their respective image, with mean pixel values $\mu_{tissue\ type}$

fibroglandular). The segmented mass region is subjected to tissue classification of benign and malignant classes, along with the randomized cropped region of fatty

Original Image	Enhanced Image with Confidence Score (CS) for Mass Detection

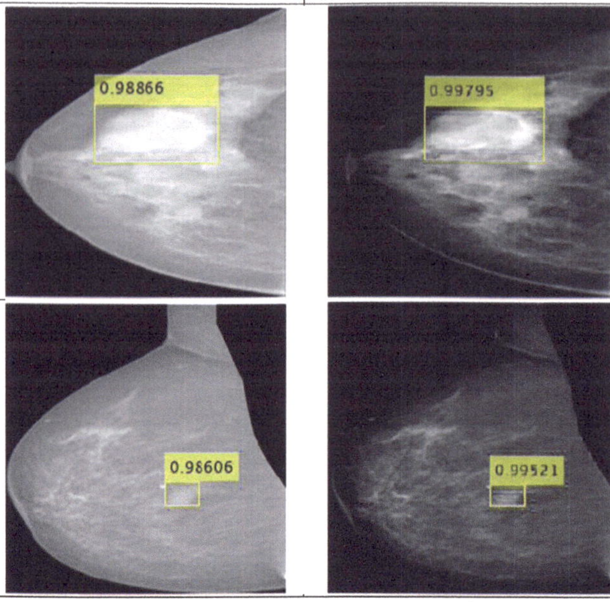

Fig. 7 Example of breast mass detection resulted from YOLO algorithm on original (left) and enhanced (right) mammogram images. The yellow box indicates the confidence score (CS) of the detected mass in bounding-box coordinates

and fibroglandular of the training image as discussed in the previous section. Table 3 summarizes the difference in the accuracy of the trained model using single or dual features on different trainer or classifier frameworks.

For comparison of a single feature, automated image feature trained using CNN architecture in Model 1 shows higher accuracy for the overall four class segmentation of 88.30%, compared to manual HC feature training on SVM in Model 2 of only 64.30%. As can be seen in Table 4, the low overall classification result is due to the low accuracy of fatty vs. fibroglandular classification, with only 57.42%. The result shows that the manual HC feature assigned on differentiation of benign and malignant features having higher accuracy at 88.43% is incompatible with being used together as a feature representing fatty and fibroglandular tissue well in a single feature input network of SVM classifier. However, for dual feature training in Model 3, combining both CNN feature and HC feature from single features of Model 1 and 2 for training shows an increase in the overall accuracy performance for all class tests, with four-class classification resulting in 93.10%. Using the same HC features as trained in Model 2 significantly increases the accuracy to 91.33% for fatty vs. fibroglandular while also increasing benign vs. malignant accuracy to 98.51%. Meanwhile, the combination of CNN and SVM optimizes both the features

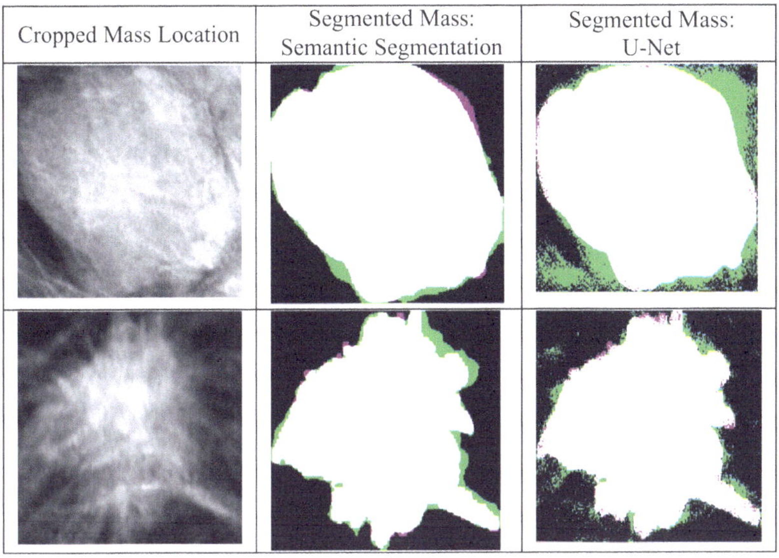

Fig. 8 Example of breast mass region segmentation from original localized cropped mass image (left) through CNN semantic segmentation (middle) and U-net segmentation (right). Color legend: White: Ground-truth, Green: FP mass, Magenta: FP background

Table 2 Evaluation metrics for mass segmentation using different segmentation techniques

Segmentation technique	Mean accuracy	Mean IoU	IoU	
			Mass	Background
U-net	0.9011	0.8155	0.8132	0.8178
ResNet50 semantic segmentation	**0.9441**	**0.8917**	**0.8878**	**0.8984**

in the proposed framework as it utilizes the best of both CNN and SVM models for training image and data features, respectively. Figure 9 shows the confusion matrix produced for this model using the CNN-SVM trainer and classifier.

Finally, a comparison of the study with previous authors on the same method of feature fusion are compared for benign and malignant classification are summarized in Table 4. It is to note here as direct comparison is limited since different studies uses different datasets and different classification schemes (benign vs malignant, dense vs. non-dense, normal vs. abnormal tissue, etc.).

From the comparison, both Mahmoud H. A. H. et al. (2022) [34] and Wang Z. et al. (2019) [18] used features from CNN and HC features to train using their best model of CNN-ELM hybrid architecture. The use of dual-feature in training the system especially for malignant vs. benign shows higher accuracy using the proposed method at 98.51%. Although the overall accuracy is lower than study by Mahmoud H. A. H. et al. (2022) [34] that uses three class classification (normal,

Table 3 Accuracy performance of the trained system using single feature (Models 1 & 2) vs. dual feature (Model 3) on CNN, SVM, and CNN-SVM training scheme for two-class and four-class model

Model	Trainer/classifier	Training feature	Test accuracy (%)		
			Two-class		Four-class
			Benign vs malignant	Fatty vs fibroglandular background	All
1	CNN	Automated (Single)	92.04	87.21	88.30
2	SVM	Manual HC (Single)	88.43	57.42	64.30
3	**CNN-SVM**	**Automated + manual HC (Dual)**	**98.51**	**91.33**	**93.10**

Table 4 Accuracy and AUC-ROC performance of the best model trained system using dual-feature for different studies on benign vs. malignant

Study	Accuracy (two-classes)	Overall accuracy	AUC
[34]		Three classes:	
(CNN-ELM)	97.06	97.43	–
(CNN-SVM)	93.06	92.93	–
[18]			
(CNN-ELM)	86.50	–	0.923
(CNN-SVM)	83.75	–	0.911
Proposed method		Four classes:	
(CNN-SVM with density)	98.51	93.10	0.980

benign, and malignant) this is indirectly caused by the increase in number of classes during classification of different tissue lowers the system's performance to differentiate between the additional classes. Both studies also applied the use of CNN-SVM, and direct comparison for our proposed method also shows higher accuracy performance showing that preparing the training image based on its density level could increase the system's robustness in detecting and classifying the tissue, even using simpler SVM model for architecture model consideration. Indeed, low accuracy resulted from lower performance to class fatty and fibroglandular tissue features as discussed from Table 3 could be overcome by specifying HC that can benefit the feature representing these classes, as well as designing algorithm with the use of more sophisticated hardware for training, where it is expected to further improve the current proposed framework for mass detection in the future.

Fig. 9 Confusion matrix on testing data for overall tissue classification using dual feature input on SVM classifier

5 Conclusion

Developing a CAD system to classify tissue types in mammograms is essential to ensure that only specific features are used to represent these tissue types to produce a good classifier against the various mammogram input encountered throughout the model development process. Combining the advantages of both features from images and radiomics HC data and using them with model training capabilities from CNN and traditional ML can produce a significantly robust model. This work also shows that the pre-trained CNN model is sufficient in extracting automated feature vectors to be used in all training models without having to train the image from scratch CNN models. Future development of a similar model should include using a second dataset to signify its ability to be used given another manufacturer's dataset and reducing the ML features through the optimization process to ensure more reliable features are included during training. An additional class for tissue classification could be considered that separates extremely dense to fibroglandular regions in the future to improve the system's robustness further by differentiating all tissue types. Practical limitation in implementation of this study in real-time could be lessened using single training-testing network using multiple datasets and more annotated training data during its development. The initial process of mass detection and segmentation could also be included in the same single network to lessen training time by using a more sophisticated detection algorithm such as YOLOv5 with more advanced computing hardware.

Acknowledgements This research work is financially supported by the Ministry of Higher Education Grant Scheme (FRGS) (Ref: FRGS/1/2022/SKK06/UITM/02/3). The authors would like to acknowledge the Breast Research Group, INESC Porto, Portugal, for providing the updated INbreast dataset and its accompanying ground-truth for this work. The authors would also like to express their gratitude to members of the Advanced Control System and Computing Research Group (ACSCRG), Advanced Rehabilitation Engineering in Diagnostic and Monitoring Research Group (AREDiM), Integrative Pharmacogenomics Institute (iPROMISE), and Centre for Electrical Engineering Studies, Universiti Teknologi MARA, Cawangan Pulau Pinang for their assistance and administrative support.

References

1. Sung H et al (2021) Global cancer statistics 2020: GLOBOCAN estimates of incidence and mortality worldwide for 36 cancers in 185 countries. CA Cancer J Clin 71(3):209–249. https://doi.org/10.3322/caac.21660
2. Nur Alham O, Balqis B, Azizah AM, Hashimah B, Nirmal K, Siti Zubaidah AR, Puteri NA, Nabihah A, Sukumaran R, Azlina AA, Nadia SMR, Sharifah SSS, Rahayu O (2019) Malaysia National Cancer Registry report (MNCR) 2012–2016. Ministry of Health Malaysia, National Cancer Registry, Putrajaya
3. Byng JW, Boyd NF, Fishell E, Jong RA, Yaffe MJ (1994) The quantitative analysis of mammographic densities. Phys Med Biol 39(10):1629–1638. https://doi.org/10.1088/0031-9155/39/10/008
4. Dembrower K et al (2020) Comparison of a deep learning risk score and standard mammographic density score for breast cancer risk prediction. Radiology 294(2):265–272. https://doi.org/10.1148/radiol.2019190872
5. Samulski M, Hupse R, Boetes C, Mus R, Heeten G, Karssemeijer N (2010) Using computer-aided detection in mammography as a decision support. Eur Radiol 20:2323–2330. https://doi.org/10.1007/s00330-010-1821-8
6. Tripathy S, Swarnkar T (2020) Performance observation of mammograms using an improved dynamic window based adaptive median filter. J Discret Math Sci Cryptogr 23(1):167–175. https://doi.org/10.1080/09720529.2020.1721881
7. Iranmakani S et al (2020) A review of various modalities in breast imaging: technical aspects and clinical outcomes. Egypt J Radiol Nucl Med 51:57. https://doi.org/10.1186/s43055-020-00175-5
8. Tsochatzidis L, Costaridou L, Pratikakis I (2019) Deep learning for breast cancer diagnosis from mammograms-a comparative study. J Imaging 5(3):37. https://doi.org/10.3390/jimaging5030037
9. Rahim SB (2021) Mamogram: Mitos Dan Fakta. Universiti Putra Malaysia. https://care.upm.edu.my/artikel/mamogram_mitos_dan_fakta-63702. Accessed 24 Jan 2022
10. Nelson HD, Fu R, Cantor A, Pappas M, Daeges M, Humphrey L (2016) Effectiveness of breast cancer screening: systematic review and meta-analysis to update the 2009 U.S.

Preventive Services Task Force recommendation. Ann Intern Med 164(4):244–255. https://doi.org/10.7326/M15-0969
11. Suh YJ, Jung J, Cho B-J (2020) Automated breast cancer detection in digital mammograms of various densities via deep learning. J Pers Med 10(4):211. https://doi.org/10.3390/jpm10040211
12. Li S et al (2017) Computer-aided assessment of breast density: comparison of supervised deep learning and feature based statistical learning. Phys Med Biol 63:025005. https://doi.org/10.1088/1361-6560/aa9f87
13. Bhat GM, Mustafa M (2010) Field programmable gate array (FPGA) implementation of novel complex PN-code-generator-based data scrambler and descrambler. Maejo Int J Sci Technol 4(1):125–135
14. Parsa S et al (2021) A security management framework for big data in smart healthcare. Big Data Res 25:100225
15. Parah SA et al (2012) On the realization of a secure, high capacity data embedding technique using joint top-down and down-top embedding approach. Comput Sci Eng 49:10141–10146
16. Parah SA, Rashid M, Vijaykumar V (2022) Artificial intelligence for innovative healthcare informatics. Springer. ISBN: 978-3-030-96568-6
17. Nazeer S, Sarosh P, Parah SA (2022) Automatic classification of COVID-19 infected *patients* using convolution neural network models. In: Parah SA, Rashid M, Varadarajan V (eds) Artificial intelligence for innovative healthcare informatics. Springer, Cham. https://doi.org/10.1007/978-3-030-96569-3_6
18. Wang Z et al (2019) Breast cancer detection using extreme learning machine based on feature fusion with CNN deep features. IEEE Access 7:105146–105158. https://doi.org/10.1109/ACCESS.2019.2892795
19. Nguyen TL et al (2021) Novel mammogram-based measures improve breast cancer risk prediction beyond an established mammographic density measure. Int J Cancer 148(9):2193–2202. https://doi.org/10.1002/ijc.33396
20. Tchou PM et al (2010) Interpretation time of computer-aided detection at screening mammography. Radiology 257(1):40–46. https://doi.org/10.1148/radiol.10092170
21. Oakden-Rayner L (2019) The rebirth of CAD: how is modern AI different from the CAD we know? Radiol Artif Intell 1(3):e180089. https://doi.org/10.1148/ryai.2019180089
22. McKinney SM et al (2020) International evaluation of an AI system for breast cancer screening. Nature 577(7788):89–94. https://doi.org/10.1038/s41586-019-1799-6
23. Pérez-Benito FJ et al (2020) A deep learning system to obtain the optimal parameters for a threshold-based breast and dense tissue segmentation. Comput Methods Prog Biomed 195:105668. https://doi.org/10.1016/j.cmpb.2020.105668
24. Huang M-L, Lin T-Y (2021) Considering breast density for the classification of benign and malignant mammograms. Biomed Signal Process Control 67:102564. https://doi.org/10.1016/j.bspc.2021.102564
25. Nazari SS, Mukherjee P (2018) An overview of mammographic density and its association with breast cancer. Breast Cancer 25(3):259–267. https://doi.org/10.1007/s12282-018-0857-5
26. Harvey H, Glocker B (2019) A standardised approach for preparing imaging data for machine learning tasks in radiology: opportunities, applications and risks. In: Artificial intelligence in medical imaging: opportunities, applications and risks. Springer, Cham, pp 61–72. https://doi.org/10.1007/978-3-319-94878-2_6
27. Mohamed AA, Berg WA, Peng H, Luo Y, Jankowitz RC, Wu S (2018) A deep learning method for classifying mammographic breast density categories. Med Phys 45(1):314–321. https://doi.org/10.1002/mp.12683
28. Saffari N et al (2020) Fully automated breast density segmentation and classification using deep learning. Diagnostics (Basel) 10(11):988. https://doi.org/10.3390/diagnostics10110988
29. Sadad T et al (2021) Internet of medical things embedding deep learning with data augmentation for mammogram density classification. Microsc Res Tech 84(9):2186–2194. https://doi.org/10.1002/jemt.23773

30. Trivizakis E et al (2019) A novel deep learning architecture outperforming "off-the-shelf" transfer learning and feature-based methods in the automated assessment of mammographic breast density. Oncol Rep 42(5):2009–2015. https://doi.org/10.3892/or.2019.7312
31. Shrivastava N, Bharti J (2020) Breast tumor detection and classification based on density. Multimed Tools Appl 79:26467–26487. https://doi.org/10.1007/s11042-020-09220-x
32. Ragab DA, Sharkas M, Marshall S, Ren J (2019) Breast cancer detection using deep convolutional neural networks and support vector machines. PeerJ 7:e6201. https://doi.org/10.7717/peerj.6201
33. Alkhaleefah M, Wu C-C (2018) A hybrid CNN and RBF-based SVM approach for breast cancer classification in mammograms. In: 2018 IEEE international conference on systems, man, and cybernetics (SMC). IEEE, pp 894–899. https://doi.org/10.1109/SMC.2018.00159
34. Mahmoud HAH, Alharbi AH, Alghamdi NS (2022) Breast cancer detection through feature clustering and deep learning. Intell Autom Soft Comput 31(2):1273–1286. https://doi.org/10.32604/iasc.2022.020662
35. Belhaj Soulami K, Kaabouch N, Saidi M, Tamtaoui A (2021) Breast cancer: one-stage automated detection, segmentation, and classification of digital mammograms using UNet model based-semantic segmentation. Biomed Signal Process Control 66:102481. https://doi.org/10.1016/j.bspc.2021.102481
36. Kaur A, Rashid M, Bashir AK, Parah SA (2022) Detection of breast cancer masses in mammogram images with watershed segmentation and machine learning approach. In: Parah SA, Rashid M, Varadarajan V (eds) Artificial intelligence for innovative healthcare informatics. Springer, Cham, pp 35–60. https://doi.org/10.1007/978-3-030-96569-3_2
37. Sha Z, Hu L, Rouyendegh D (2020) Deep learning and optimization algorithms for automatic breast cancer detection. Int J Imaging Syst Technol 30:495–506. https://doi.org/10.1002/ima.22400
38. Moreira IC, Amaral I, Domingues I, Cardoso A, Cardoso MJ, Cardoso JS (2012) INbreast: toward a full-field digital mammographic database. Acad Radiol 19(2):236–248. https://doi.org/10.1016/j.acra.2011.09.014
39. Al-Antari MA, Al-Masni MA, Kim T-S (2020) Deep learning computer-aided diagnosis for breast lesion in digital mammogram. Adv Exp Med Biol 1213:59–72. https://doi.org/10.1007/978-3-030-33128-3_4
40. Afzal I, Parah SA, Hurrah NN, Song OY (2020) Secure patient data transmission on resource constrained platform. Multimed Tools Appl:1–26. https://doi.org/10.1007/s11042-020-09139-3
41. Redmon J, Farhadi A (2018) YOLOv3: an incremental improvement. arXiv:1804.02767. https://doi.org/10.48550/arXiv.1804.02767
42. Mohiyuddin A et al (2022) Breast tumor detection and classification in mammogram images using modified YOLOv5 network. Comput Math Methods Med 2022:1359019. https://doi.org/10.1155/2022/1359019
43. Hamed G, Marey M, Amin SE, Tolba MF (2021) Automated breast cancer detection and classification in full field digital mammograms using two full and cropped detection paths approach. IEEE Access 9:116898–116913. https://doi.org/10.1109/ACCESS.2021.3105924
44. Zhao J, Chen T, Cai B (2021) A computer-aided diagnostic system for mammograms based on YOLOv3, vol 81. Multimed Tools Appl, pp 19257–19281. https://doi.org/10.1007/s11042-021-10505-y
45. Djebbar K, Mimi M, Berradja K, Taleb-Ahmed A (2019) Deep convolutional neural networks for detection and classification of tumors in mammograms. In: 2019 6th International conference on image and signal processing and their applications (ISPA). IEEE. https://doi.org/10.1109/ISPA48434.2019.8966895
46. Aly GH, Marey MAE-R, El-Sayed Amin S, Tolba MF (2021) YOLO V3 and YOLO V4 for masses detection in mammograms with ResNet and inception for masses classification. Adv Intell Syst Comput 1339:145–153. https://doi.org/10.1007/978-3-030-69717-4_15

47. He K, Zhang X, Ren S, Sun J (2016) Deep residual learning for image recognition. In: 2016 IEEE conference on computer vision and pattern recognition (CVPR). IEEE, pp 770–778. https://doi.org/10.1109/CVPR.2016.90
48. Hazarika M, Mahanta LB (2018) A new breast border extraction and contrast enhancement technique with digital mammogram images for improved detection of breast cancer. Asian Pac J Cancer Prev 19(8):2141–2148. https://doi.org/10.22034/APJCP.2018.19.8.2141
49. Mehraj U et al (2021) The tumor microenvironment as driver of stemness and therapeutic resistance in breast cancer: new challenges and therapeutic opportunities Springer Nature (IF-7.051). Cell Oncol 44:1209–1229. https://doi.org/10.1007/s13402-021-00634-9
50. Zeiser F et al (2020) Segmentation of masses on mammograms using data augmentation and deep learning. J Digit Imaging 33:858–868. https://doi.org/10.1007/s10278-020-00330-4
51. Wang J, Gou C, Shen T, Wang F-Y (2019) Global segmentation-aided local masses detection in X-ray breast images. In: Proceedings 2018 Chinese automation congress, CAC 2018. IEEE, pp 3655–3660. https://doi.org/10.1109/CAC.2018.8623575

Deterministic SEIR Mathematical Model for Infectious Diseases Like COVID-19

Khalid Nazir and Aasma Rafiq

1 Introduction

The Severe Acute Respiratory Syndrome Coronavirus 2 (SARS-COV-2) is one of the types of the Coronavirus family also called COVID-19. The illness started in Wuhan (China) in late December 2019 and expanded at a fast rate step by step which lead to significant worry for the Chinese government. The Chinese government has embraced drastic actions to moderate the outbreak. On 23 January 2020, the administration of Wuhan suspended all public deals inside the city, and shut all inbound and outbound transportation for just about a month and a half. Not long after Wuhan, China secured 15 different urban areas. According to reports and media data, China has bit by bit assumed responsibility for the sickness and on 18 March 2020, China detailed no new neighbourhood cases interestingly since the outbreak started. The number of deaths related with COVID-19 incredibly surpasses the other two Coronaviruses (Severe Acute Respiratory Syndrome Coronavirus (SARS-CoV) and Middle East Respiratory Syndrome Coronavirus (MERS-CoV), and the episode is as yet continuous, which represented a colossal danger to the worldwide general wellbeing and financial aspects areas. The WHO [16] has authoritatively pronounced the episode a pandemic and has approached äll nations to avoid potential risk and gave warnings to proceed with endeavors that have been viable in restricting the quantity of cases and easing back the infection spreading". Keeping in view the blended population the vast majority of the nations on the planet received lockdown and social distancing as a device to decrease the spread of the disease. The states proclaimed the date of lockdown after they found that there were instances of the illness in their nations, additionally keeping in view the monetary and social design of the country, the date and the length of lockdown

K. Nazir · A. Rafiq (✉)
Department of School Education J&K, Srinagar, India

S. A. Parah et al. (eds.), *Intelligent Multimedia Signal Processing for Smart Ecosystems*, https://doi.org/10.1007/978-3-031-34873-0_10

shift from the country to country. This destructive infection didn't spare the super powers including USA and European nations. Despite the fact that Wuhan is the focal point of this destructive infection yet USA got most noticeably awful hit by COVID-19 both regarding infected people just as number of deaths. UK went into full Coronavirus lockdown on 23 March 2020, Russia on 30 March 2020. India which has world's biggest population has forced a cross country lockdown trying to moderate the spread of COVID-19 on 22 March 2020 for 1 day and later on March 24 it was chosen to close all public level foundations for a time of 21 days. Notwithstanding in the lockdown, there were still expansion in the number of cases [17], which might be on the grounds that every one of the tainted people were either not screened by the specialists or were not kept in isolation and that lead to the spread of the disease. A lot of research work has been going on various techniques for the study and detection of COVID-19 disease [9]. A cohort study on COVID-19 was done by [11] on the patients discharged from the hospitals. Similarly [12] discussed the priorities to be taken in consideration during COVID-19 pandemic. Many researchers including [13, 14] framed various mathematical models during the recent years for the study of dynamics of COVID-19. The authors [6] discussed mathematical model for noval coronavirus and its control. A mathematical model was also formulated on case study in Ethiopia on COVID-19 [15]. In this work, we plan to examine the job of quarantine/social distance of individuals in lessening the pace of outbreak of this destructive disease COVID-19.

The leftover show of the present work is depicted into following sections: In Sect. 2, the numerical model with respect to the elements of the infection is shaped. In Sect. 3, numerical examination for non-negativity and boundedness are researched. Section 4 has been argued whether or not endemic and disease-free equilibria exist and are stable. The numerical simulations of the model for various parameters and the importance of our findings were covered in Sect. 5. Section 6 summarises the conclusion in the end.

2 Mathematical Formulation

The objective to examine the study of disease transmission through numerical models is to see how irresistible sicknesses are sent in the general public and furthermore to track down the most ideal systems to control the spread of an infection or to annihilate it. The first breakthrough in the modelling framework came in 1927 by Kermack and McKendrick [4]. They developed an epidemiological susceptible-infectious-recovered (SIR) model with the following assumptions:

1. When there are no births or deaths and a stable population, the disease spreads in a closed environment.
2. The total number of susceptibles S with the proportional (transmission) coefficient is proportionate to the number of susceptibles who are infected by an infected person I per unit of time.

3. The recovery rate coefficient γ determines the number of persons R who are recovered from the contaminated compartment per unit of time, and these individuals develop a lifelong immunity to the infectious disease.

Although the SIR model was precise for portraying the viral illness transmission like flu, measles, and chickenpox yet the model was improper for managing bacterial illnesses like encephalitis and gonorrhea as in these infections the recovered people acquired no immunity and might contract the virus again at a future time-frame. Following 5 years, Kermack and McKendrick [5] proposed the susceptible-infectious-susceptible (SIS) model to examine bacterial infections' dynamics. Both the old-style SIR and SIS models make the assumption that there is little time for infection before susceptible S individuals recover and develop a lifelong immunity. Afterward, Anderson and May [1] showed that these models have been valuable for a wide range of diseases in different regions everywhere in the world. A few diseases like flu and tuberculosis, on having sufficient touch with a susceptible person exposes them for a while, making them infected but not yet contagious. Hence, it is sensible to present a dormant compartment, prompting a susceptible-exposed-infectious-recovered (SEIR) model. In epidemology, SEIR models with indispensable elements and lasting or transitory resistance gives a numerical depiction of the irresistible sicknesses and comparing spread in the science. It takes a specific amount of time for the infection to multiply inside the bodies of the susceptible S persons and reach the necessary level for them to become infected I individuals during such a latent phase. The exposed E individuals quickly develop into infected I individuals after the disease's incubation period. Every one of these models expect that the illness spreads in a shut climate and having homogeneous blending of the people of the population which is additionally thought to be steady. Such models have been generally examined in the writing.

Most mathematical epidemic models typically exhibit threshold behaviour known as basic reproduction number $\mathbf{R_0}$ that measures how many secondary infections an infectious disease typically causes. If it is less than one, the disease will go extinct; if it is greater than one, an endemic state will result [2, 3]. This is one of the fundamental and important findings in mathematical epidemiology. The best course of action when infectious diseases like COVID-19 are present is to completely eradicate them by taking either preventive measures or isolating the sick people. Individuals who are infected I are identified for quarantine and subsequently isolated, receiving the local medication which will boost the immunity and head straight for the R compartment that was found. By lowering the fundamental reproduction value $\mathbf{R_0}$ to less than unity, quarantine contributes to the eventual extinction of the infectious disease. Figure 1 represents the block schematic of the SEIR model includes compartments made up of individuals from the overall population N as well as a quarantine subcompartment Q between I and R.

In the present work, the SEIR model is rescaled utilizing quarantine compartment and analysed to determine its equilibrium points as well as the related local stabilities. To test the SEIR model, mathematical simulations have been utilized along including a bunch of discretionarily characterized parameters for even propagation

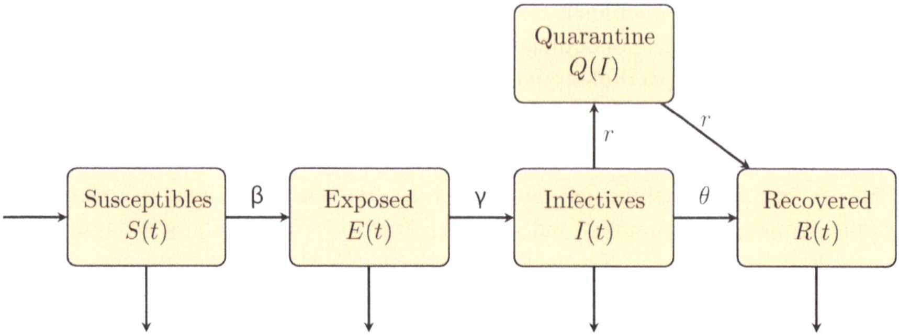

Fig. 1 Flow chart of disease progression in SEIR along with Quarantine compartment

of the contagious illness in the simulation. Table 1 sums up the understandings of the distinctive positive parameters implanted in the SEIR model. The model is progressed and joins indispensable elements by including the quarantine for depicting the spread of infectious diseases. The model depends on the following assumptions:

1. It is assumed that only those infected people which have been detected by the authorities are quarantined. Some of the infected people may remain in the society due to the negligence of the authorities or by the individuals themselves.
2. No separate dynamical equation for the quarantined individuals has been introduced, the class has been considered as a subclass of the infective class.
3. It is assumed that all those people who have been shifted for quarantine after getting the local treatments like medicine to boost their immunity are shifted to the recovery class at the same rate.

The following ordinary system of equations govern the behavior of the SEIR model.

$$
\begin{aligned}
\frac{dS}{dt} &= \delta N - \beta IS - \lambda S \\
\frac{dE}{dt} &= \beta IS - \gamma E - \lambda E \\
\frac{dI}{dt} &= \gamma E - \mu I - \lambda I - Q(I) \\
\frac{dR}{dt} &= \mu I - \lambda R + Q(I)
\end{aligned}
$$

where all the parameters are defined in the Table 1.

The term $Q(I)$ represents those infected individuals who have been detected by the authorities and have been put under quarantine. Here, we have assumed that the rate of quarantined individuals is proportional to the number of the infected people, that is:

Table 1 Description of model parameters and their inverses

Parameter	Name	Meaning
N	Total population (constant)	Total population of the region in a closed environment.
δ	Constant birth rate from the whole population	Rate at which there is constant recruitment from the population to the susceptible class.
β	Transmission rate from susceptibles to infectives	Rate of exposure of susceptible people to infected people, leaving the susceptible class, and entering the exposed class.
γ	Disease outcome rate from exposed to infectives	Rate at which exposed people feel ill from an incubating infection, then depart exposure in one class, infection in another.
μ	Recovery rate from infectives to recovery class	Rate at which infected people enter the recovered class after leaving the infected class.
λ	Death rate, includes natural and disease death	Death rate of S, E, I and R
$\frac{1}{\beta}$	Mean susceptibility period (Pre-exposed individuals)	Time for susceptibles to become exposed ones.
$\frac{1}{\gamma}$	Average latency time	When exposed people will eventually become infected people.
$\frac{1}{\mu}$	Average infectious time	When infected people will eventually become recovered people.

$$Q(I) = rI; \qquad 0 \le r \le 1$$

where r is the parameter depends upon the rate of incidence of the infected people who are going under quarantine.

Thus, the model equations are reduced to the following equations

$$\left.\begin{aligned} \frac{dS}{dt} &= \delta N - \beta IS - \lambda S; & S_0 &= S(0) \ge 0 \\ \frac{dE}{dt} &= \beta IS - \gamma E - \lambda E; & E_0 &= E(0) \ge 0 \\ \frac{dI}{dt} &= \gamma E - \mu I - \lambda I - rI; & I_0 &= I(0) \ge 0 \\ \frac{dR}{dt} &= \mu I - \lambda R + rI; & R_0 &= R(0) \ge 0 \end{aligned}\right\} \tag{1}$$

where S_0, E_0, I_0 and R_0 represents the initial susceptibles, exposed, infectives and recovered respectively.

3 Mathematical Solution

To show that the model is mathematically and epidemologically relevant, we need to find the answer of some basic questions. The first question that arise in mind for solving the system of ordinary differential equations is about its existence and uniqueness. Since we are dealing with the population, the next question is about the positive and bounded solution. For this purpose, we will make use of some fundamental results from differential equations [8]. Also, our main motive is find the disease free environment about which we will answer by making use of stability analysis.

It is sufficient to mathematically analyse the following reduced system of equations.

$$\left.\begin{aligned}
\frac{dS}{dt} &= \delta N - \beta I S - \lambda S; \qquad & S_0 = S(0) \geq 0 \\
\frac{dE}{dt} &= \beta I S + \gamma E - \lambda E; \qquad & E_0 = E(0) \geq 0 \\
\frac{dI}{dt} &= \gamma E - \mu I - \lambda I - r I; \qquad & I_0 = I(0) \geq 0
\end{aligned}\right\} \tag{2}$$

We must demonstrate that the model's solutions are both unique and exist for all positive times. Additionally, for given positive initial circumstances, the solutions to the given model continue to be positive.

Theorem 3.1 (Existence and Uniqueness) *Let* $S_0, E_0, I_0 \in \mathbb{R}$ *be given, then* $\exists\, t_0 > 0$ *and continuously differentiable functions* $\{S, E, I : [0, t_0) \to \mathbb{R}\}$ *such that S,E,I satisfies Eq.* (2) *and* $(S, E, I)(0) = (S_0, E_0, I_0)$

Proof To prove this result, we will make use of Picard-Lindelöf Theorem. Choosing $S = y_1$; $E = y_2$; $I = y_3$, we have:

$$f(y) = \begin{pmatrix} \delta N - \beta y_1 y_3 - \lambda y_1 \\ \beta y_1 y_3 - (\gamma + \lambda) y_2 \\ \gamma y_2 - (\mu + \lambda + r) y_3 \end{pmatrix}$$

$$\text{Jacobian matrix} = \nabla f(y) = \begin{pmatrix} -\beta y_3 - \lambda & 0 & -\beta y_1 \\ \beta y_3 & -(\gamma + \lambda) & \beta y_1 \\ 0 & \gamma & -(\mu + \lambda + r) \end{pmatrix}$$

which is linear in y and therefore locally bounded for every $y \in \mathbb{R}^3$, hence f has continuous bounded derivative on any compact subset of $\mathbb{R}^3$. Hence f is locally Lipschitz, therefore by Picard Lindläof theorem, there exists a unique solution $y(t)$ to $y'(t) = f(y(t))$ wit $y(0) = y_0$ on $[0, t_0]$ for some $t > 0$ and $\exists\, t_0 > 0$ and thus, $\exists$ continuously differentiable functions $\{S, E, I : [0, t_0) \to \mathbb{R}\}$ such that S,E,I satisfies Eq. (2) and $(S, E, I)(0) = (S_0, E_0, I_0)$

Theorem 3.2 (Non Negativity and Boundedness) *Assume that* $S_0 > 0$, $E_0 > 0$ *and* $I_0 > 0$ *and if unique solution to the system* (2) *exists in the interval* $[0, t_0]$ *for some* t_0, *then the functions* $S(t)$, $E(t)$ *and* $I(t)$ *remain bounded and positive* $\forall\, t \in [0, t_0]$

Proof Since $S(t)$, $E(t)$ (and) I(t) have positive initial values. Then by Theorem 3.1, $\exists t^*$ such that solution to system (2) exists on $\left[0, t^*\right]$.

Let $T^* = \sup\left\{t \in [0, t^*] : S(s), E(s), I(s) > 0 \;\forall\, s \in [0, t]\right\}$, then on the interval $[0, T^*]$ we can estimate the population values.

We can estimate the bounds on the solution as well.

Since

$$\frac{dE}{dt} = \beta SI - (\gamma + \lambda)E \geq -(\gamma + \lambda)E;$$

implies

$$E(t) \geq E(0)\exp\{-(\gamma + \lambda)t\} > 0 \;\forall t \;\in [0, T^*].$$

Also

$$\frac{dI}{dt} = \gamma E - (\mu + \lambda + r)I \geq -(\mu + \lambda + r)I;$$

implies

$$I(t) \geq I(0)\exp\{-(\mu + \lambda + r)t\} > 0 \;\forall t \;\in [0, T^*].$$

And

$$\frac{dS}{dt} = \delta N - \beta SI - \lambda S \leq \delta N;$$

implies

$$S(t) \leq c_1(1 + t); \qquad c_1 = \max\{S_0, \delta N\}$$

Since

$$\frac{d(E + I)}{dt} \leq \beta SI - \lambda E \leq c_2(1 + t)(E + I); \qquad c_2 = \max\{\lambda, \beta c_1\}$$

implies

$$(E + I) \leq c_3\exp(t^2);$$

And

$$\frac{dS}{dt} = \delta N - \beta SI - \lambda S \geq -(\lambda + \beta c_3 \exp(t^2))S \geq -c_4(1 + \beta \exp(t^2))S;$$
$$c_4 = \max\{\beta c_3, \lambda\}$$

implies

$$S(t) \geq S_0 \exp\Big\{ - c_4 \int_0^t (1 + \exp(\tau^2) d\tau)\Big\} > 0$$

4 Equilibrium and Stability Analysis

4.1 Equillibrium Analysis

For the analysis of equilibrium points of the non linear system, we have adopted the method of nullcline [8]. Setting each equation of the system (2) to zero.

$$\delta N - \beta IS - \lambda S = 0 \tag{3}$$

$$\beta IS - \gamma E - \lambda E = 0 \tag{4}$$

$$\gamma E - \mu I - \lambda I - rI = 0 \tag{5}$$

From Eqs. (3) and (4), we have

$$S = \frac{\delta N - (\gamma + \lambda)E}{\lambda} \tag{6}$$

From Eq. (5), we have

$$I = \frac{\gamma E}{(\mu + \lambda + r)} \tag{7}$$

From Eqs. (4), (6) and (7), we have

$$E = 0 \text{ or } E = \frac{\delta N}{(\gamma + \lambda)} - \frac{\lambda}{\beta\gamma}(\mu + \lambda + r) \tag{8}$$

Now if $E = 0$, then from Eqs. (6) and (7), we have,

$$I = 0 \text{ and } S = \frac{\delta N}{\lambda}$$

Hence, we obtain the Disease-Free Equilibrium (DFE) given by $Q_0 = \left(\frac{\delta N}{\lambda}, 0, 0\right)$.

And, for $E \neq 0$, from Eq. (7), we have

$$I = I^* = \frac{\gamma \delta N}{(\gamma+\lambda)(\mu+\lambda+r)} - \frac{\lambda}{\beta} = \frac{\lambda}{\beta}\left[\frac{\beta\gamma\delta N}{(\gamma+\lambda)(\mu+\lambda+r)} - 1\right] = \frac{\lambda}{\beta}(\mathbf{R_0} - 1)$$

$$\text{where } \mathbf{R_0} = \frac{\beta\gamma\delta N}{\lambda(\gamma+\lambda)(\mu+\lambda+r)} \tag{9}$$

is called Basic Reproduction Number $\mathbf{R_0}$.

The most crucial factor to consider when analysing any epidemiological model is the fundamental reproduction value, or $\mathbf{R_0}$. It establishes whether an outbreak of infectious illnesses happens or not. Since estimated secondary infection rates caused by one sick person throughout the progression of the mean infection time in a population that is totally susceptible is $\mathbf{R_0}$. If $\mathbf{R_0} < 1$, it suggests that the average number of new infections caused by one affected person is less than unity, indicating that the infectious disease will eventually go away. And, if $\mathbf{R_0} > 1$, the infectious disease continues to spread as an epidemic because, on average, more new infections are caused by one diseased person than by a healthy one. If the value of $\mathbf{R_0}$ for an infectious disease is one, then the statement ($\mathbf{R_0} = 1$) means that each infection that already exists results in one new infection. The disease will continue to exist and be stable in this instance, but it won't cause an outbreak or an epidemic[10].

Clearly, the value of $\mathbf{R_0}$ directly depends upon the contact rate of infected individuals and susceptibles, β and inversely proportional to the rate at which the infected people are quarantined, r. Therefore in order to reduce the value of epidemiological parameter $\mathbf{R_0}$, we need to reduce the value of β or increase the value of r. The value of E and S in terms of $\mathbf{R_0}$ is given by

$$E = E^* = \frac{\beta\gamma\delta N - \lambda(\gamma+\lambda)(\mu+\lambda+r)}{\beta\gamma(\gamma+\lambda)} = \frac{\lambda(\mu+\lambda+r)}{\beta\gamma}(\mathbf{R_0} - 1)$$

$$S = S^* = \frac{(\lambda+\gamma)(\mu+\lambda+r)}{\beta\gamma} = \frac{\delta N}{\lambda\mathbf{R_0}}$$

Summarizing the above facts, the dynamical system has the Endemic Equilibrium (EE) given by

$$Q^* = \left(S^*, E^*, I^*\right) = \left\{\frac{\delta N}{\lambda\mathbf{R_0}}, \frac{\lambda(\mu+\lambda+r)}{\beta\gamma}(\mathbf{R_0} - 1), \frac{\lambda}{\beta}(\mathbf{R_0} - 1)\right\}$$

4.2 Stability Analysis

The stability properties of linear ODEs depends only on the eigenvalues of the system. We approximate the nonlinear system of differential Eq. (2), with a linear system at the points Q_0 and Q^*, in order to look into the local stability characteristics of the equilibria. This can be accomplished by using the Jacobian for system (2) to linearize the system about each equilibrium.

$$\mathrm{J} = \begin{pmatrix} -\beta I - \lambda & 0 & -\beta S \\ \beta I & -(\gamma + \lambda) & \beta S \\ 0 & \gamma & -(\mu + \lambda + r) \end{pmatrix}$$

Then studying the linearized system

$$\dot{\zeta(t)} = J(Q)\zeta(t) \tag{10}$$

Theorem 4.1 *The DFE Q_0 is locally asymptotically stable, if* $\mathbf{R_0} < 1$.

Proof Evaluating Jacobian matrix J at Q_0, we have

$$J(Q_0) = \begin{pmatrix} -\lambda & 0 & -\frac{\beta\delta N}{\lambda} \\ 0 & -(\gamma + \lambda) & \frac{\beta\delta N}{\lambda} \\ 0 & \gamma & -(\mu + \lambda + r) \end{pmatrix}$$

The characteristic equation of the matrix is given by

$$\Delta(x) = (x + \lambda)\left[(x + \gamma + \lambda)(x + \mu + \lambda + r) - \frac{\gamma\beta\delta N}{\lambda}\right] = 0$$

One of the eigenvalue is $x = -\lambda < 0$. The other two eigenvalues satisfy the equation

$$x^2 + a_1 x + a_2 = 0$$

where $a_1 = \gamma + 2\lambda + \mu + r$: and $a_2 = (\gamma + \lambda)(\mu + \lambda + r) - \dfrac{\gamma\beta\delta N}{\lambda}$

The other two eigenvalues are given by

$$x_{\pm} = \frac{-(\gamma+2\lambda+\mu+r) \pm \sqrt{(\gamma+2\lambda+\mu+r)^2 - 4\left\{(\gamma + \lambda)(\mu + \lambda + r) - \frac{\gamma\beta\delta N}{\lambda}\right\}}}{2}$$

These eigenvalue will be negative , if

$$\text{if} \qquad 4\left[(\gamma+\lambda)(\mu+\lambda+r)\left(1-\frac{\gamma\beta\delta N}{\lambda(\gamma+\lambda)(\mu+\lambda+r)}\right)\right] > 0$$

$$\text{implies} \qquad (\gamma+\lambda)(\mu+\lambda+r)(1-\mathbf{R_0}) > 0$$

Given that all parameters are positive, we can demonstrate that, if and only if, $\mathbf{R_0} < 1$, all eigenvalues have a negative real part. Therefore, in this instance, the origin is an equilibrium point that is locally asymptotically stable. The asymptotic behaviour of the system is equivalent to that of this linear system for local perturbations according to the Hartman-Grobman Theorem, and the conclusion follows.

Theorem 4.2 *The EE Q^* is locally asymptotically stable if* $\mathbf{R_0} > 1$

Proof Evaluating Jacobian matrix J at Q^*, we have

$$J_{(Q^*)} = \begin{pmatrix} -\lambda\mathbf{R_0} & 0 & -\dfrac{\beta\delta N}{\lambda\mathbf{R_0}} \\ \lambda(\mathbf{R_0}-1) & -(\gamma+\lambda) & \dfrac{\beta\delta N}{\lambda\mathbf{R_0}} \\ 0 & \gamma & -(\mu+\lambda+r) \end{pmatrix}$$

$$= \begin{pmatrix} -\lambda\mathbf{R_0} & 0 & -m \\ \lambda(\mathbf{R_0}-1) & -(\gamma+\lambda) & m \\ 0 & \gamma & -(u+\lambda) \end{pmatrix}$$

$$\text{where } m = \frac{\beta\delta N}{\lambda\mathbf{R_0}} \text{ and } u = \mu + r$$

The characteristic equation of the matrix is given by

$$x^3 + (2\lambda+u+\gamma+\lambda\mathbf{R_0})x^2 + \Big\{(\lambda+\gamma)(u+\lambda) - m\gamma$$

$$+\lambda\mathbf{R_0}(2\lambda+u+\gamma)\Big\}x + \lambda\mathbf{R_0}\Big\{(\lambda+\gamma)(u+\lambda) - m\gamma\Big\} + m\lambda\gamma(\mathbf{R_0}-\mathbf{1}) = 0$$

$$\text{Since } \mathbf{R_0} = \frac{\beta\gamma\delta N}{\lambda(\gamma+\lambda)(\mu+\lambda+r)} = \frac{\beta\gamma\delta N}{\lambda(\gamma+\lambda)(u+\lambda)} \text{ and } m = \frac{\beta\delta N}{\lambda\mathbf{R_0}}$$

Therefore, the characteristic equation reduces to

$$x^3 + (2\lambda + u + \gamma + \lambda \mathbf{R_0})x^2 + \lambda \mathbf{R_0}(2\lambda + u + \gamma)x + m\lambda\gamma(\mathbf{R_0} - 1) = 0$$

$$\text{or, } x^3 + b_1x^2 + b_2x + b_3 = 0$$

$$\text{where } b_1 = (2\lambda + u + \gamma + \lambda \mathbf{R_0});$$

$$b_2 = \lambda \mathbf{R_0}(2\lambda + u + \gamma);$$

$$b_3 = m\lambda\gamma(\mathbf{R_0} - 1).$$

All roots of this cubic equation have a negative real part, using the Routh-Hurwitz criteria, iff $b_1, b_2, b_3 > 0$ and $b_1b_2 > b_3$. Hence, it is sufficient to show that $\mathbf{R_0} > 1$ if and only if the Routh-Hurwitz criteria are satisfied. Clearly, $b_1, b_2, b_3 > 0$ if $\mathbf{R_0} > 1$.

$$\text{Now, } b_1b_2 - b_3 = \lambda \mathbf{R_0}(2\lambda + u + \gamma + \lambda \mathbf{R_0})(2\lambda + u + \gamma) - m\lambda\gamma(\mathbf{R_0} - 1)$$

$$= \lambda \mathbf{R_0}\Big\{4\lambda^2 + 2\lambda u + 2\lambda\gamma + 2\lambda u + u^2 + u\gamma + 2\lambda\gamma + u\gamma + \gamma^2 + 2\lambda^2\mathbf{R_0}$$

$$+ u\lambda\mathbf{R_0} + \gamma\lambda\mathbf{R_0}\Big\} - \frac{\beta\delta N\gamma}{\lambda\mathbf{R_0}}\lambda(\mathbf{R_0} - 1)$$

$$= \lambda \mathbf{R_0}\Big\{4\lambda^2 + 2\lambda u + 2\lambda\gamma + 2\lambda u + u^2 + u\gamma + 2\lambda\gamma + u\gamma + \gamma^2 + 2\lambda^2\mathbf{R_0}$$

$$+ u\lambda\mathbf{R_0} + \gamma\lambda\mathbf{R_0} - (\gamma + \lambda)(u + \lambda)\Big(1 - \frac{1}{\mathbf{R_0}}\Big)\Big\}$$

$$= \lambda \mathbf{R_0}\Big\{4\lambda^2 + 2\lambda u + 2\lambda\gamma + 2\lambda u + u^2 + u\gamma + 2\lambda\gamma + u\gamma + \gamma^2 + 2\lambda^2\mathbf{R_0}$$

$$+ u\lambda\mathbf{R_0} + \gamma\lambda\mathbf{R_0} - \gamma u - \gamma\lambda - \lambda u - \gamma^2 + \frac{\gamma u}{\mathbf{R_0}} + \frac{\gamma\lambda}{\mathbf{R_0}} + \frac{\lambda u}{\mathbf{R_0}} + \frac{\lambda^2}{\mathbf{R_0}}\Big\}$$

$$= \lambda \mathbf{R_0}\Big\{3\lambda^2 + 3\lambda u + 3\lambda\gamma + u^2 + u\gamma + \gamma^2 + 2\lambda^2\mathbf{R_0} + u\lambda\mathbf{R_0}$$

$$+ \gamma\lambda\mathbf{R_0} + \frac{\gamma u}{\mathbf{R_0}} + \frac{\gamma\lambda}{\mathbf{R_0}} + \frac{\lambda u}{\mathbf{R_0}} + \frac{\lambda^2}{\mathbf{R_0}}\Big\}$$

$$> 0$$

Therefore, by Routh-Herwitz criteria, it follows that all the eigenvalues of $J(Q^*)$ are negative when $\mathbf{R_0} > 1$.

Theorem 4.3 *If* $\mathbf{R_0} < 1$, *the DFE* Q_0 *is globally asymptotically stable.*

Proof Consider the Lyapunov function

$$
\begin{aligned}
V(S,E,I) &= \gamma E + (\gamma+\lambda)I \\
\text{implies } \frac{dV}{dt} &= \gamma\frac{dE}{dt} + (\gamma+\lambda)\frac{dI}{dt} \\
&= \Big(\gamma\beta S - (\gamma+\lambda)(\mu+\lambda+r)\Big)I \\
&\le (\gamma+\lambda)(\mu+\lambda+r)(\mathbf{R_0}-1)I; \qquad \text{at } Q_0
\end{aligned} \tag{11}
$$

Therefore, the maximum invariant set in $\{(S,E,I)\in\Omega : \frac{dV}{dt}\le 0\}$ is a singleton set$\{Q_0\}$. Hence by LaSalle's Principal[7] the DFE Q_0 is globally asymptotically stable if $\mathbf{R_0} > 1$.

Theorem 4.4 *If* $\mathbf{R_0} > 1$, *then EE* Q^* *is globally asymptotically stable.*

Proof Consider the Lyapunov function

$$
\begin{aligned}
V(S,E,I) &= \left(S - S^* - S^*\ln\frac{S}{S^*}\right) + \left(E - E^* - E^*\ln\frac{E}{E^*}\right) \\
&\quad + \left(\frac{\gamma+\lambda}{\gamma}\right)\left(I - I^* - I^*\ln\frac{I}{I^*}\right) \\
\text{implies, } \frac{dV}{dt} &= \left(1-\frac{S^*}{S}\right)\frac{dS}{dt} + \left(1-\frac{E^*}{E}\right)\frac{dE}{dt} + \left(\frac{\gamma+\lambda}{\gamma}\right)\left(1-\frac{I^*}{I}\right)\frac{dI}{dt}
\end{aligned}
$$

Substituting the values of derivatives from the system (2) and using $\delta N = \beta S^* I^* + \lambda S^*$, we have

$$
\begin{aligned}
\frac{dV}{dt} &= \left(1-\frac{S^*}{S}\right)\Big(\beta I^* S^* + \lambda S^* - \beta I S - \lambda S\Big) + \left(1-\frac{E^*}{E}\right)\Big(\beta I S - (\gamma+\lambda)E\Big) \\
&\quad + \left(\frac{\gamma+\lambda}{\gamma}\right)\left(1-\frac{I^*}{I}\right)\Big(\gamma E - (\mu+\lambda+r)I\Big) \\
&= -\frac{\lambda}{S}(S-S^*)^2 + \beta S^* I^* - \beta S^* I^* \frac{S^*}{S} + \beta S^* I - \beta I S\frac{E^*}{E} + (\gamma+\lambda)E^* \\
&\quad - (\gamma+\lambda)E\frac{I^*}{I} - \frac{(\gamma+\lambda)}{\gamma}(\mu+\lambda+r)I + \frac{(\gamma+\lambda)}{\gamma}(\mu+\lambda+r)I^*
\end{aligned} \tag{12}
$$

Note that $\gamma E^* = \Big(\mu+\lambda+r\Big)I^*$ and hence $\beta S^* I - \left(\frac{\gamma+\lambda}{\gamma}\right)\Big(\mu+\lambda+r\Big)I = 0$

Hence, Eq. (12) reduces to

$$
\frac{dV}{dt} = -\frac{\lambda}{S}(S-S^*)^2 - \beta S^* I^* \frac{S^*}{S} - \beta S I\frac{E^*}{E} - (\gamma+\lambda)E\frac{I^*}{I}
$$

$$+\beta S^*I^* + (\gamma+\lambda)E^* + \frac{(\gamma+\lambda)}{\gamma}(\mu+\lambda+r)I^*$$

$$= -\frac{\lambda}{S}(S-S^*)^2 + (\gamma+\lambda)E^*\left[3 - \frac{S^*}{S} - \frac{SE^*I}{S^*EI^*} - \frac{EI^*}{E^*I}\right] \leq 0 \qquad (13)$$

Therefore, the maximum invariant set in $\{(S, E, I) \in \Omega : \frac{dV}{dt} \leq 0\}$ is a singleton set $\{Q^*\}$. Hence, by LaSalle's Principal [7] the EE Q^* is globally asymptotically stable if $\mathbf{R_0} > 0$.

5 Results and Discussion

We simulate the SEIR model to outline the dynamics of the framework. The mathematical simulation was done utilizing Matlab(@Ode45). There is a DFE in every epidemiological model where the population is kept free of the disease. The basic reproductive number $\mathbf{R_0}$, which is the normal number of secondary infectious cases that an infectious person produces in a population that is fully susceptible over the whole infectious period, is another threshold parameter present in these models. $\mathbf{R_0}$, which is also linked to long-term behaviours and the amount of intercession necessary for destruction, controls the spread of diseases.

5.1 *Case I: Non Epidemic Situation*

If $\mathbf{R_0} < 1$, the disease gradually disappears from the community because, on average, each afflicted person cannot ensure that the infectious agent will be transmitted to one vulnerable person. Therefore, the populations cannot be infected by the disease, and the DFE are asymptotically stable.

Considering the hypothetical values of the parameters $\lambda = 0.08, N = 1, \gamma = 0.76, \mu = 0.048, \delta = 0.08, r = 0$(which indicates none of the infected persons are kept in quarantine) and $\beta = 0.09$ (0.09 susceptible people who are exposed to infected people leave the susceptible class and join the exposed class per day, which means there is less social contact) for the model. For these values, $\mathbf{R_0}$ was calculated as 0.6362, less than unity and indicates no epidemic for the infectious illness. Figure 2 uses a variety of unique initial conditions to show the time-series of the proportionate populations for the state variables against time for the period of 90 days. From Fig. 2, we observe that the number of infected and exposed individuals sharply decreases to zero. The disease appears to vanish from the host population after 70–75 days. According to our numerical simulations, the model's DFE is globally stable. Since 90 days of time period for the infected population to reduce to zero is a long duration, so in order to reduce this time period we have

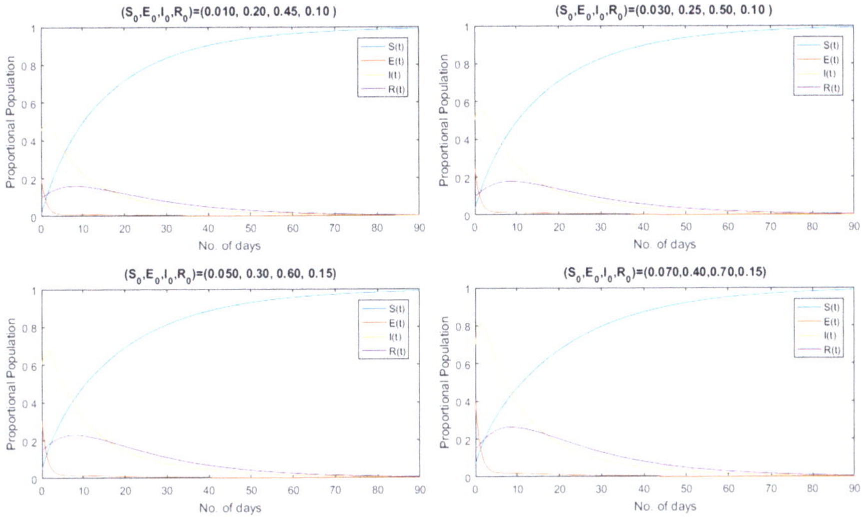

Fig. 2 Proportional Population $r = 0$, $\beta = 0.09$, $\mathbf{R_0} = 0.6362$ (No Epidemic)

taken the following case which incorporates the quarantine. By considering $r = 0.1$ (only 0.1 of the infected persons are kept in quarantine) and $\beta = 0.09$ for the model, and $\mathbf{R_0}$ was calculated as $\mathbf{R_0} = 0.3571$, is less than unity and suggests that there is no infectious disease outbreak. The time-series of the proportional populations for the state variables plotted against time over a 90 days period with the same initial conditions is shown in Fig. 3. Figure 3 shows that the number of exposed people and infected people rapidly declines to zero. The sickness seemed to have disappeared from the host population after 60 days.

5.2 *Case II: Epidemic Situation*

If $\mathbf{R_0} > 1$, each infected person creates on average more than one new infection, which allows the disease to spread across the population. As a result, the balance of a disease-free environment is unstable and invasion is always a possibility. Consider the hypothetical values of the parameters $\lambda = 0.08, N = 1, \gamma = 0.90, \mu = 0.062, \delta = 0.08, r = 0$ (implying none of the infected persons are kept in quarantine) and $\beta = 0.3$ (less social contact for the model is caused by susceptible individuals who are exposed by infected individuals, leave the susceptible class, and enter the exposed class each day). $\mathbf{R_0}$ was calculated as 1.94 for these values, which is greater than unity and indicates the infectious disease is experiencing an epidemic. Using different individual initial conditions, Fig. 4 illustrates time-series of the proportional populations for the state variables against time for a 90-days time period. From Fig. 4, we see that the number of infected people does not go to zero,

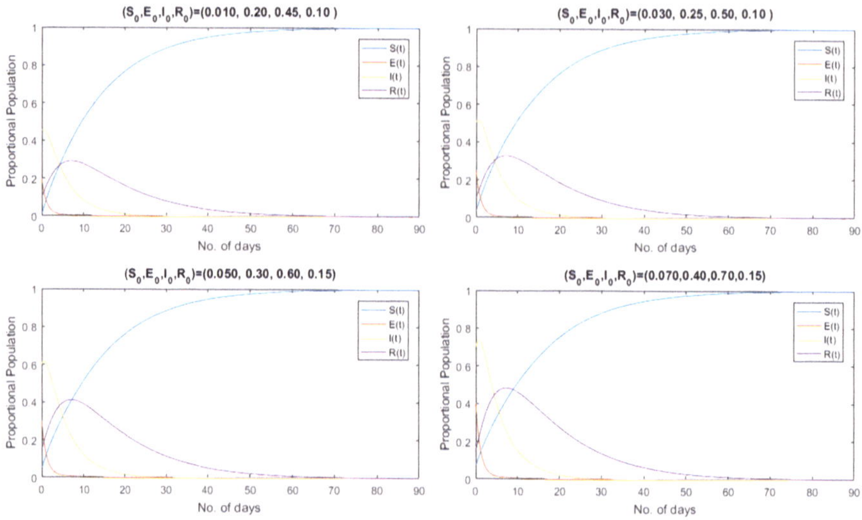

Fig. 3 **Proportional Population** $r = 0.1, \ \beta = 0.09, \ \mathbf{R_0} = 0.35, \ $ (No Epidemic)

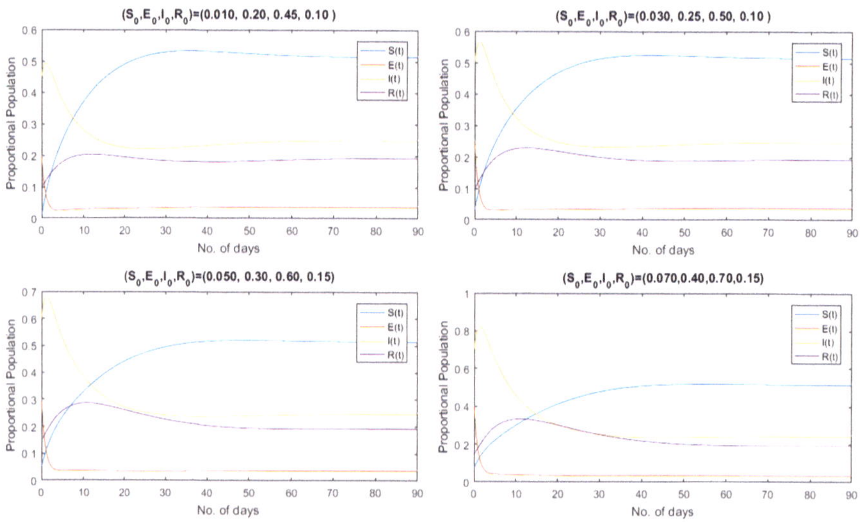

Fig. 4 **Proportional Population** $r = 0, \ \beta = 0.3, \ \mathbf{R_0} = 1.94, \ $ (Epidemic)

meaning that as time goes on, infected people are still existing among the hosts. This suggests that an endemic disease is affecting the population. Thus, it appears that the DFE will change throughout time.

From Fig. 4, it is clearly understood that if we have to reduce or irradicate the epidemic we must have to quarantine the infected people. We consider $0.0 \leq r \leq 1.0$ for the epidemic case of the model. $\mathbf{R_0}$ was calculated, which comes less than

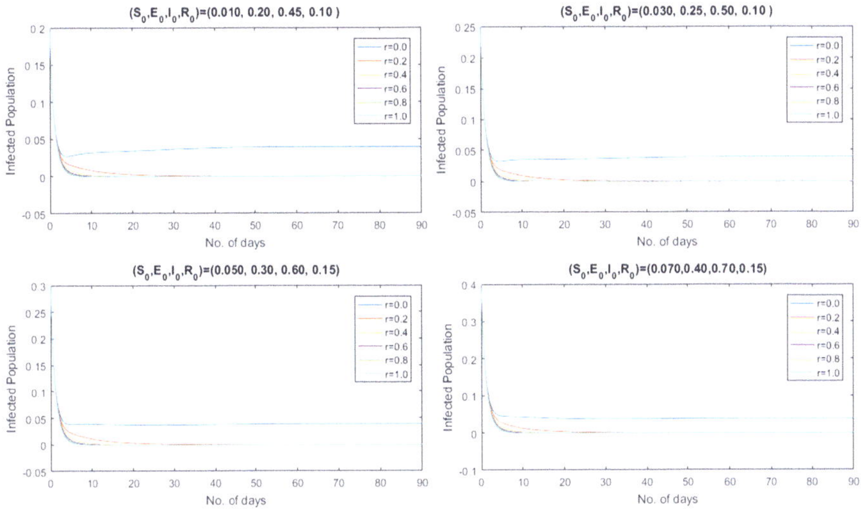

Fig. 5 **Proportional Population** $r = (0.0 - 1.0)$, $\beta = 0.3$, (No Epidemic)

unity for each r in all cases except for $0.0 \leq r < 0.2$, implies no epidemic for the infectious disease. Figure 5 uses a variety of unique initial conditions to show the time-series of the proportional infected populations for the state variables against time over a 90-days period. Figure 5 shows that the number of infected people rapidly drops to zero. The sickness seemed to have disappeared from the host population after 20 days. The MATLAB codes of each non epidemic and epidemic cases are shown in the appendix.

6 Conclusion

The work is devoted towards the basic epidemiological model along with the quarantine conditions. A mathematical model based on basic SEIR model with appropriate initial conditions has been constructed to estimate the effect of diseases like COVID-19 at various conditions. The parameters of population dynamics have been involved in the model for their significant contribution towards the determination of the spread of disease. The model has been solved and the numerical values have been obtained using MATLAB(@ode45).

The main focus was to improve the measures like quarantine in order to reduce the spread of the disease. The local stability of the DFE and EE equilibrium points was evaluated for the scenarios of no epidemic ($\mathbf{R_0} < 1$) and epidemic ($\mathbf{R_0} > 1$) from the experimental findings of the model by using the time-series of the S, E, I and R individuals. Eventually, the I individuals who were affected vanished after approximately 70–75 days in non epidemic case and the infected I

individuals reached their steady-state in approximately 30–60 days (epidemic). The time duration for the epidemic cases was also reduced to 20 days approximately by incorporating the quarantine facilities for the infected individuals I. S, E, I and R individuals arrived at the respective equilibrium points DFE and EE for local stability in both the epidemic and no epidemic situations. The model incorporates the quarantine factor in consideration. The model can be helpful in such diseases which spread due to the viruses or such agents whose vaccinations have not been yet developed. In order to irradicate or overcome the epidemic situation, we vary the quarantine parameter so that $\mathbf{R_0} < 1$. In case of the epidemic situation, the probability is that more number of people may be in contact with the infected persons, so it is better to detect those persons and put in quarantine. For future work, age structure could be included to the SEIR model with quarantine parameter along with proper treatment to develop mathematical models of epidemics that are increasingly more realistic.

Appendix

(No Epidemic, $r = 0$, $\mathbf{R_0} = 0.6362$*)*

```
function noepidemic = noepidemic(t, x);
lambda = 0.08; delta = 0.08; beta = 0.09; gamma = 0.76; mu = 0.048;
r = 0; N = 1; [t, x] = ode45(@rhs, [0 90], [0.010 0.20 0.45 0.10]);
subplot(2, 2, 1); plot(t, x); title('(S0, E0, I0, R0) = (0.010, 0.20, 0.45, 0.10)');
xlabel('No.ofdays'); ylabel('ProportionalPopulation'); legend('S(t)',
'E(t)',' I(t)',' R(t)'); [t, x] = ode45(@rhs, [0 90], [0.030 0.25 0.50 0.10]);
subplot(2, 2, 2); plot(t, x); title('(S0, E0, I0, R0) = (0.030, 0.25, 0.50, 0.10)');
xlabel('No.ofdays'); ylabel('ProportionalPopulation'); legend('S(t)',
'E(t)',' I(t)',' R(t)'); [t, x] = ode45(@rhs, [0 90], [0.050 0.30 0.60 0.15]);
subplot(2, 2, 3); plot(t, x); title('(S0, E0, I0, R0) = (0.050, 0.30, 0.60, 0.15)');
xlabel('No.ofdays'); ylabel('ProportionalPopulation'); legend('S(t)',
'E(t)',' I(t)',' R(t)'); [t, x] = ode45(@rhs, [0 90], [0.070 0.40 0.70 0.15]);
subplot(2, 2, 4); plot(t, x); title('(S0, E0, I0, R0) = (0.070, 0.40, 0.70, 0.15)');
xlabel('No.ofdays'); ylabel('ProportionalPopulation'); legend('S(t)',
'E(t)',' I(t)',' R(t)'); functiondxdt = rhs(t, x);
ds = delta * N - beta * x(1) * x(3) - lambda * x(1);
de = beta * x(1) * x(3) - gamma * x(2) - lambda * x(2);
di = gamma * x(2) - mu * x(3) - lambda * x(3) - r * x(3);
dr = mu * x(3) - lambda * x(4) + r * x(3);
dxdt = [ds; de; di; dr];
end;
end;
```

(Epidemic,r = 0, R_0 *= 1.4)*

```
function epidemic = epidemic(t, x);
lambda = 0.08; delta = 0.08;   beta = 0.3; gamma = 0.90;   mu = 0.062;
r = 0; N = 1; [t, x] = ode45(@rhs, [0 90], [0.010 0.20 0.45 0.10]);
subplot(2, 2, 1); plot(t, x); title('(S0, E0, I0, R0) = (0.010, 0.20, 0.45, 0.10)');
xlabel('No.ofdays'); ylabel('ProportionalPopulation'); legend('S(t)',
'E(t)',' I(t)',' R(t)'); [t, x] = ode45(@rhs, [0 90], [0.030 0.25 0.50 0.10]);
subplot(2, 2, 2); plot(t, x); title('(S0, E0, I0, R0) = (0.030, 0.25, 0.50, 0.10)');
xlabel('No.ofdays'); ylabel('ProportionalPopulation'); legend('S(t)',
'E(t)',' I(t)',' R(t)'); [t, x] = ode45(@rhs, [0 90], [0.050 0.30 0.60 0.15]);
subplot(2, 2, 3); plot(t, x); title('(S0, E0, I0, R0) = (0.050, 0.30, 0.60, 0.15)');
xlabel('No.ofdays'); ylabel('ProportionalPopulation'); legend('S(t)',
'E(t)',' I(t)',' R(t)'); [t, x] = ode45(@rhs, [0 90], [0.070 0.40 0.70 0.15]);
subplot(2, 2, 4); plot(t, x); title('(S0, E0, I0, R0) = (0.070, 0.40, 0.70, 0.15)');
xlabel('No.ofdays'); ylabel('ProportionalPopulation'); legend('S(t)',
'E(t)',' I(t)',' R(t)'); functiondxdt = rhs(t, x);
ds = delta * N - beta * x(1) * x(3) - lambda * x(1);
de = beta * x(1) * x(3) - gamma * x(2) - lambda * x(2);
di = gamma * x(2) - mu * x(3) - lambda * x(3) - r * x(3);
dr = mu * x(3) - lambda * x(4) + r * x(3);
dxdt = [ds; de; di; dr];
end;
end;
```

References

1. Anderson RM, May RM (1991) Infectious disease of humans,dynamics and control. Oxford University Press, Oxford
2. Brauer F, Van den Driessche P, Wu J (2008) Mathematical epidemiology. Springer, New York, pp 415.
3. Lu H (2020) Drug treatment options for the 2019-new coronavirus (2019-nCoV). Biosci Trends 14(1):69–71
4. Kermack WO, McKendrick AG (1927) A contribution to the mathematical theory of epidemics. Proc R Soc A 115:700–721
5. Kermack, WO, McKendrick AG (1932) Contributions to the mathematical theory of epidemics II. the problem of endemicity. Proc. R. Soc., 138; 55–83
6. Alqarni MS, Alghamdi M, Muhammad T, Alshomrani AS and Khan MA (2020) Mathematical modeling for novel coronavirus (COVID-19) and control. Numer Methods Partial Differ Equ. 38(4):760–776
7. Lasalle J (1976) The stability of dynamical systems. SIAM, Philadelphia
8. Morris WH, Stephen S, Robert LD (2013) Differential equations. Dynamical equations, and introduction to chaos, 3rd edn. Elsevier, Oxford
9. Nazeer S, Sarosh P, Arah SA (2022) Automatic classification of COVID-19 infected patients using convolution neural network models. Artificial intelligence for innovative healthcare informatics. Springer International Publishing, New York, pp 119–131

10. Shohel A, Sumaiya R, Arifuzzaman (2017) A mathematical modeling and dynamical features in the spreading of virus infectious diseases. Int J Adv Appl Mech Mech 5(1):88–98
11. Huang C, Huang L, Wang Y, et al. (2021/2022) 6-month consequences of COVID-19 in patients discharged from hospital: a cohort study. Lancet 397(10270):220–232
12. Holmes EA, O'Connor RC, Perry VH, et al. (2021) Multidisciplinary research priorities for the COVID-19 pandemic: a call for action for mental health science. Lancet Psych 8(4)
13. Ali A, Alshammari FS, Islam S, Khan MA, Ullah S (2021) Modeling and analysis of the dynamics of novel coronavirus (COVID-19) with Caputo fractional derivative. Results Phys 20:103669
14. Sene N (2020) SIR epidemic model with Mittag-Leffler fractional derivative. Chaos Solut Fractals 137:109833
15. Kifle ZS, Obsu LL (2022) Mathematical modeling for COVID-19 transmission dynamics: a case study in Ethiopia. Results Phys 34:105191
16. WHO. Coronavirus disease (COVID-19) pandemic, available at : https://www.who.int/emergencies/diseases/novel-coronavirus-2019
17. Worldmeter, Coronavirus cases per day, available at: https://www.worldometers.info/Coronavirus

Part III
Intelligent Signal Processing for Smart Industry

Detection of Power Distribution Fault in Thermal Images Using CNN

Nurul Huda Ishak, Muhammad Ameer Izzat Halim, and Iza Sazanita Isa

1 Introduction

A failure in an electrical circuit causes electricity to be diverted from its intended route, resulting in a malfunction in electrical equipment or apparatus. Faults in power systems are growing increasingly prevalent as the systems' structure and operations become more complex. Years of study and development into fault-diagnosis technology for power systems have resulted [1]. Thermal imaging technology is used in diverse range of applications in many industrial sectors. The technology is typically used to examine the invisible defects in various fields such as industrial, electrical, structural, and medical. Thermal image processing is therefore utilized to detect hotspots in electrical systems and highlight potential problem areas [2]. Algorithms based on artificial intelligence (AI) now have a significant advantage: high accuracy. Study on AI is closely related to smart technologies such as Industrial Forgery Detection [3], High Capacity and Secure Electronic Patient Record [4] and Reversible Data Hiding for Electronic Patient Information Security for Telemedicine Applications [5]. Because of this characteristic, AI-based fault diagnostic technology is more dependable than the prior approach. CNN have been applied in distribution system fault localization and detection. The CNN is a transfer-learning method that identifies the underlying relationship between input data and output labels, resulting in a more robust and universal model [6].

Fault detection is recognized as a crucial component of an electrical system that enables ongoing cost reduction, and elimination, enhanced productivity, and equipment availability [2]. Because it is necessary to monitor the whole electrical network at regular intervals in order to minimize electrical losses and related

N. H. Ishak · M. A. I. Halim (✉) · I. S. Isa
Centre for Electrical Engineering Studies, School of Electrical Engineering, Universiti Teknologi MARA, Cawangan Pulau Pinang, Permatang Pauh, Malaysia
e-mail: izasazanita@uitm.edu.my

S. A. Parah et al. (eds.), *Intelligent Multimedia Signal Processing for Smart Ecosystems*, https://doi.org/10.1007/978-3-031-34873-0_11

problems, automatic fault detection is commonly utilized [6]. Many studies have focused on power fault detection, but thermal imaging detection is also a popular approach due to its efficacy, dependability, and non-contact nature. As a result, there is a requirement for thermal imaging to identify power problem distribution systems automatically. Deep neural networks will be utilized in this research to detect and assess power failure distribution systems automatically. Because this study employs CNN architecture, it will aid in the resolution of an important job in power defect detection: thermal image detection. CNN is the most successful image detection technology because of its capacity to recognize patterns in images, and it has been used in power failure analysis. The CNN is not only useful in image detection, but it may also aid in enhancing the computer's accuracy and defect detection ability.

This research aims to detect power fault distribution systems using a convolution neural network by improving the performance of pre-trained CNN model using ResNet-50 which contains 50 layers of convolutional layer, and data augmentation which is image augmentation to generate new images to train the neural network. Generally, the scope of this study is to analyse the performance of CNN by using MATLAB to detect power faults based on thermal imaging.

2 Literature Review

Recently, in power preventive maintenance at substation, manual thermographic scanning has widely been used for detecting faulty on electrical power system as routine task. Since heat dissipation in high voltage electrical equipment can be recognized using thermal image technology [7], thermal imaging approach has been used in parallel with visual detection approach to predict line faulty at electrical power system [8]. To avoid the overheating that can cause electrical equipment failure, the heat dissipation can be detected in the early stage by efficient screening through investigating the thermal image features induced in the images [9]. The main issue in processing the thermographic images is difficult to interpret the results since it requires a certain experience and knowledge [10]. Moreover, obtaining high accuracy can be difficult due to varying emissivity of the different materials, reflections from other surfaces and other characteristics [11]. At present, image fusion techniques have been widely applied to the target recognition as a type of image pre-processing methods, especially for infrared-visible image fusion [12, 13]. Therefore, it is necessary to use thermal imaging for the purpose of automatically detecting power fault distribution systems. Deep neural networks will be used for the purpose of this research to automatically identify faults in power distribution systems and evaluate them.

Thermal imaging is one of the approaches used to detect and predict faults in the transmission line during preventive maintenance work. The process involves capturing images of expected equipment or objects using a thermal camera that creates an image by using infrared radiation. This image is converted into a temperature-based format and displayed as temperature distribution. A thermographic image can be used to identify any potentially problematic areas by detecting hotspots in the electrical system image. Based on past research, thermal imaging has widely been used to detect faults in power system [14]. Most of the fault inspection method is since most components in a system show an increase in temperature when malfunctioning. Therefore, infrared image recognition technology is found reliable for fast prediction methods due to contactless monitoring of electrical equipment without shutting down the equipment or system. Ideally this detection should be computationally efficient and the decision should be highly reliable [15, 16].

2.1 *Power Fault at Distribution Line*

Since distribution system is responsible for distributing electrical power to individual consumers from the high-voltage (HV) transmission lines, the distribution system is an essential component of the electric power system. Electrical power is sent to end consumers across significant distances (in the form of kilometres), since the distribution system serves a large geographical area that is comprised of both urban and remote rural locations [17]. These lines are susceptible to experiencing short-circuit or shunt failures because of severe weather conditions, mechanical breakage, or insulation deterioration. These are the three main reasons for line failure. Faults in distribution lines may arise when electrical power networks are operating because of unanticipated natural disasters or accidents. These faults can disrupt the flow of electricity. These conditions might result in interruptions to the delivery of goods, the risk of fires that put people in danger, and significant financial losses [18]. Protective relays and other control devices were commonly installed to assist circuit breakers in taking appropriate measures, thereby minimising damage and the duration of interruption while maintaining the maximum desirable resilience level [19]. This was done to reduce the risks and losses associated with the situation.

In order to serve an efficient and avoidable interruptions in power distribution, the utmost importance to ensure that the reliability of power supply in power transmission and distribution networks is highly maintained within specific time range. Distribution systems are substantially more prone to faults than transmission systems due to the fact that distribution systems are significantly larger in size. More than 80% of all problems can be traced back to distribution systems [20]. So that end users can get good, reliable energy, it's important to find faults in energy distribution lines as quickly and accurately as possible and figure out where the faults lie.

When applied to power distribution lines, the use of deep learning or a deep neural network for fault detection leads to better energy system security, quicker restoration time, cost savings, and a reduction in the number of shunt faults and weak regions across the system.

2.2 Convolution Neural Network

The procedure of fault diagnosis is expected to be intelligent enough to automatically detect and recognize the health states of machines [21, 22] using AI. The integration of thermal imaging and automated computer-based approaches have been proposed to detect faults in power transmission lines such as fault identification [13], fault classification [23] and fault localization [24]. Conventionally, Artificial Neural Network (ANN) was mostly applied in fault analysis [24, 25] and showed effectiveness in designing the fault diagnosis models through proposed fault analysis algorithms. However, ANN is only dominant for problems where datasets are limited, and image inputs are not necessary. To the latest, a special class of deep learning namely the CNNs model has gained popularity among researchers to be applied in the detection and classification of faulty insulator components at overhead transmission lines [23]. Figure 1 shows the architecture of the CNNs model. CNNs recognize an input image and hence assigns the image to various aspects of features and differentiate one from the other.

In the field of computer vision, image recognition is one of the most common applications for a subclass of the deep learning algorithms known as CNNs. The convolutional layer, the pooling layer, and the fully connected layer are the three basic layers that make up a CNN architecture. In addition to these levels, there are various more parameters and layers that serve specialised purposes. An ideal feature map of the input data is generated by the 2D convolution operation, which makes use of a large number of previously learned filter kernels (as opposed to handcrafted or engineered filters typical in classical signal or image processing). The pooling layer performs the function of a down-sampling layer, which reduces the size of the data features. In conclusion, a completely connected layer is employed for the categorization process. It helps in learning patterns from time-frequency representations and lowers the requirement for hand-crafted expert features because the convolution and pooling procedures are in charge of feature extraction. Error-back propagation is a method of optimization that is used to train CNN. This method allows a CNN to minimise classification errors across network parameters such as

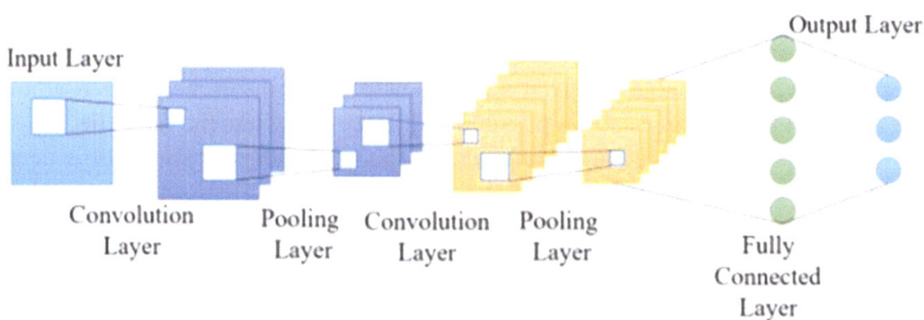

Fig. 1 The basic architecture of CNN [26]

filter kernels or weights. To put it another way, the network is trained to learn or mine discriminative features from the underlying data, which can then be used to classify the data using a classifier such as SVM or CNN's SoftMax layer. During training, the network is exposed to the data and is instructed to learn these features. As a result, you will need a significant number of labelled data samples in order to train a CNN from scratch.

A CNN network that has been pre-trained on a large benchmark dataset that is freely accessible, such as the ImageNet dataset for object recognition, has the potential to be used as an off-the-shelf feature extractor for a completely different domain. This is because the network has been trained on a large dataset that has been used as a standard. In the realm of computer vision, a number of well-known CNN architectures, like AlexNet, VGGNet, GoogLeNet, and others, have been proposed, and their pre-trained versions have been made available to the general public [22]. The practical limitations of the discussed model in real-time applications include model complexity when power is interrupted or hardware parameter tuning.

2.3 Power Fault Detection Using Convolution Neural Network

This development has been accelerating in recent years. In recent years, neural networks, also known as "deep mastering," have begun to beat standard AI in tasks like characterizing pictures, detecting voice, and creating meaningful, readable text. Deep learning techniques are a kind of ANN that is more complex than traditional ones. They consist of several layers and are designed to gain high-order abilities from the input before outputting the predicted fee at the network's peak. CNN, which is classified as one of the deep learning algorithms, has seen widespread use in creative computer work and predictive analysis. The results of CNN have shown its robustness in object recognition and localization across a range of visual variations. For several years, the detection of power issues has benefited from the use of deep learning. Most of the exciting deep learning research presented to improve the overall performance of fault imaging did so by either modifying the design of already established deep learning networks or proposing new ones. Deep learning has the potential not only to carry out the essential task but also to improve the overall performance of the computer's fault imaging system when it comes to classifying and locating issues.

Based on the past studied, [13, 23] had conducted research to classify the thermal condition of electrical equipment to predict the coordinates, orientation angle, and class type of each equipment part. Furthermore, CNN was also used by [23] to recognize and detect a fault at the insulator. There are a few advantages of using CNNs such as the ability to eliminate the need for manual feature extraction [27], capable to produce highly accurate recognition results [28] and can be retrained for new recognition tasks [27]. In addition, the proposed research in image fusion decomposition [29] also shown promising results for better detection in thermal imaging by applying CNN-based models. Table 1 shows the previous research

Table 1 Convolution Neural Network for fault detection

Study	Dataset	Method	Finding	Performance measure
Aziz et al., 2020 [22]	Total 3456 data samples and labels were collected. Each class (No Fault, LL, OC, PS, Fault in PS and arc fault) has 576 instances	Applying two convolution neural network: (1) few layers of pre-trained AlexNet are fine-tuned to yield 6-way output. (2) fc7 layer of a pre-trained AlexNet and then used conjunction with classical ML methods for classification	Results showing that AlexNet not only has roughly half the execution time but provides comparable.	73.53% performance accuracy
Tao et al., 2020 [25]	Chinese power line insulator dataset (CPLID)	Localizing insulator and detecting defects with CNN using aerial images analysed	Datasets are normally too small to train CNN. Augmenting the data was necessary in order to avoid overfitting. Data augmentation generates new datasets using existing data	91% based on proposed method and 96% if using the standard insulator dataset
M. Gowrishankar, P. Nagaveni and P. Balakrishnan, 2016 [30]	MATLAB is used to simulate the fault condition on the 220 kV transmission line	Detect and classify fault in transmission line based on Discrete Wavelet Transform and Artificial Neural Network	A power system protection plan can be developed using the proposed method, according to simulation findings, which demonstrate that it produces adequate results.	High accuracy rate of 90.6% in determine the correct fault types
Sowah et al., 2018 [31]	Experimental data was collected with the device and then processed to determine the criteria for detection, location and classification of faults in the network	Machine Learning	Decision Tree is found to be better compared to kNN and SVM with respect to accuracy	High accuracy rate of 99.42% using the Decision Tree compared to others machine language
Huang et al., 2020 [32]	Aerial insulator image dataset is built to train and testing. 200 each for normal, broken, dropped, crack and noisy insulators	Deep Learning	The multi-feature fusion based deep learning network shows great performance compared to DCNN and BPNN in classification	The proposed method had achieved performance 82.4% with less training time compared with others method

has been conducting the study in analysing or detecting the fault by using CNN on the power system. Aziz et al. [22] conducted a study on fault classification in photovoltaic arrays with CNN, this research is applying two CNNs', one of which is from the fc7 layer of a pre-trained AlexNet and then used in conjunction with classical ML methods for classification. According to the findings of this investigation, CNN obtained 73.53% performance accuracy.

Tao et al. [25] proposed a convolution neural network to identify power line insulator problems in aerial images. The goal of this project is to use CNN to both localise insulators and identify faults. This study offered data augmentation before combining insulator localisation and defect detection with cascading CNN. According to this study, the performance accuracy is 91% when utilising the suggested approach and 96% when using the standard insulator dataset. M. Gowrishankar et al. [30] did research on deep learning method to identify and classify fault in transmission line based on Discrete Wavelet Transform and ANN. The fault condition on the 220 kV transmission line was simulated in this research using MATLAB. This research obtained a detection rate of 90.6%.

Sowah et al. [31] in their study investigated design power distribution defect detection, location, and classification using machine learning. This research employed three machine learning algorithms to forecast power defect, location, and categorization inside a single-phase distribution network: K-Nearest Neighbor (KNN), Decision Trees, and Support Vector Machines (SVM). When compared to other machine languages, this research demonstrated a high accuracy rate of 99.42% when employing the Decision Tree.

Following that, Haung et al. [32] performed multi-feature fusion-based deep learning for insulator picture recognition and flaw detection. The aerial insulator picture collection was utilised for training and testing in this work, comprising five types of insulator conditions (broken, normal, dropped, broken, cracked and noisy). Back Propagation neural network and deep convolution neural network were compared in this research, as was the suggested approach, the multi-feature fusion based deep learning network. This investigation on the suggested strategy produced 82.4% performance with less training time than previous methods.

3 Methodology

This research used a deep learning architecture that has been pre-trained for 1000 times with different types of image categories known as ResNet-50. Residual Network (ResNet) is a sort of neural network that uses identity mapping, which implies that input is transferred directly or indirectly to another layer. With these techniques, a network with a high number of layers may be built and trained quickly without raising the training error percentages. ResNet can also help with the vanishing gradient problem by applying identity mapping. The ResNet model utilized in this work has 50 layers, thus the name ResNet-50. At the same time, a data augmented, or called image augmented also used in the model to get a better

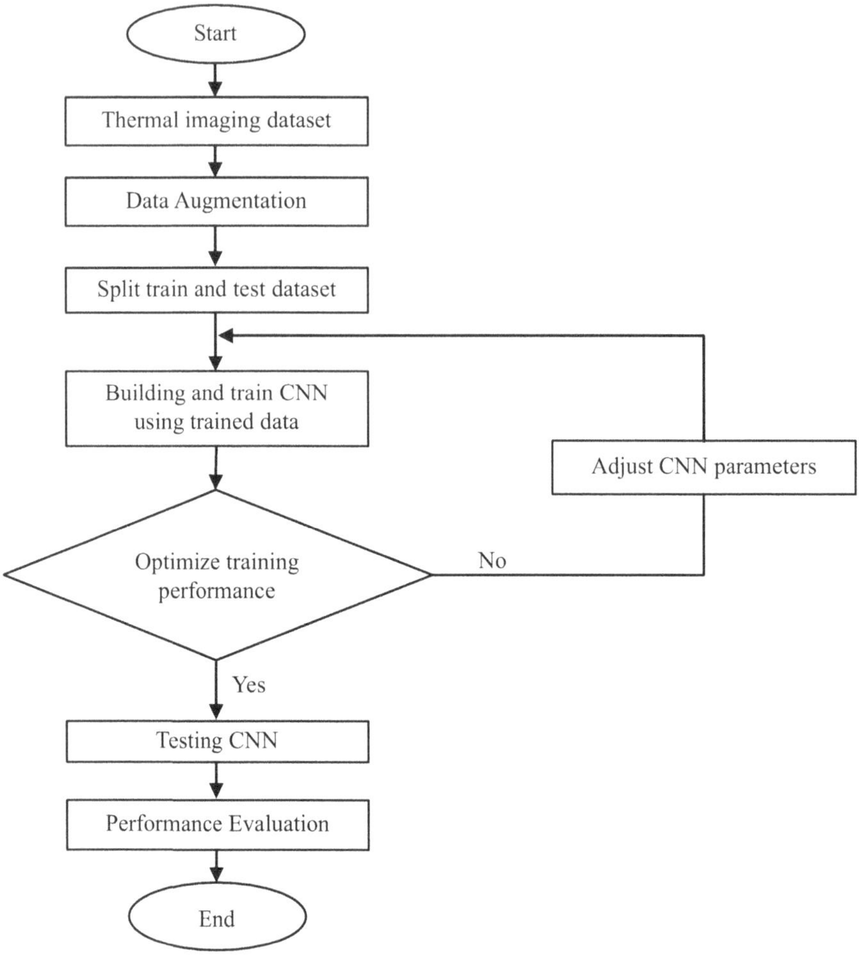

Fig. 2 Proposed power fault detection based thermal imaging using CNN

performance in detecting power faults by using thermal images. Overall proposed work for power fault detection based thermal imaging using CNN is presented in Fig. 2.

3.1 Data Acquisition

Data acquisition of fault rating system for cable temperature had been collected from the Ti Thermal Imaging LTD thermal imaging report. The rating of cable temperature reference has been taken from the British Standard BS7671 which structurally resembles IEC 60364 used for Malaysia's Standard electrical installations of

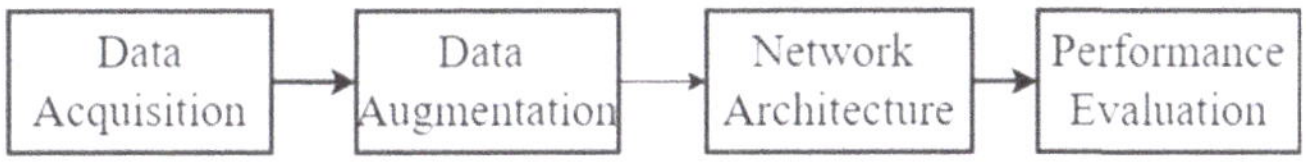

Fig. 3 Data acquisition of power fault detection using convolution neural network based on thermal images

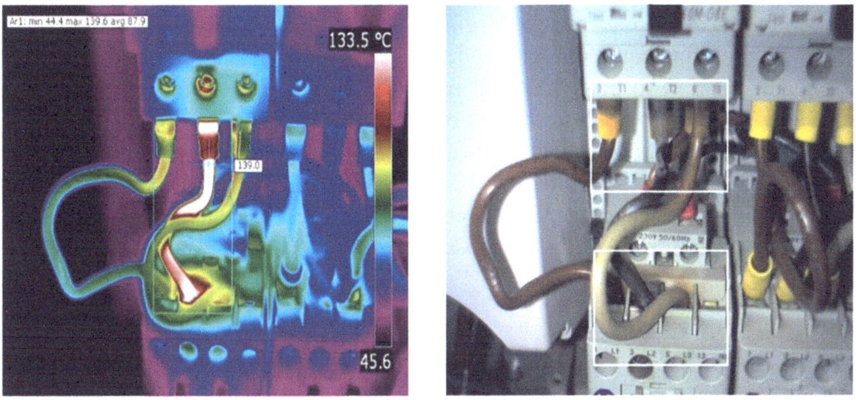

Fig. 4 Data acquisition (**a**) Thermal Images and (**b**) standard images of electrical equipment capture and compare side by side [33]

buildings to detect whether the distribution system is in a normal or faulty condition. Data acquisition including a step of image pre-processing which is done together with data augmentation to double the amount data from the original amount. After attaining the image pre-processing done together with data augmentation, the research proceeds with the network architecture of CNN framework development to train and test the network capability. Finally, the performance evaluation in terms of accuracy of the convolution neural network will be recorded as the output as shown in Fig. 3.

The data acquisition is collected from Ti Thermal Imaging LTD thermal imaging report to uncover anomalous component temperature. In this study, the data of fault detection at power distribution consisted of total 250 thermal images together with the trending data of the electrical equipment. Each thermal image for normal condition and faulty condition distribution system are 125 images respectively. Figure 4 shows the example of thermal images and their respective original images which are collected as attributes for training and testing in the proposed CNN.

3.2 Deep Learning

Deep learning is a machine learning technique that interprets data using several layers of the neural network algorithm. A neural network is a complicated structure that generates artificial neurons that can transform many inputs into a single output. The neurons in this network will interact and be linked to one another. This neural network can recognize and observing any feature in the datasets. Focus of this research is on the CNN in a form of deep neural network that is extensively used in deep learning to analyze visual pictures that utilizes the folding method [7].

The CNN detects power faults in the distribution system. The thermal imaging of distribution system as input data will be supplemented first, then separated into training and testing datasets. The divided datasets for train and test will be fed into the neural network to train the CNN using the trained datasets. If the train CNN is not performing optimally in terms of accuracy. The train CNN must fine-tune various parameters, such as the proportion of splitting train tests, until it is optimal in terms of accuracy. After the CNN has been optimized and trained, the testing datasets will pass through the CNN model architecture and generate the output that will be utilized as the model performance accuracy.

3.3 Convolution Neural Network

CNN is a neural network design that uses spatial correlation of data inputs to locate power outages using thermal imaging. Weight sharing in CNN design increases learning features, particularly in images, while also decreasing computation when compared to fully connected ANN. The CNN algorithm is exceptionally good at spotting patterns in photographs. CNN is made up of neurons that can be programmed to have biases and to weigh. CNN architecture has two major learning components: feature extraction and classification. Convolution is conducted on the input dataset during the feature learning stage to generate the feature map using a kernel-based filter. Then CNN will compute the possibility of the picture specification to a particular class, which in this case consists of two classes: fault and non-fault of the distribution system. The CNN-processed image architecture consists of two stages namely feature learning and classification as presented in the Fig. 5. The goal of the CNN, or ConvNet, is to compress pictures into a format that is easier to process while retaining key components for producing a successful prediction. This is critical for developing an architecture that is not just capable of learning features but also scalable to large datasets.

Data availability enhances the performance of deep learning neural networks regularly. Data augmentation is a method of creating new training data from old training data. This is accomplished by applying domain-specific approaches to training data examples, yielding new and distinct training instances. Picture data augmentation is the most extensive type since it entails changing copies of images in the training dataset that belong to the same class as the original image. Shifts, flips, zooms, and other picture alteration operations are included in transforms. The goal is to add more trustworthy cases to the training collection. CNN has been widely used in signal processing as one of the most powerful deep learning technologies. CNN

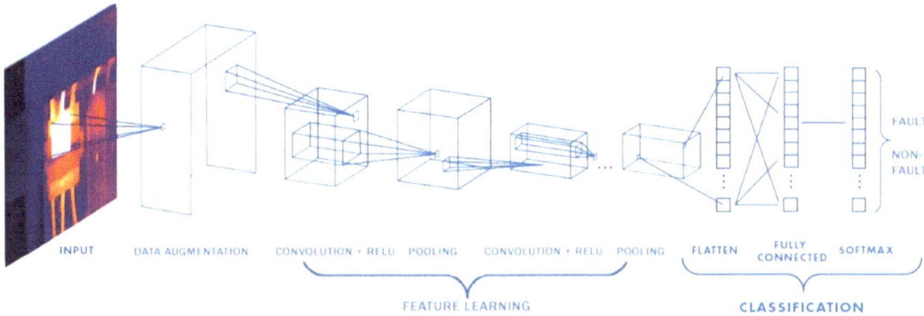

Fig. 5 A framework of CNN architecture consists of feature learning and classification for network training [34]

comprises convolution, pooling, and a fully connected layer (FC), using the rectified linear unit (ReLU) as the activation function [8]. Convolutional layers convolute the input and deliver the result to the next layer. Convolution is a procedure that converts all pixels in its receptive region into a single value. When a convolution layer is applied to an image, it decreases the image size while simultaneously combining all of the information in the field into a single pixel, with the ultimate result being a vector. Non-linearity is necessary for any neural network to be effective. This is accomplished by passing the weighted sum of its input through an activation 16 function. At this stage, the outcome of the convolution process is delivered via the ReLu activation function.

A pooling layer is a separate component of a CNN. Its purpose is to gradually lower the spatial dimension of the representation, hence reducing the number of parameters and computation in the network. The pooling layer functions independently on each feature map. Then, towards the end of CNN, there is a layer of neurons that are totally linked, or FC. Neurons in a totally connected layer are fully linked to all activations in the previous layer, much as neurons in a conventional Neural Network. This layer is created from the output of the preceding layers, which incorporate the image's properties to categorize it. The number of totally connected layers in this research corresponds to the number of picture classes; the number of image classes is separated into two fault and non-fault power distribution classes. The output of the fully connected layers was then normalized before the classification layers used SoftMax to determine the classification probabilities as shown in Fig. 5.

Three critical parameters must be changed to change the behavior of DCNN's convolution layers: filter size, stride, and zero padding. The size of the filter is essential to recognize the relevant attributes. The most appropriate filter size should be chosen, and it should not be too large or too small, because if the filter size is too large, it may overlook the picture's features, and if the filter size is too small, it may cause confusion since it gives more information about the image. Stride is used to control the number of steps required to move the filter over the input picture, as well as the amount of stride, which may be used to change the size of the output volume of the 17 convolution layers. Then there's zero padding, which is utilized to define the spatial size of the output volume by padding the input volumes with zero around the border.

The output size of the CNN layer was calculated using the Eq. (1), where W represents the input height or length value, K represents the filter size, P represents padding, and S represents stride.

$$Output_{height\ or\ length} = \frac{(W - K + 2P)}{S} + 1 \tag{1}$$

3.4 Data Augmentation

In order to properly train the deep ANN, a vast corpus of training data is required. However, collecting such a considerable corpus of training data is often both time-consuming and labor-intensive. This problem may be solved by data augmentation, which involves artificially expanding the size of the training set using transformations that preserve labels. Recently, there has been a significant increase in the usage of generic data augmentation to enhance CNN task performance.

Data Augmentation, a regularization strategy that artificially inflates the dataset by employing label-preserving transformations to add invariant instances, might be one solution to the problem. Data Augmentation is a regularization scheme. As a result, data augmentation is a strategy that may further improve CNN performance while preventing dataset from over-fitting.

3.5 Performance Evaluation

Testing and validation are two aspects of performance evaluation. The thermal image dataset will be utilized for training to evaluate the CNN model architecture's accuracy. The accuracy of the CNN model design will be evaluated using the method provided in Eq. (2), where TP denotes True Positive, TN denotes True Negative, FP denotes False Positive, and FN denotes False Negative. The accuracy measurement is used to forecast and judge how well the system worked.

$$Accuracy = \frac{TP + TN}{TP + TN + FP + FN} \tag{2}$$

4 Result and Discussion

4.1 Image Augmentation Analysis on ResNet-50 Performance

ResNet-50 was chosen as the CNN to detect the thermal images because the architecture is smaller in parameter size without taking into consideration the estimated size of the objects [35]. The thermal image data will be divided into training and testing datasets. In this study, 30% of the data thermal images were tested and 70% of the thermal images were trained, while all the datasets are picked randomly for the ResNet-50 model.

Without using any additional method to improve the CNN, it can achieve a performance accuracy of 75%. Various methods can be used to increase the accuracy. However, for this study, two methods have been used: a pre-trained CNN model, ResNet-50, which contains 50 layers of convolutional layer, and data augmentation which is image augmentation to generate new images to train the neural network.

Those new images are generated using the existing training images and not collecting the images manually. In addition, data that had been augmented will be resized, and color preprocessing to make the input image compatible with the ResNet-50, which can help prevent the network from overfitting and memorizing the exact details of the training images.

This study used to avoid overfitting that may cause poor performance in the model total of 250 thermal images. With the randomized pick, thermal images, training, and testing at 30% and 70% for each normal condition and fault condition for the CNN, as shown in Table 2.

Figures 6 and 7 shows the ResNet-50 can detect the thermal images for normal and fault condition at different total number of thermal images. The accuracy of the system with ResNet-50 is obtained and presented in Table 2. With the obtained average accuracy of ResNet-50 in Table 2 the neural network is a good model in term of performance and able to detect the power fault thermal images since the average accuracy is above 75% which the minimum accuracy for a Convolution Neural Network.

When power fault thermal image was detected using pre-trained CNN (ResNet-50) and image augmentation, the average accuracy increased as the data increased, reaching a maximum of 83.3% for normal condition detection and 82.8% for a fault condition, as shown in Table 3. The accuracy of CNNs' increases as the dataset grows. Table 3 demonstrates that the normal and fault detection accuracy is 87.1%

Table 2 Description of dataset for ResNet-50 model

Dataset	Normal	Fault
Training	38	38
Testing	87	87
Total	250	

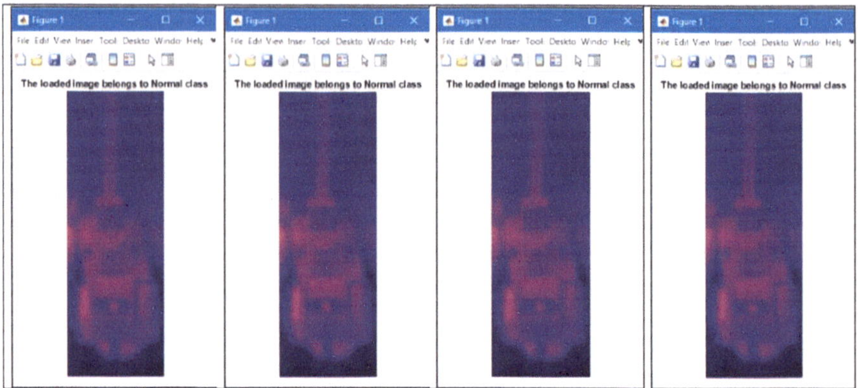

a) ResNet-50 detecting normal condition at100 thermal images
b) ResNet-50 detecting normal condition at 150 thermal images
c) ResNet-50 detecting normal condition at 200 thermal images
d) ResNet-50 detecting normal condition at 250 thermal images

Fig. 6 **(a–d)** Illustrate the ResNet-50 detection for a normal condition at various number of thermal images

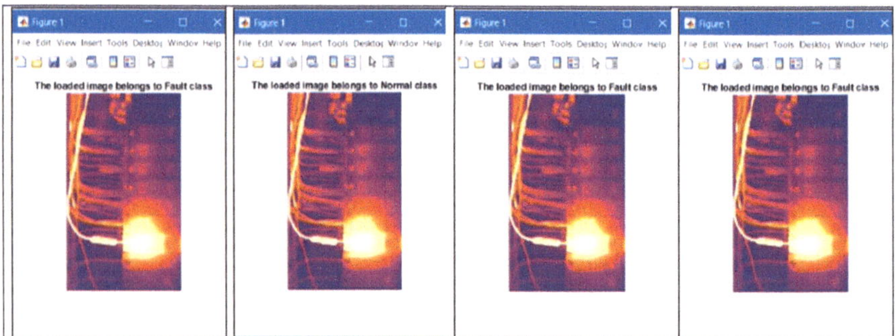

a) ResNet-50 detecting normal condition at 100 thermal images
b) ResNet-50 detecting normal condition at 150 thermal images
c) ResNet-50 detecting normal condition at 200 thermal images
d) ResNet-50 detecting normal condition at 250 thermal images

Fig. 7 **(a–d)** Illustrate the ResNet-50 detection for a fault condition at various number of thermal images

Table 3 Total data of thermal images affecting the accuracy of CNN

ResNet-50		
	Accuracy (%)	
Total data of thermal images	Normal	Fault
100	87.1	87.1
150	83.7	91.3
200	86.4	77.7
250	83.3	82.8

when the total data is 100. The accuracy of normal detection at 150 total data is 83.7%. Fault detection, on the other hand, is 91.3%.

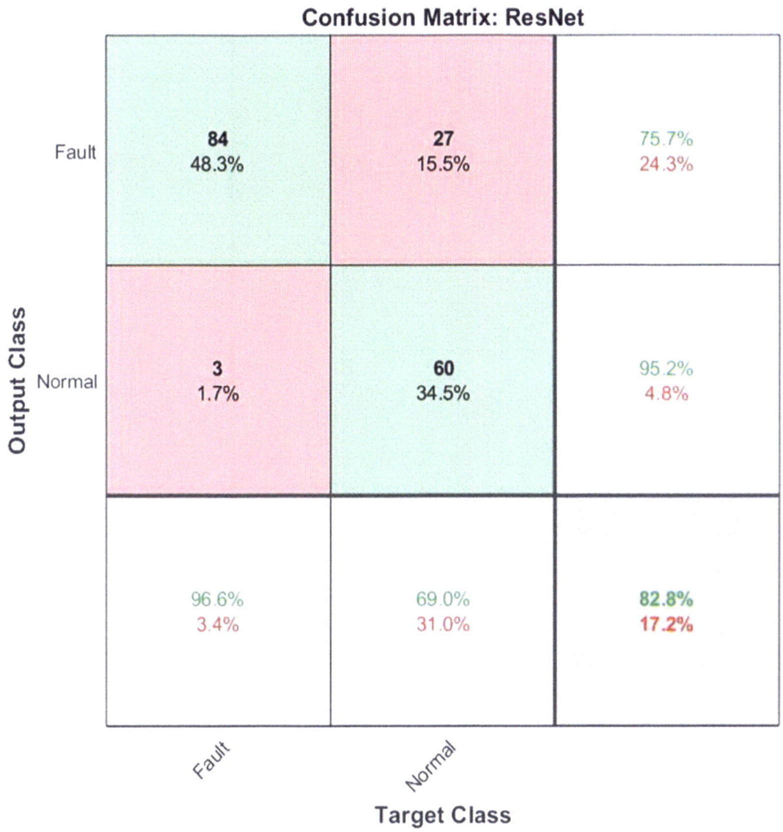

Fig. 8 Confusion Matrix prediction of accuracy for Fault condition detection

However, it cannot detect the fault condition due to the underfitting of the dataset. Nonetheless, at total data 200, the C can detect normal and fault conditions even though the accuracy for normal and fault condition detection is 86.4% and 77.7%, respectively. Finally, at the target data of 250 images, the accuracy detection is at 83.3% and 82.8% for normal and fault conditions, respectively.

Testing was performed on 70% of dataset and average accuracy and loss at random runs for fault condition are observed to be 82.8% and 17.2% respectively. The confusion matrix is presented in Fig. 8 which portrays the detection accuracy. To compute accuracy from a confusion matrix, firstly compute the four parameters TP, FP, TN, FN from it with respect to a certain class.

The confusion matrix in Fig. 9 represented an analysis of normal condition detection. TP (number of normal conditions correctly classified as normal) is found to be 75. FP (number of irrelevant instances mistakenly classified as normal) is found to be 17. TN (number of irrelevant instances correctly classified not belonging to normal) is found as 70. FN (number of normal instances mistakenly classified as

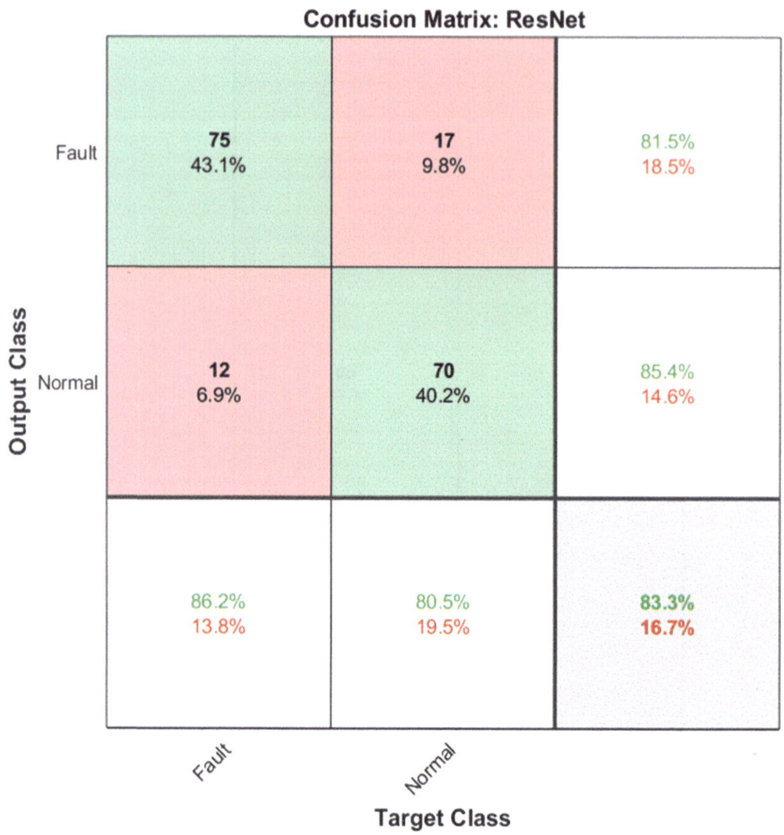

Fig. 9 Confusion Matrix prediction of accuracy for Normal condition detection

other conditions, which in this case fault) is found as 12. Based on the discussion, accuracy for the normal condition is 0.833. The values for accuracy are presented in Table 2 based on averaging the results for a total of 250 thermal images [7].

4.2 Testing Performance of Improved ResNet-50

Linear regression is one of the statistical methods that is used for predictive analysis. Linear regression makes predictions for continuous/real or numeric variables. Linear regression algorithm shows a linear relationship between a dependent (y) and one or more independent (y) variables, hence called as linear regression. Since linear regression shows the linear relationship, which means it finds how the value of the dependent variable is changing according to the value of the independent variable. In this study, a simple linear regression is used to make a comparison

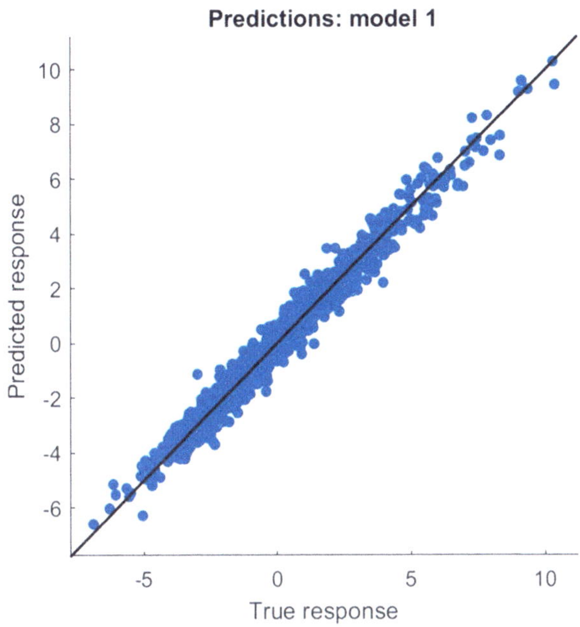

Fig. 10 Linear regression for the thermal image's detection of Fault conditions

for ResNet-50 model in normal and fault condition between the total numbers of thermal images, 100, 150, 200 and 250. By doing so it can validate the ResNet-50 to detect the thermal images for normal and fault condition.

Figure 10 shows the Linear Regression for thermal pictures produced by the network, one with a fault condition. Linear regression is sufficient because the model identifies the best-fitting linear line between the independent and dependent variables. Take note that the regression applies to both fault and normal detection. The performance of proposed ResNet-50 has been evaluated for 250 thermals images and the model is able to forecast detection since the regression is virtually linear to the linear line where all the probabilities of correct and wrong classification are presented in the confusion matrix of Figs. 8 and 9. Considering the small size of the dataset utilized, the overall model performance is adequate.

Figure 10 shows the linear regression in which the regression points is much closer to the linear line which is prediction line compared to Fig. 11. This indicate that the accuracy of normal condition with 250 thermal images is higher than fault condition as refer to result in Table 2.

Figure 12 demonstrates comparison between 100, 150, 200 and 250 total number of thermal images. During normal condition the dots is closed to the regression line it is observed that the dots position is almost similar between 100, 150, 200 and 250 thermal images. This is because the accuracy is in the interval of average accuracy between 83% and 87%. Hence, this ResNet-50 is a good model and able to detect the thermal images in normal conditions with high accuracy.

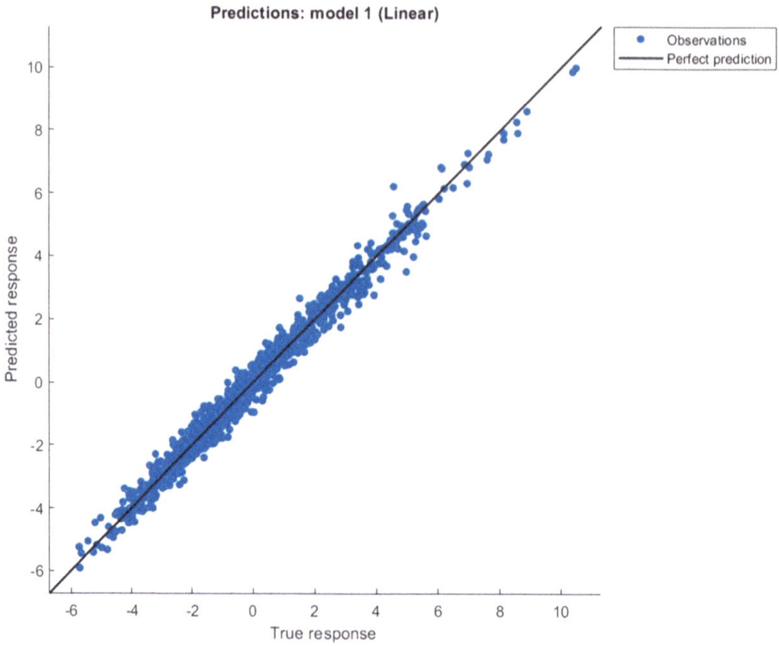

Fig. 11 Linear Regression for the thermal image's detection of normal conditions

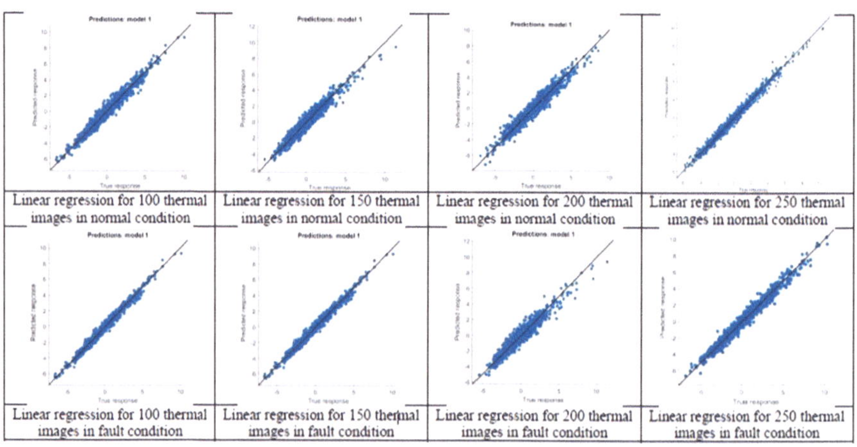

Fig. 12 Linear Regression comparison for normal and fault condition at different total number of thermal images

Figure 12 demonstrate comparison between 100, 150, 200 and 250 total number of thermal images. During normal condition the dots are close to the regression line it is observed that the position of the dot is almost similar between 100, 150, 200, and 250 thermal images. This is because the accuracy is in the interval of average accuracy between 83% and 87%. Hence, this ResNet-50 is a good model and able to detect the thermal images in normal conditions with high accuracy.

Meanwhile, the best average accuracy was achieved during a fault condition at 100 and 150 thermal images. But at a total of 150 thermal images, the ResNet-50 was unable to recognize the thermal image due to underfitting in the amount of training data. The lowest accuracy that can be seen with linear regression is at 200 thermal images. ResNet-50, however, is able to identify the thermal image at the fault condition from the regression line.

5 Conclusion

Fault detection and analysis for power distribution are critical methods to improve power system security and dependability. Until now, the fault statistics of power distribution have varied due to the high fault rate, but they have been manually documented or recorded. Based on the strong performance of the convolution neural network, this research applies it to the identification of power faults using thermal images. A power fault detection technique is provided, which can efficiently identify faults in electrical equipment such as power distribution. Furthermore, a pre-trained model CNN and picture augmentation are introduced in this research to improve detection. In the simulation experiment, thermal images are preprocessed, and the result of image processing, including grayscale and resize, are analyzed. A linear classifier is then used to detect the feature information that CNN has retrieved together with the thermal images, and the detection performance is assessed. Using linear for detection, the accuracy rate on detecting thermal pictures in two conditions, normal and fault, was 83.3% and 82.8%, respectively, at the required thermal images out of a total of 250 thermal images. Finally, this study shows that the suggested CNN-based detection technique is highly significant with 75% classification accuracy between normal and fault conditions. It is recommended that future research on various types of faults be conducted to produce an efficient and mode generalized model.

References

1. Han J, Miao S, Li Y, Yang W, Yin H (2021) Fault diagnosis of power systems using visualized similarity images and improved convolution neural networks. IEEE Syst J. https://doi.org/10.1109/JSYST.2021.3056536
2. Haider M, Doegar A, Verma RK (2018) Fault identification in electrical equipment using thermal image processing. In: 2018 International conference on computing, power and communication technologies (GUCON), pp 853–858. https://doi.org/10.1109/GUCON.2018.8675
3. Hurrah NN, Loan NA, Parah SA, Sheikh JA, Muhammad K, de Macedo ARL, de Albuquerque VHC (2021) INDFORG: industrial forgery detection using automatic rotation angle detection and correction. IEEE Trans Industr Inform 17:3630–3639

4. Parah SA, Sheikh JA, Ahad F, Bhat GM (2017) High capacity and secure electronic patient record (EPR) embedding in color images for IoT driven healthcare systems. In: Internet of things and big data analysis toward next-generation intelligence. Springer Link, pp 409–437
5. Mansour RF, Parah SA (2021) Reversible data hiding for electronic patient information security for telemedicine applications. Arab J Sci Eng 46:9129–9144
6. Olivatti Y, Penteado C, Aquino PT, Maia RF (2018) Analysis of artificial intelligence techniques applied to thermographic inspection for automatic detection of electrical problems. In: 2018 IEEE international smart cities conference (ISC2). https://doi.org/10.1109/ISC2.2018.8656724
7. Maduako I et al (2021) Deep learning for component fault detection in electricity transmission lines. https://doi.org/10.21203/rs.3.rs-1028973/v1
8. Caetano DG et al (2018) Design and implementation of an automatic vehicle for thermographic inspections in. In: 3rd International conference on computational intelligence and applications, ICCIA 2018, pp 145–150. https://doi.org/10.1109/ICCIA.2018.00034
9. Woolery B (2021) Infrared thermal imaging-based thermal diffusivity measurement using angstrom method. University of Oklahoma
10. Shao H et al (2020) Intelligent fault diagnosis of rotor-bearing system under varying working conditions with modified transfer convolutional neural network and thermal images. IEEE Trans Ind 17:3488–3496
11. Liu CH, Qi Y, Ding WR (2017) Infrared and visible image fusion method based on saliency detection in sparse domain. Infrared Phys Technol 83:94–102
12. Li H, Li X, Yu Z, Mao C (2016) Multifocus image fusion by combining with mixed-order structure tensors and multiscale neighborhood. Inf Sci 349–350:25–49
13. Naik VV, Gharge S (2016) Satellite image resolution enhancement using DTCWT and DTCWT based fusion. In: Proceedings of the international conference on advances in computing, Jaipur, pp 1957–1962
14. Palimkar NH (2016) Fault prediction in electrical equipment using thermographic inspection. Int J Eng Res Technol 5:685–687
15. Hurrah NN, Parah SA, Sheikh JA (2020) Embedding in medical images: an efficient scheme for authentication and tamper localization. Multimed Tools Appl 79:21441–21470
16. Bhat GM, Mustafa M, Ahmad S, Ahmad J (2009) VHDL modeling and simulation of data scrambler and descrambler for secure data communication. Indian J Sci Technol 2(10):41–44
17. Kazim M, Khawaja AH, Zabit U, Huang Q (2020) Fault detection and localization for overhead 11-kV distribution lines with magnetic measurements. IEEE Trans Instrum Meas 69(5):2028–2038. https://doi.org/10.1109/TIM.2019.2920184
18. Zhao M, Barati M (2021) A real-time fault localization in power distribution grid for wildfire detection through deep convolutional neural networks. IEEE Trans Ind Appl 57(4):4316–4326. https://doi.org/10.1109/TIA.2021.3083645
19. Alashter MA, Mrehel OG, Shamekh AS (2020) Design and evaluation a distance relay model based on artificial neural networks (ANN). In: 6th IEEE international energy conference (ENERGYCON). IEEE, Piscataway
20. Sun K, Chen Q, Gao Z (2016) An automatic faulted line section location method for electric power distribution systems based on multisource information. IEEE Trans Power Deliv 31(4):1542–1551. https://doi.org/10.1109/TPWRD.2015.2473681
21. Zeng F, Zhang Q, Wang J, Chen Y, Qi G, Zhu Z (2018) Morphology-based visible-infrared image fusion framework for smart city. Int J Simul Process Model 13(6):523. https://doi.org/10.1504/IJSPM.2018.10016917
22. Aziz F, Ul Haq A, Ahmad S, Mahmoud Y, Jalal M, Ali U (2020) A novel convolutional neural network-based approach for fault classification in photovoltaic arrays. IEEE Access 8:41889–41904. https://doi.org/10.1109/ACCESS.2020.2977116
23. Fahim SR, Sarker SK, Muyeen SM, Das SK, Kamwa I (2021) A deep learning based intelligent approach in detection and classification of transmission line faults. Int J Electr Power Energy Syst 133. https://doi.org/10.1016/j.ijepes.2021.107102

24. Ren H, Hou ZJ, Vyakaranam B, Wang H, Etingov P (2020) Power system event classification and localization using a convolutional neural network. Front Energy Res 8. https://doi.org/10.3389/fenrg.2020.607826
25. Tao X, Zhang D, Wang Z, Liu X, Zhang H, Xu D (2020) Detection of power line insulator defects using aerial images analyzed with convolutional neural networks. IEEE Trans Syst Man Cybern: Syst 50(4):1486–1498. https://doi.org/10.1109/TSMC.2018.2871750
26. Gu H, Wang Y, Hong S, Gui G (2019) Blind channel identification aided generalized automatic modulation recognition based on deep learning. IEEE Access 7:110722–110729
27. Mukherjee A, Kundu PK, Das A (2021) Transmission line faults in power system and the different algorithms for identification, classification and localization: a brief review of methods. J Inst Eng India Ser B 102:855–877. https://doi.org/10.1007/s40031-020-00530-0
28. Siriwithtayathanakun P, Sriyanyong P (2018) Effect of faults on electrical equipment in power substation: a case study of metropolitan electricity authority's power system. In: 2018 15th International conference on electrical engineering/electronics, computer, telecommunications and information technology (ECTI-CON), pp 176–179. https://doi.org/10.1109/ECTICon.2018.8619878
29. Wei H et al (2019) A novel precise decomposition method for infrared and visible image fusion. In: 2019 Chinese Control Conference (CCC), pp 3341–3345. https://doi.org/10.23919/ChiCC.2019.8865921
30. Gowrishankar M, Nagaveni P, Balakrishnan P (2016) Transmission line fault detection and classification using discrete wavelet transform and artificial neural network. Middle-East J Sci Res 24:1112–1121
31. Sowah RA et al (2018) Design of power distribution network fault data collector for fault detection, location and classification using machine learning. In: 2018 IEEE 7th international conference on adaptive science & technology (ICAST), pp 1–8. https://doi.org/10.1109/ICASTECH.2018.8506774
32. Huang X, Shang E, Xue J, Ding H, Li P (2020) A multi-feature fusion-based deep learning for insulator image identification and fault detection. In: 2020 IEEE 4th information technology, networking, electronic and automation control conference (ITNEC), pp 1957–1960. https://doi.org/10.1109/ITNEC48623.2020.9085037
33. TICOR (2012) Thermal imaging report. Thermal Imaging Ltd
34. Prabhu (2018) Understanding of convolutional neural network (CNN) – deep learning. 4 Mar 2018
35. Zahisham Z, Lee CP, Lim KM (2020) Food recognition with ResNet-50. In: 2020 IEEE 2nd international conference on artificial intelligence in engineering and technology (IICAIET), pp 1–5. https://doi.org/10.1109/IICAIET49801.2020.9257825

Text to Speech Synthesis Using Deep Learning

Rabbia Mahum, Aun Irtaza, and Ali Javed

1 Introduction

The best ways to communicate among human beings are speaking and writing. They communicate through natural languages, and the written form of language is called text. The vocal communication is done using speech that is easy to understand for the listeners. Speech or natural language processing techniques aim to develop a system that can analyze, generate, and understand human languages. Before the computer systems were able to understand only machine languages, however, with the advancement in machine learning and deep learning algorithms, the systems have become so powerful that they can process the statement in written or spoken form such as Google Search Engine [1].

Advancements in Internet of Things (IOTs) have made ways for the researchers to work for development of smart systems that aid people suffering from different health issues. Various health-care applications have been developed [2–5] to aid the smart technologies. One of the most important applications is artificial speech synthesis based-on various machine learning [6–8] techniques.

The computational models for human language processing utilize the human cognition in machines i.e. how humans store, and process the text. Due to enhancement in models, computer systems have become able to speak just like humans [9]. The artificial intelligent systems generate waveforms from an input text to synthesize speech. The input can be in any natural language form and the system may comprise of several phases for the conversion from text to speech. There exist various

R. Mahum (✉) · A. Irtaza
Computer Science Department, University of Engineering and Technology, Taxila, Pakistan
e-mail: rabbia.mahum@uettaxila.edu.pk; aun.irtaza@uettaxila.edu.pk

A. Javed
Software Engineering Department, University of Engineering and Technology, Taxila, Pakistan
e-mail: ali.javed@uettaxila.edu.pk

S. A. Parah et al. (eds.), *Intelligent Multimedia Signal Processing for Smart Ecosystems*, https://doi.org/10.1007/978-3-031-34873-0_12

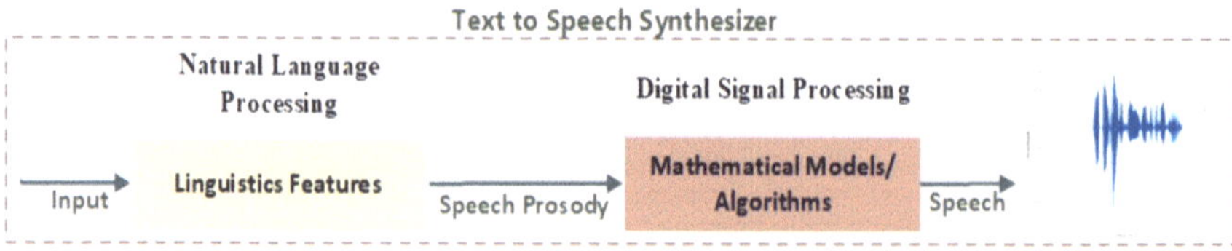

Fig. 1 The basic architecture of TTS synthesizer

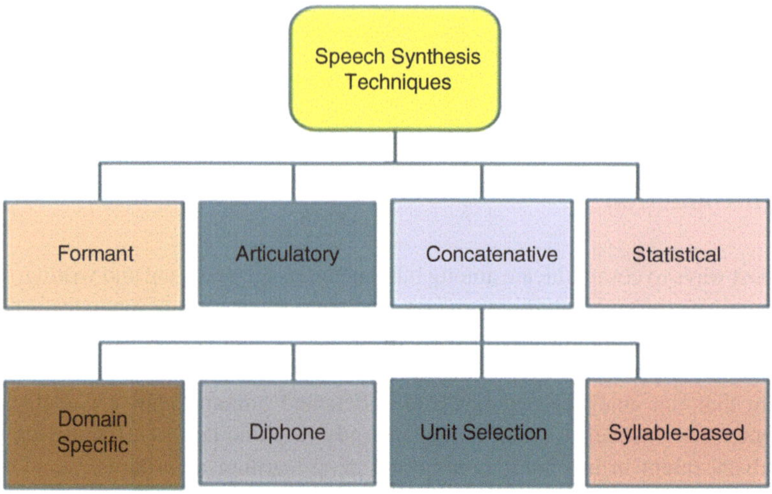

Fig. 2 Types of speech synthesis

applications of TTS synthesizers such as people with speaking difficulties are able to communicate better with those listeners who may not be able to understand sign language. Moreover, in educational background, TTS synthesis based systems can teach 365 days a year without any break. Furthermore, the applications can be programmed for any specific task i.e. for spell correction and pronunciation teaching. TTS synthesizers may be installed in security systems such as in warning or alarm systems, synthesized speech can give accurate information. Besides the positive uses of TTS synthesis systems, one drawback of these systems is to get used by hackers. The hackers may generate spoofed speech to fool the Automatic Spoofing Verification systems (ASV), which create a big threat for the community.

The basic architecture of TTS synthesizer is shown in Fig. 1.

There exist various types of speech synthesis methods as shown in Fig. 2.

The remaining chapter is organized as: Sect. 2 demonstrates some existing techniques for TTS synthesis and detection, Sect. 3 demonstrates the applications of TTS synthesizers, Sect. 4 refers to our proposed TTS synthesizer based on reinforcement block using encoder-vocoder. Furthermore, Sect. 5 demonstrates some experiments and in the end, conclusion of the chapter is presented in Sect. 6.

2 TTS Synthesis and Detection

There exist three types of modalities i.e., replay attack (RA), TTS, and voice cloning (VC) to synthesize speech. TTS and VC comprise of regenerated content and more similar to natural speech than RA. In ASV spoof 2019 competition, logical access (LA) and physical access (PA) tasks were introduced for synthesized speech, and RA for the development of ASV systems respectively. Various researchers have proposed different approaches [10–12] for the spoofing detection including all modalities. Some algorithms exist based on machine learning techniques to discern the audios based on data driven and knowledge focused countermeasures [13]. However, in traditional machine learning algorithms, hand crafted features extraction is performed which is time consuming task, moreover they may ignore the deep features underlying the audios spectrograms [14]. With the improvement in domain of convolutional neural networks (CNNs), some methods have been proposed based on deep layers such as in [15], an end to end algorithm employing raw waveforms as input is developed. Chintha et al. proposed a recurrent CNN structure to detect spoofed (fake) audios [16]. Moreover, a lightweight CNN has been employed by [17], namely LCNN utilizing softmax loss function to detect anti spoofed attacks. Furthermore, various combinations of detecting systems have been tested along with ResNet [18], and explored with other classifiers as well for better performance [19, 20] in spoofed speech detection. In [21], a model was employed based on an end-to-end ensemble method to learn the fusions of various detection systems. Even though, the performance of these proposed algorithms was satisfactory, however there exist an issue of generalization for unseen attacks in the models, and requirement of high computational resources such as time and memory, therefore it is required to introduce an efficient and robust system that can carry out detection of fake audios from any source.

From last two decades text to speech systems have become so powerful that are capable to generate a realistic voice after training of limited audio samples from target speakers [22]. Therefore, it is a huge threat for ASV systems as they may be attacked by the naturalness of the speech generated [23]. The applications that can protect the ASV systems from the spoofed audio attacks are called deepfake speech detectors. Thus, various machine learning and deep learning based works have been proposed in various domains [6–8] including the detection of forged speech. In [24], a SVM based classifier has been utilized as ASV employing Gaussian Mixture Model (GMM). They attained equal error rate (EER) as 4.92%, and 7.78% on the 2006 NIST for speaker identification core test. The authors have proposed the GMM, and a Relative Phase shift with Support Vector Machine (SVM) for the synthetic speech detection to minimize the weaknesses of speaker verification systems. Moreover, a detailed comparison of Hidden Markov Model (HMM), and DNN has been performed for the detection of spoofed speech [25]. In [26], proposed model employs the spectrograms in image form as input to CNN, thus forming a base of audio processing using images. In [27], various features descriptors have been used such as Mel Frequency Cepstral Coefficient (MFCC), spectrogram, etc.

and the effect of GMM-UBM on the accuracy has been analyzed. It is concluded that the combination of different feature descriptors gives better result in terms of EER. Chao et al. [24] utilized SVM to discern the real speeches from the fake recordings of the claimed man. Similarly, in [28] Chao has employed two core methods such as Kernel Fisher Discriminant (KFD) and SVM to verify speakers and attained better results as compared to their previous work based on GBM and UBM method. Moreover, to decline the computational cost of the polynomial kernel SVM by exchanging the dot product among two utterances with two i-vectors [29] was used. Furthermore, authors applied feature selection technique attaining 64% dimensionality reduction in features having EER of 1.7% [29]. Whereas, Loughran et al. [30] overcame the issue of imbalanced data (where the one class samples are greater than the other) utilizing Genetic algorithm (GA) with adjusted cost function. Malik et al. [31], developed a system for audio forgery detection based on acoustic signatures of environment by investigating the integrity of audio. However, these proposed models failed to address synthesized audio content with high precision.

3 Applications of the TTS Synthesis

The intelligent speech synthesizers have widespread domain of usage in development of human-machine interaction systems [32, 33]. Moreover, the systems based on TTS synthesizers are becoming more affordable to common people for daily use [34]. Some important applications of TTS synthesizers are given below.

3.1 Speaker

TTS synthesizers are widely used for the people who have speech difficulty due to disability, therefore they use speech synthesizers as a speaker to communicate. The implication of TTS synthesizers in small devices help them to improve their lifestyle without using sign language. Moreover, the English language is utilized as a medium in the most of these devices by service providers.

3.2 Screen Reader

The TTS synthesizers may be employed as screen readers for the people who have reading difficulties. Moreover, the people with visual impairment are not able to read screens from a specific distance, therefore TTS synthesizers work as an aid for these type of people. These screen readers are developed for various languages such as Urdu, Hindi, and English. They are very helpful for the people having dyslexia.

The TTS based screen reader may be used as listening any document while working, cooking or driving. The old age people can also use them if they have vision issues without any dependency. Various screen readers are available developed by different companies as web portals such as the IBM TTS (https://www.ibm.com/watson/services/text-to-speech/), and VoiceReader (https://www.linguatec.de/en/text-to-speech/voice-reader-home-15/). However, these screen recorders are mostly based on foreign languages, and lack Urdu language as a medium. Therefore, an effort is required to be done for Urdu screen recorder.

3.3 Language Teaching

The TTS synthesizers may be combined with a learning algorithm to develop helpful systems to learn new language by listening the words pronunciations. For example Capti-Voice is a TTS synthesizer system based on cloud platform supporting various foreign languages. However, there is still room for general language learning and teaching that may be able to teach and educate all existing natural languages.

3.4 Multimedia Applications and Telecommunication

The combination of speech synthesis and speech identification provide a significant user interface for mobile devices. The latest applications of TTS synthesizers include multimedia domain and internet search engines where a person is able to interact in his native language. The synthesized speeches are employed in call inquiry systems as well. Moreover, people who are unable to understand English language become able to understand the scenario in their native language using various TTS synthesized applications.

3.5 Entertainment

The synthesized speech is widely used in various games and talking robots or toys. Initially, the voice was not of good quality in talking calculators, however it has been improved in various 3D applications i.e. talking heads. Aiming to improve the learning process in kids, TTS synthesis can be incorporated in kid's toys easily. The IBM TTS [35] also provides the services for interactive toy development for kids. However, those toys mostly understand or speak English language which is a big challenge for illiterate people who are unable to understand English language. Therefore, the work is required for TTS synthesizers which can generate audios of different languages in toy industry.

3.6 Human-Machine Interaction

Various computer-machine interactive systems such as kiosks and automated tellers employ TTS synthesizers. Moreover, TTS systems may be used in alarms due to more accurate information delivery through TTS synthesis. The response due to warning speech is faster instead of warning light or signal from different room or place. Additionally, the activity log of printer devices can be heard on connected computer. However, all these applications mostly employ English language only.

3.7 Other Applications

Other than above-mentioned applications, the TTS synthesizers have been used in browsing, as reader in mails and SMS, dictionary, pronunciation, and PDF readers. Although, all of these applications use English language, however the work is yet to be done for other natural languages as well. The browser plug-ins and SMS reader application on android exist for Indian languages as well. During the last decade, the communication methods have been formed for 3 dimensional audio-visuals techniques. The vast applications of speech synthesis in different domains bring more funds and ideas. TTS synthesizers may also be employed in language interpreters, or talking mobile devices.

4 The Proposed Solution

Deep learning architectures are composed of various layers such as input, hidden, and classification layer. These hidden layers have various types i.e. convolutional, batch normalization, pooling, activation etc. The deep learning models extracts features utilizing various filters convolving over the input images. Moreover, when the filters are convolved over all the data then feature map is formed. These feature maps are reduced in dimensions employing pooling layers minimizing the computational power of the system. These feature maps can be fed again to convolutional layers repeating the above steps again. Numerous applications exist for various purposes such as facial feature recognition [36], speech identification [37], and emotion detection [38]. The basic architecture of deep network is shown in Fig. 3.

There exist various possibilities for flow of data in TTS synthesizers such as shown in Fig. 4.

Our proposed TTS synthesizer, consists of four different trained CNNs such as: (1) An encoder that generates a feature vector of speech from dataset [39], (2) A synthesizer that predicts a mel-spectrogram from grapheme sequences, conditioned on the vector from speaker embedding [34], (3) A Vocoder based on autoregressive

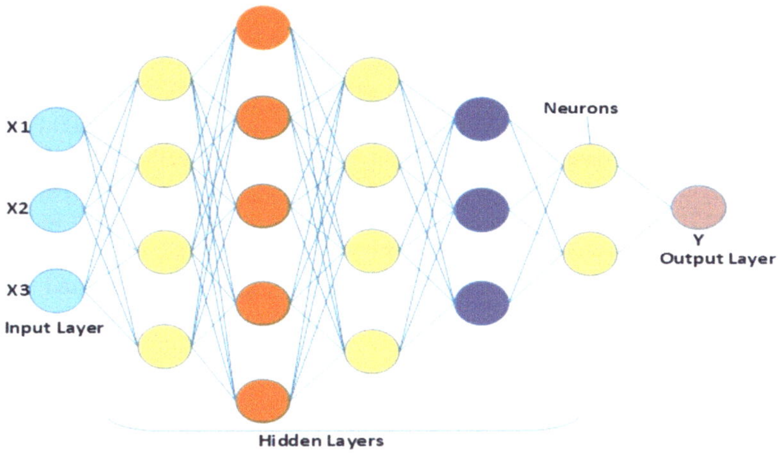

Fig. 3 The architecture of Deep network

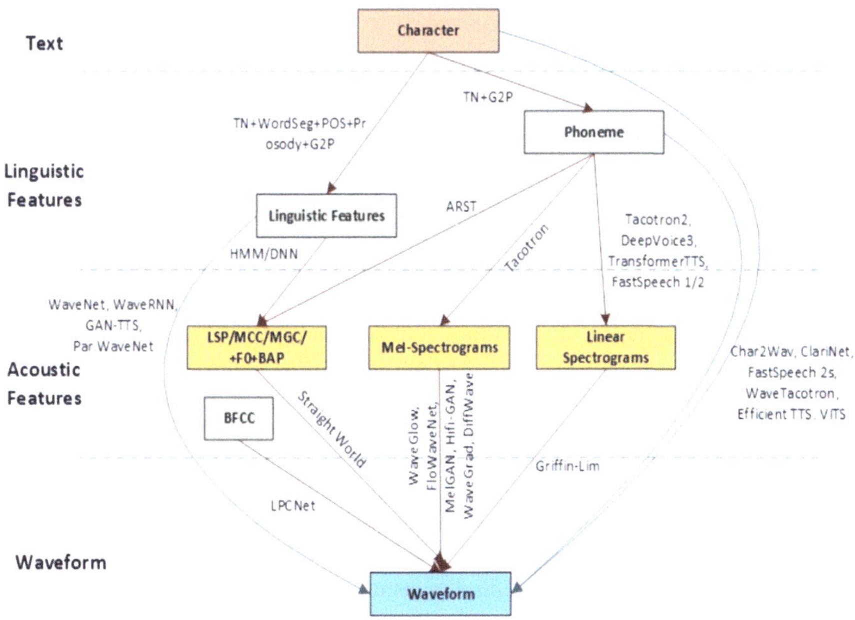

Fig. 4 A taxonomy of data flow from character to waveforms

WaveNet that generates the waveforms from spectrograms [40], (4) an additional RL block. The RL block consists of a neural network and make our proposed TTS model an agent which learns on the basis of reward. The accuracy of the network is computed and considered as a reward to give feedback to the main model comprising of Encoder, Synthesizer, and Vocoder.

4.1 Dataset

In the first phase of our proposed idea, we gathered data for English language from [41] i.e. LibriSpeech. The dataset comprises of 1000 hours having 16,000 Hz read English speech. The data was gathered from the LibriVox project's audio books. For French language based synthesis, we employed the SIWIS French speech synthesis database [42] having speech recordings of high quality along with text files. The database has been designed mainly for TTS synthesis systems. There exist 9750 utterances attained from numerous sources i.e. parliament speeches and speech talent by French professionals. The database is available freely, and comprised of 10 hours of speech in total. Moreover, for Chinese TTS synthesis, we employed Mandarin Chinese Speech Corpus [43]. It is comprised of 30 hours of annotated speeches. The speeches were collected using a single carbon microphone, and mostly participants were young speaking Mandarin with fluency.

4.2 The Proposed Encoder

The encoder is employed to form a condition for synthesizer block on a speech input by dataset. For the good generalization of the model, feature representations from various speakers are essential to capture without considering the background noise and phonetic content. These requirements are attained utilizing a speaker-discriminative technique that is trained employing a speaker verification task independent of text. We utilized [39], a very reliable neural network for the speaker verification. This network matches log mel-spectrograms sequentially attained from the speeches having an arbitrary length to an embedding vector of fixed dimensions, known as *d-vector* [44, 45]. The network's training is optimized based on generalized end-to-end loss for speaker verifications, so the same speaker's embedding exhibit the high similarity between cosine, whereas varying speaker utterances are far apart in embedding space. The training data utilized for the network comprises of speech that are segmented as 1.6 seconds audios having corresponding speaker identity labels. The mel-spectrograms of 40-channel are fed to the network comprising of a 3 LSTM layers with 768 cells, each individual is followed through a projection to 256 dimensions. The last embedding is produced using L_2-normalization of the output at top layer in the final frame. Moreover, through the inference, an utterance of arbitrary length is divided into 800 ms windows, overlapping by half (50%).

The network get trains for each window, and the outputs are combined to compute average, and normalized for final utterance embedding formation. However, the encoder network was not directly optimized during training that could extract features relevant to speaker identification.

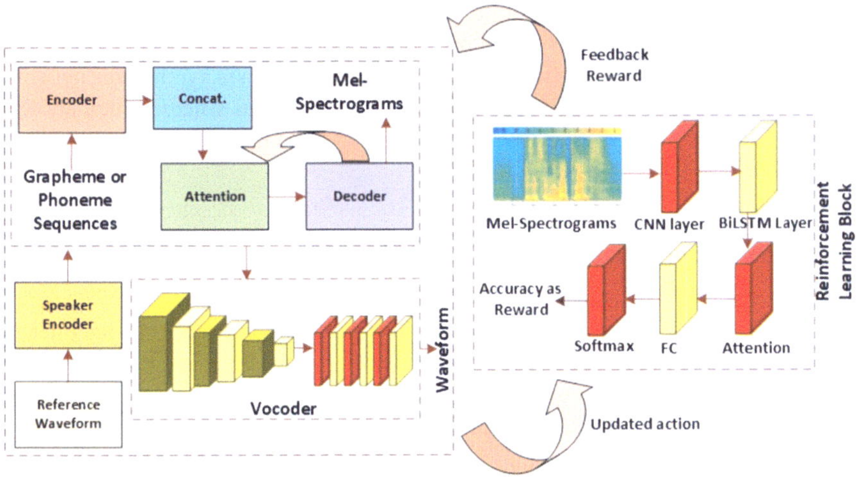

Fig. 5 The proposed model's architecture

4.3 The Proposed Synthesizer

In this section, we discuss the synthesizer's network which is based on Tacotron 2 architecture [34] using recurrent sequence to sequence model for supporting multiple speakers as described in [46]. At each time step, the synthesizer encoder's output is combined with an embedding vector. We analyzed that passing embedding to an attention layer as shown in Fig. 5, is better and network converges significantly across multiple speakers. We trained synthesizer over target audios and pairs of text transcripts. We mapped the text to phonemes at the input, which lead to improved pronunciation of difficult words and fast convergence. The proposed network is trained in a transfer learning manner, utilizing a pre-trained encoder (frozen parameters) to extract embedding from the target audios such as the target speech should be same to the speaker's reference signal. The features from target spectrograms are extracted from 50 ms windows along with 12.5 ms step, passed through 80 channel mel-scale filter bank followed by logs compression of dynamic range. We improved [34] by applying augmentation of the L_2 loss on the computed spectrogram with an additional L_1 loss. We concluded that the combined loss is more robust towards noisy training data. Moreover, we didn't employ an additional loss term for speaker embedding as in [47].

4.4 The Proposed Vocoder

We employed autoregressive WaveNet [40] vocoder sample-by-sample to transform the mel-spectrograms generated by synthesized network into time-domain waveforms. The architecture applied for the proposed vocoder is same as [34] comprising 30 dilated convolutional layers. The proposed network is not directly conditioned on the encoder's output. The mel-spectrograms generated via synthesizer network extract all of the details for high quality synthesis of various voices, which allow a vocoder to act as multi-speaker voice generator based on training data.

4.5 The Proposed RL Block

The RL block and our proposed model works together as an agent. The parameters of the proposed model are known as policy. Furthermore, the purpose of the policy is to predict the audio features at each instant. The features of speech describe the actions of agent. After the features prediction, the reward value is computed which gives feedback on accuracy. We employed policy gradient method for back-propagation that optimizes the model for maximum reward gain. As shown in Fig. 5, left block is TTS synthesizer based on Tacotron. Moreover, we utilized Griffin-Lim [48] technique for rebuilding the speech signals. Our proposed text editor is based on same structure as Tacortron2 that take input of text embedding and form output for encoder embedding. Whereas, the decoder based on attention module takes the output from encoder and embedding as input. Then, decoder predicts frame by frame speech features.

Reward The reward function is utilized for the computation of more natural speech that is based on a network as shown in Fig. 5. The network consists of a CNN layer, an attention layer, BiLSTM layer, a fully connected layer (FC) and softmax layer. The extracted features proved better efficiency for the natural speech synthesis using a neural network. The network takes the mel-spectrogram first as input, and then give output in the form of constant size latent output. Then, BiLSTM takes the temporal information and transform it into modified form. Whereas, an attention layer gets training for each frame's weight.

Let us suppose an input is x_i and the speech is $\grave{y}_i$ having speaker type as t_i, and the goal is to form a natural speech y_i having desired emotion l_{yi}. To update the model, we employed accuracy as the reward value. To form the end of sequence (EOS) token, the proposed RL network is employed to evaluate how accurate a mel-spectrogram feature y_i matches with the original speech label. The probability p_i for target speech with emotion l_{yi} of y_i is computed as below:

$$p_i = \mathrm{RL}\ (l_{yi} \mid \mathrm{yi}\ ;\ \vartheta), \tag{1}$$

Here, ϑ represents the parameters of RL network, and the value of probability varies from 0 to 1. The RL reward is computed as:

$$R = \frac{N}{K} = \frac{\sum_{i=1}^{K} 1\,(p_i > \Omega)}{K} \tag{2}$$

Here, R is reward of spectrum features y_i. The reward value also varies from 0 to 1. K refers the size of sample, and Ω is a threshold number that is set to 0.5. N represents the total samples.

5 Experimental Evaluation

The generated speeches from our proposed TTS synthesizer are evaluated using Mean Opinion Score (MOS) for similarity with real speech and naturalness, based on subjective listening tests. The MOS score varies according to Absolute Category Rating scale [49] from 1 to 5 such as Excellent, Good, Fair, Poor, and Bad respectively. In total, 100 participants were included to assign MOS for naturalness and similarity, separately from 1 to 5 including 80 males and 20 females. 100 audios for each language, ranging from 20 to 35 seconds were generated and reviewed by participants. Then, the average score was computed to evaluate the quality of the synthesized audios. We attained 4.67 average MOS for similarity and 4.70 for naturalness for synthesized speech through our proposed model that is a significant number. More particularly, we achieved best MOS for naturalness as 4.75 and MOS for similarity as 4.71 for Chinese language as shown in Table 2.

5.1 Environmental Setup

We performed the experiments using a GPU NVIDIA card i.e. GEFORCE GTX with 4 GB memory. The details of employed hardware are shown in Table 1. The operating system was Windows 10 having a RAM of 16 GB. The experiment was performed on the Matlab 2021a.

Table 1 System specifications for the employed model

Hardware	Specifications
Computer	GPU Server
CPU	Intel Core i5
RAM	16 GB
GPU	NVIDIA GEFORCE GTX $\times$ 4

Table 2 Results over French, Chinese, and English datasets

Dataset	MOS Similarity	MOS Naturalness
French	4.6	4.64
Chinese	4.71	4.75
English	4.7	4.72
Average	4.67	4.70

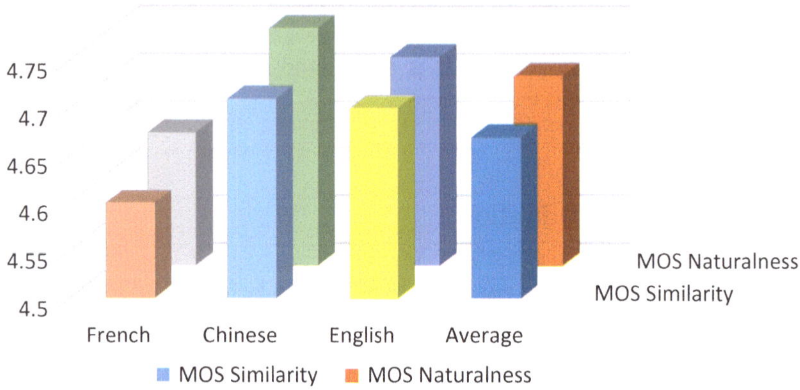

Fig. 6 The comparison plot for French, Chinese, and English Dataset

5.2 Assessment Over All Datasets

As mentioned in section of datasets, we have described the details of three datasets such as for English language: LibriSpeech, French language: SIWIS French speech synthesis database, and Chinese language: Mandarin Chinese Speech Corpus. In this section, we assess the individual results for our proposed TTS synthesizer as shown in Table 2. We employed the similar protocol for all three datasets i.e. average MOS for naturalness and similarity from group of 100 people. We achieved MOS similarity as 4.7 for English, 4.54 for Chinese, and 4.6 for French language dataset and average MOS similarity is 4.67. Whereas, MOS naturalness for French dataset is 4.64, for Chinese 4.75, and for English 4.72. Moreover, average MOS naturalness is 4.70. The comparison plot for all three datasets is shown in Fig. 6.

5.3 Time Complexity of Existing Techniques

In this section, we analyse various existing TTS synthesizers in terms of time complexity for training and inference time. In Table 3, T refers to the number of steps or iterations in the algorithm. Whereas, N represents the sequence length. Our proposed model attained $O(N)$ computational time as it is based on Tacotron 2 and inference time is also $O(N)$. Although, some algorithms take constant time such

Table 3 Comparative analysis of existing TTS models in context of time complexity

Model	Training	Inference
Tacotron 2	O(N)	O(N)
Deep Voice 3	O(1)	O(N)
FastSpeech 2	O(1)	O(1)
WaveGlow	O(T)	O(T)
WaveNet	O(1)	O(1)
Flowtron	O(1)	O(1)
Our proposed TTS	O(N)	O(N)

Table 4 Comparison with existing TTS synthesizers

Model	MOS	Inference Time(ms)
Tacotron 2	4.46	538
Fast Speech 2	3.83	374
VARA-TTS	3.88	33
BVAE-TTS	4.1	19.1
Our proposed TTS	4.67	13.5

as *O(1)*, however the resulting MOS is not considerable for these techniques as reported in Table 4. The details of complexities for training and inference time are reported in Table 3.

5.4 Comparison with Existing Techniques

In this section, we perform an experiment to compare our proposed TTS synthesizers with some existing models such as Tacotron2 [50], FastSpeech 2 [51], VARA-TTS [52], and BVAE-TTS [53] based on MOS and Inference Time. We utilized LibriSpeech dataset for assessment of all the mentioned systems in Table 4. It is exhibited from the table that MOS for Tacotron 2 is 4.46 that is maximum after our proposed TTS synthesizer. However, inference time for Tacotron 2 is maximum than other existing models i.e. 538 ms. Furthermore, Fast Speech 2 achieved 3.83 MOS, VARA-TTS attained 3.88 MOS, and BVAE-TTS obtained 4.1 MOS that is significant than former models. Moreover, the minimum inference time is attained by our proposed TTS synthesizer i.e. 13.5 ms. The Fast Speech 2 utilized 374 ms, VARA-TTS used 33 ms, and BVAE-TTS produced speech in 19.1 ms. Therefore, we believe that our proposed TTS synthesizer, achieved maximum MOS than existing models exhibiting significant naturalness. A comparison plot with existing TTS synthesizers is shown in Fig. 7.

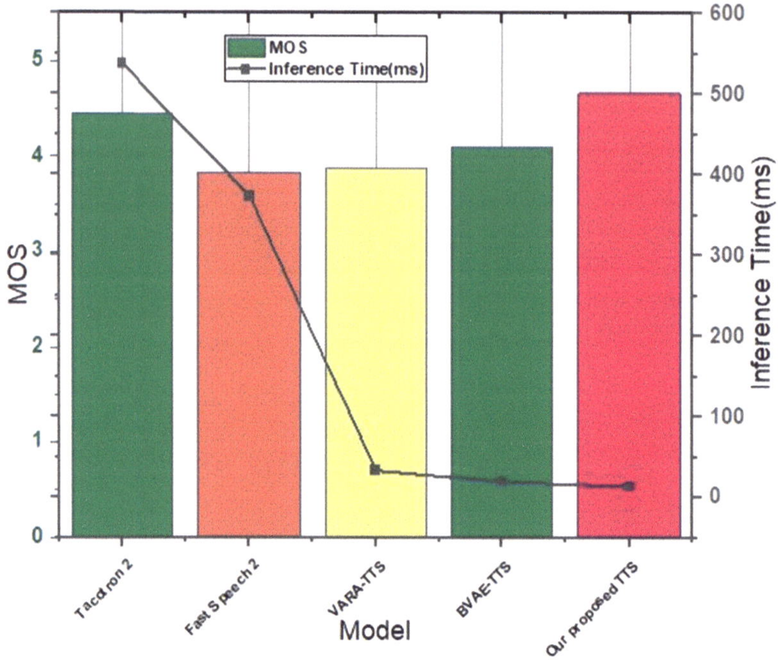

Fig. 7 A comparison plot of proposed model with existing TTS synthesizers

6 Conclusion

In this chapter, we have presented an efficient TTS synthesizer employing an encoder, synthesizer, decoder, and an additional reinforcement learning-based network. We utilized an existing structure of Tacotron2 for the synthesis of speech from text. More specifically, our proposed TTS synthesizer is comprised of three main stages such as (1) An encoder that generates a feature vector of speech from the dataset, (2) A synthesizer that predicts a Mel-spectrogram from grapheme sequences, conditioned on the vector from speaker embedding, and (3) A Vocoder based on autoregressive WaveNet that generates the waveforms from spectrograms. Furthermore, the RL block is attached consisting of CNN, BiLSTM, attention, FC, and softmax layer. Then, the accuracy of the RL network is considered as a reward value for the training. The proposed TTS synthesizer is based on a reference audio signal being fed to the encoder. Additionally, the quality of the synthesized speech varies according to the length, therefore short utterances provide better naturalness. We assessed our proposed algorithm using MOS, which vary from 1 to 5 such as Excellent, Good, Fair, Poor, and Bad respectively. In total, 100 participants were assigned the task to listen the audio and assign MOS to a generated speech by our proposed synthesizer. We attained 4.67 MOS overall which is a significant number than achieved from existing TTS synthesizers.

Although, our proposed synthesizer is significant for three languages i.e. English, Chinese, and French, however, we aim in the future to make it more generalized TTS synthesizer which can generate maximum natural languages exhibiting naturalness and similarity to human speech.

Acknowledgement The researchers want to thank University of Engineering and Technology Taxila to provide research environment.

References

1. Mishra R, Tripathi SP (2021) Deep learning based search engine for biomedical images using convolutional neural networks. Multimed Tools Appl 80(10):15057–15065
2. Parah SA, Sheikh JA, Ahad F, Bhat GM (2018) High capacity and secure electronic patient record (EPR) embedding in color images for IoT driven healthcare systems. In: Internet of things and big data analytics toward next-generation intelligence. Springer, Cham, pp 409–437
3. Hurrah NN, Parah SA, Sheikh JA (2020) Embedding in medical images: an efficient scheme for authentication and tamper localization. Multimed Tools Appl 79:21441–21470
4. Sarosh P, Heidari AA, Muhammad K (2021) Secret sharing-based personal health records management for the internet of health things. Sustain Cities Soc 74:103129
5. Ahad F, Bhat GM (2015) On the realization of robust watermarking system for medical images. In: 2015 Annual IEEE India conference (INDICON), New Delhi, pp 1–5. https://doi.org/10.1109/INDICON.2015.7443363
6. Mahum R et al (2022) A novel framework for potato leaf disease detection using an efficient deep learning model. Hum Ecol Risk Assess: Int J 29:1–24
7. Mahum R et al (2021) A novel hybrid approach based on deep CNN features to detect knee osteoarthritis. Sensors 21(18):6189
8. Mahum R et al (2021) A novel hybrid approach based on deep CNN to detect glaucoma using fundus imaging. Electronics 11(1):26
9. Korzekwa D et al (2022) Computer-assisted pronunciation training—speech synthesis is almost all you need. Speech Comm 142:22–33
10. Korshunov P et al (2016) Overview of BTAS 2016 speaker anti-spoofing competition. In: 2016 IEEE 8th international conference on biometrics theory, applications and systems (BTAS). IEEE, New York
11. Wu H et al (2020) Defense against adversarial attacks on spoofing countermeasures of ASV. In: ICASSP 2020-2020 IEEE international conference on acoustics, speech and signal processing (ICASSP). IEEE, Piscataway
12. Wu D (2019) An audio classification approach based on machine learning. In: 2019 International conference on intelligent transportation, big data & smart city (ICITBS). IEEE, Los Alamitos
13. Todisco M et al (2019) ASVspoof 2019: future horizons in spoofed and fake audio detection. arXiv preprint arXiv:1904.05441
14. Dinkel H, Qian Y, Yu K (2018) Investigating raw wave deep neural networks for end-to-end speaker spoofing detection. IEEE/ACM Trans Audio Speech Lang Process 26(11):2002–2014
15. Chintha A et al (2020) Recurrent convolutional structures for audio spoof and video deepfake detection. IEEE J Sel Top Signal Process 14(5):1024–1037
16. Lavrentyeva G et al (2019) STC antispoofing systems for the ASVspoof2019 challenge. arXiv preprint arXiv:1904.05576
17. He K et al (2016) Identity mappings in deep residual networks. In: European conference on computer vision. Springer, Berlin

18. Alzantot M, Wang Z, Srivastava MB (2019) Deep residual neural networks for audio spoofing detection. arXiv preprint arXiv:1907.00501
19. Lai C-I et al (2019) ASSERT: anti-spoofing with squeeze-excitation and residual networks. arXiv preprint arXiv:1904.01120
20. Monteiro J, Alam J, Falk TH (2020) An ensemble based approach for generalized detection of spoofing attacks to automatic speaker recognizers. In: ICASSP 2020-2020 IEEE international conference on acoustics, speech and signal processing (ICASSP). IEEE, Piscataway
21. Verma NK et al (2015) Intelligent condition based monitoring using acoustic signals for air compressors. IEEE Trans Reliab 65(1):291–309
22. Wu Z et al (2015) Spoofing and countermeasures for speaker verification: a survey. Speech Commun 66:130–153
23. Wu Z et al (2016) Anti-spoofing for text-independent speaker verification: an initial database, comparison of countermeasures, and human performance. IEEE/ACM Trans Audio Speech Lang Process 24(4):768–783
24. Chao Y-H et al (2008) Using kernel discriminant analysis to improve the characterization of the alternative hypothesis for speaker verification. IEEE Trans Audio Speech Lang Process 16(8):1675–1684
25. Ze H, Senior A, Schuster M (2013) Statistical parametric speech synthesis using deep neural networks. In: 2013 IEEE international conference on acoustics, speech and signal processing. IEEE, Piscataway
26. Dörfler M, Bammer R, Grill T (2017) Inside the spectrogram: convolutional neural networks in audio processing. In: 2017 international conference on sampling theory and applications (SampTA). IEEE, Piscataway
27. Balamurali B et al (2019) Toward robust audio spoofing detection: a detailed comparison of traditional and learned features. IEEE Access 7:84229–84241
28. Chao Y-H (2014) Using LR-based discriminant kernel methods with applications to speaker verification. Speech Comm 57:76–86
29. Yaman S, Pelecanos J (2013) Using polynomial kernel support vector machines for speaker verification. IEEE Signal Processing Lett 20(9):901–904
30. Loughran R et al (2017) Feature selection for speaker verification using genetic programming. Evol Intel 10(1):1–21
31. Zhao H, Malik H (2013) Audio recording location identification using acoustic environment signature. IEEE Trans Inf Forensics Secur 8(11):1746–1759
32. Handley Z (2009) Is text-to-speech synthesis ready for use in computer-assisted language learning? Speech Comm 51(10):906–919
33. McCoy KF et al (2013) Speech and language processing as assistive technologies. Comput Speech Lang 27(6):1143–1146
34. Shen J et al (2018) Natural TTS synthesis by conditioning wavenet on MEL spectrogram predictions. In: 2018 IEEE international conference on acoustics, speech and signal processing (ICASSP). IEEE, Piscataway
35. Alghoul A et al (2018) Email classification using artificial neural network. Int J Acad Dev 2(11):8–14
36. Yang S et al (2015) From facial parts responses to face detection: a deep learning approach. In: Proceedings of the IEEE international conference on computer vision. IEEE
37. Dhamyal H et al (2021) Fake audio detection in resource-constrained settings using microfeatures. Proc Interspeech 2021:4149–4153
38. Ng H-W et al (2015) Deep learning for emotion recognition on small datasets using transfer learning. In: Proceedings of the 2015 ACM on international conference on multimodal interaction. ACM
39. Wan L et al (2018) Generalized end-to-end loss for speaker verification. In: 2018 IEEE International conference on acoustics, speech and signal processing (ICASSP). IEEE, Piscataway
40. Oord AVD et al (2016) Wavenet: a generative model for raw audio. arXiv preprint arXiv:1609.03499

41. Panayotov V et al (2015) Librispeech: an ASR corpus based on public domain audio books. In: 2015 IEEE international conference on acoustics, speech and signal processing (ICASSP). IEEE, Piscataway
42. Honnet P-E et al (2017) The SIWIS French speech synthesis database? Design and recording of a high quality French database for speech synthesis. Idiap
43. Wang D, Zhang X (2015) Thchs-30: a free chinese speech corpus. arXiv preprint arXiv:1512.01882
44. Variani E et al (2014) Deep neural networks for small footprint text-dependent speaker verification. In: 2014 IEEE international conference on acoustics, speech and signal processing (ICASSP). IEEE, Piscataway
45. Heigold G et al (2016) End-to-end text-dependent speaker verification. In: 2016 IEEE international conference on acoustics, speech and signal processing (ICASSP). IEEE, Piscataway
46. Arık SÖ et al (2017) Deep voice 2: multi-speaker neural text-to-speech. In: Proceedings of the 31st international conference on neural information processing systems. Curran Associates Inc., Long Beach, California, pp 2966–2974
47. Wang X et al (2020) ASVspoof 2019: a large-scale public database of synthesized, converted and replayed speech. Comput Speech Lang 64:101114
48. Griffin D, Lim J (1984) Signal estimation from modified short-time Fourier transform. IEEE Trans Acoust Speech Signal Process 32(2):236–243
49. Rec I (1996) P. 800: methods for subjective determination of transmission quality. International Telecommunication Union, Geneva, p 22
50. Elias I et al (2021) Parallel tacotron 2: a non-autoregressive neural TTS model with differentiable duration modeling. arXiv preprint arXiv:2103.14574
51. Ren Y et al (2020) Fastspeech 2: fast and high-quality end-to-end text to speech. arXiv preprint arXiv:2006.04558
52. Liu P et al (2021) VARA-TTS: non-autoregressive text-to-speech synthesis based on very deep vae with residual attention. arXiv preprint arXiv:2102.06431
53. Lee Y, Shin J, Jung K (2020) Bidirectional variational inference for non-autoregressive text-to-speech. In: International conference on learning representations

Advanced Sequence Learning Approaches for Emotion Recognition Using Speech Signals

Mustaqeem Khan, Muhammad Ishaq, Monorama Swain, and Soonil Kwon

1 Introduction

Speech is the most effective and significant form of human communication, as well as a potential medium for HCI. A new area of HCI study focuses on detecting audible emotions using sensors. This is true across various applications, including virtual reality, human-robot interaction, healthcare, behavior evaluation, and emergency call centers that analyze the user's voice to infer their emotional condition. To be practical for real-time commercial applications, the baseline speech-emotion recognition (SER) approach must be improved from its current low accuracy and high computation. The scarcity of datasets, feature choice, and model configuration are the main causes of low accuracy and high computational cost. Like this, emotion identification is a current, ongoing problem that is essential to real-time applications [1]. The SER system hasn't had real-time speech processing in the modern period. Due to its multimodality and the dearth of available data, emotion identification from oral speech signals [2] is one of the crucial and difficult issues. For modern real-world applications to effectively process and identify the responsive state of the speaker, a smart system is needed. Improving analytical skills to aid communication through a human-machine interface (HMI) [3]. Designing a trustworthy and dependable SER system is essential for accurately identifying emotions in a variety of contexts, including oral speech, photos, videos, gestures, phonemes, and text transcription. This work investigates high-level models and methods, such as 1D/2D Convolutional neural networks (CNN) models, to meet the

M. Khan (✉) · M. Ishaq · S. Kwon (✉)
Sejong University, Software Convergence, Seoul, Republic of Korea
e-mail: mustaqeem@icp.edu.pk; skwon@sejong.edu

M. Swain
Silicon Institute of Technology, Electronics and Communication, Bhubaneswar, India
e-mail: mswain@silicon.ac.in

S. A. Parah et al. (eds.), *Intelligent Multimedia Signal Processing for Smart Ecosystems*, https://doi.org/10.1007/978-3-031-34873-0_13

described issues. Long short-term memory (LSTM) gated recurrent units (GRU), self-attention mechanisms, and auto-encoders for SER are examples of sequential models. Due to this, we outperformed cutting-edge techniques in terms of precision with efficient processing.

The most common and extravagant method of transmitting information between humans and computers is SER. In real-world applications of human-machine interaction, it is also very important. In the field of ordinal signal processing, the speech signals generated by sensors for emotion recognition is a vibrant topic of research, employing speech signals, which include more information than vocal words, to assess the presenters' emotional states on a qualitative level [1]. Many academics are working in this area to develop a machine intelligent enough to distinguish the state from a speaker's voice and analyze or categorize their emotional state. One of the difficult jobs in SER is the selection and abstraction of relevant and discriminative cues [2]. Using AI and deep-learning techniques, researchers have recently been looking for strong and salient indicators for SER [3]. In order to extract hidden information, CNN features were used to train different CNN models [4, 5], which improved SER's performance and eased the burden on the computer. Due to the massive usage of social media, the low cost and excessive bandwidth of the Internet, and other factors, emotion identification has encountered many difficulties and limitations in the modern era. Semantic gaps occur because of the low cost of social media and the internet. Researchers are working to develop new strategies to close the semantic gap in this field. To properly identify the speaker's voice from audio inputs and trained models, the best salient cues must be extracted. As a result of daily technological advancements, researchers can now present fresh and adaptable platforms to test out new AI-based procedures.

The development of HCI, including the understanding of emotions, is greatly aided by the advanced knowledge, technology, and application of artificial intelligence and deep learning techniques. It's a vibrant area of HCI with many real-time applications, such as using it in call centers to gauge customer happiness and human connection. Emergency contact centers categorize a customer's emotional condition for the best possible response, and virtual reality is also used to discern human feelings. Figure 1 shows a categorization of SER techniques using the data at hand. Only acoustic-based emotion recognition is stressed in this work. To address this, we created a variety of cutting-edge Deep Learning-based techniques that effectively lowered model complexity while boosting overall accuracy.

CNN recently surpassed cutting-edge methods in several fields, including image classification, object detection, and localization. However, because an action is carried out through the successions of photos, processing a single image will almost certainly not produce superior results for fully encapsulating the activity. As a result, it can be identified in movies by scrutinizing the movements and other visual alterations of human body parts like hands and legs over the course of a series of frames. For instance, both skipping rope and jumping for a headshot in football share identical action pose in the first frame. However, a series of frames can clearly show how these behaviors differ from one another. For example, diving for swimming is one action, but every time the swimmer dives into the water, he

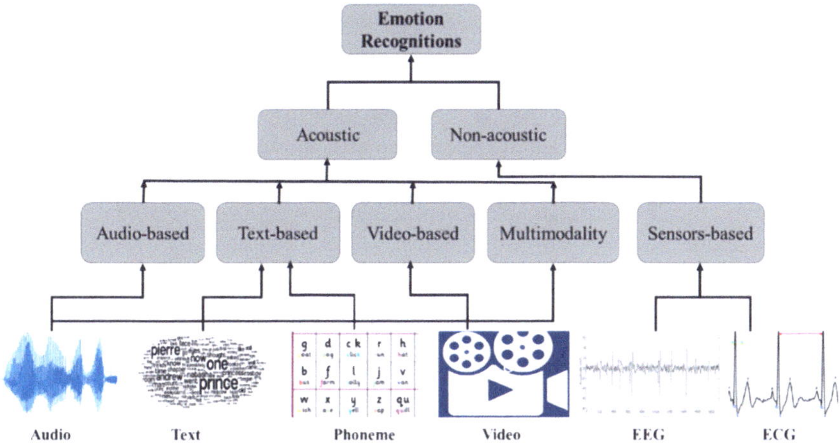

Fig. 1 Categorization of emotion recognition model and publicly available data for the acoustic and non-acoustic systems

also jumps into the air, which is another activity. This presents a barrier to analyzing long-term activities in video data. Therefore, the sequence is broken when both actions, such as jumping and diving, appear in the same order. Numerous researchers [6, 7] have developed various methods for representing sequences and learning those sequences for activity recognition in order to overcome these problems. For instance, the current literature on human activity recognition recommends some well-liked techniques for sequence representation, including support vector machine (SVM), decision tree, recurrent neural networks (RNN), and LSTM, as well as popular methods such as volume motion template [6], Spatial-temporal features [7], motion history volume [8], interest point descriptors [9], bag-of-words [10], optical flow trajectories [11], and CNN, based learned features [12, 13]. For sequence representation and learning, these techniques are insufficient. The LSTM is one of the most efficient sources of sequence learning, according to recent studies. So, for the purpose of recognizing actions and activities, we will study sequence learning concepts combined with LSTM in this chapter.

2 Recurrent Neural Networks

Because of their ability to learn time series data, RNNs are the type of artificial neural network that has gained popularity in recent years. The RNN is a unique network that, in contrast to conventional feedforward neural networks, comprises of recurrent connections. The main advantage is that it can handle any time series sequence of input data thanks to the network of its recurrent connections. For examining latent sequential patterns in both temporal sequential and spatial

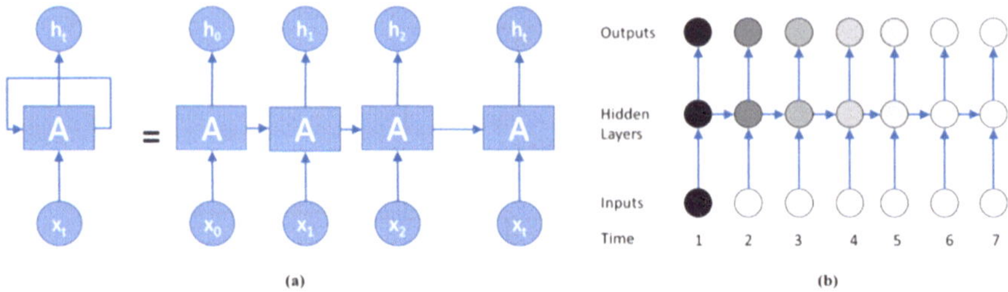

Fig. 2 (**a**) A standard unrolled RNN network (**b**) vanishing gradient problem in RNN

sequential data, RNNs are described [14]. Figure 2 displays a typical RNN (a). It processes data by receiving input over time at various time intervals. It takes two inputs at each time step: the first is the hidden state of the previous RNN, and the second is the time series data.

Figure 2a illustrates how each time step's input and output are provided by the RNN. The patterns for sequences in time series data are disturbed by the act of taking input at each recurring step. Since the neural network multiplies weights with data and adds bias factors. As a result, the effects of early sequential data in the series are forgotten when data is multiplied numerous times with various hidden states. This problem, called the vanishing gradient problem, is illustrated in Fig. 2b, where the influence of the data is depicted as gray color shades that get darker as the sequence gets longer. Standard RNNs can only access about 10 time steps of sequential data between the relevant input and the desired output in practice. Video data is also sequential, with changes in the visual contents reflected in several frames, making it easier to understand the context of an action by looking at a series of frames. In the case of long-term sequences, RNNs may interpret such sequences but forget the sequence's earlier inputs.

3 Long Short-Term Memory

An RNN variant known as LSTM can be used to tackle the vanishing gradient problem [15]. It has the capacity to learn enduring sequential dependencies. The long-term sequential pattern detection in time series data is controlled by its unique structure, which includes input, output, and forget gates. A sigmoid function attached to each unit adjusts the gates by learning during training which states the gate should be open and closed in. The processes carried out in the LSTM unit are described by Eqs. (1) to (7) [16], where x_t is the input at time t and input is fed as chunks from the long sequence. At time t, the forget gate (f_t) keeps track of the previous frame whose information needs to be erased from memory and clears information from the memory cell as needed. A recurrent unit with an activation function of "tanh" and information about the next step is stored in the output gate

o_t, which is computed using the input of the current frame and the state of the prior frame s_{t-1}. Tanh activation and memory cell c_t are used to calculate the hidden state of an RNN step. Action recognition does not require the intermediate output of the LSTM, so the final choice is made by applying the softmax classifier to the final state of the LSTM network, which provides a label and probabilities for the predicted activity. The output of each state can be used in a variety of other tasks.

$$i_t = \sigma\left((x_t + s_{t-1})\, W^i + b_i\right) \tag{1}$$

$$f_t = \sigma\left((x_t + s_{t-1})\, W^f + b_f\right) \tag{2}$$

$$o_t = \sigma\left((x_t + s_{t-1})\, W^o + b_o\right) \tag{3}$$

$$g = tanh\left((x_t + s_{t-1})\, W^g + b_g\right) \tag{4}$$

$$c_t = c_{t-1}.f_t + g.i_t \tag{5}$$

$$s_t = tanh\,(c_t)\,.o_t \tag{6}$$

$$Predictions = softmax\,(Vs_t) \tag{7}$$

Recent studies [17, 18] have demonstrated that a large amount of data boosts a machine learning task's accuracy. Consequently, it is highly difficult to learn complicated sequential patterns in video data from large-scale datasets, and a single LSTM is insufficient to do it. To learn long-term dependencies in video data, researchers recently used multi-layer ML-LSTM, which is built by stacking numerous LSTM cells.

3.1 Multi-layer LSTM

By adding more layers to the neural network models, the accuracy and performance of many deep neural networks can be improved. By adding several LSTM layers to the network, LSTM employs the same tactic. It captures a higher level of sequence information by adding deep layers [19]. In a typical RNN, input is fed into a single layer for processing and activation before being output, but to capture hierarchical patterns in time sequence problems, we need to process the data on many layers. Each layer in the network is a hierarchy that receives the hidden state of the preceding layer as input thanks to the stacking of LSTM layers. A two-layer LSTM is depicted in Fig. 3. Layer 2's input comes from its previous time step, s_{t-1} (2), while Layer 1's output comes from the current time step, s_t. Layer 1 receives input from data x_t (1). The computation of the LSTM cell is the same as before; the only

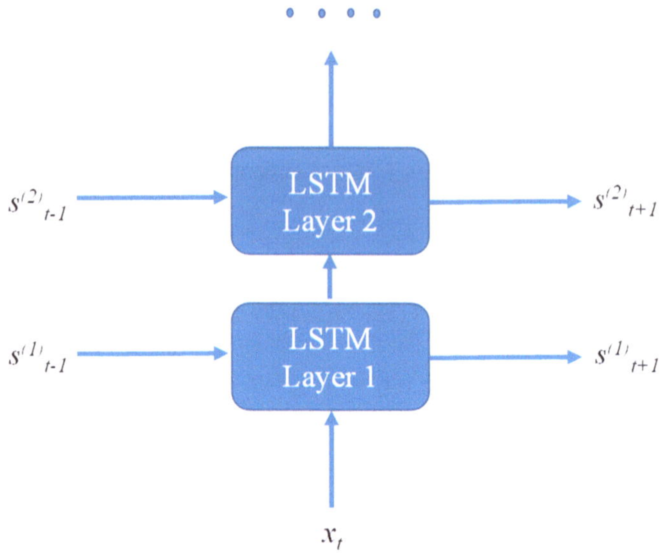

Fig. 3 Multi-layer long short term memory networks [19]

difference is that the superscript of each of it, f_t, o_t, c_t, and s_t now includes the layer's information. Below is a description of how to calculate the layer's status.

$$s_t^l = tanh\left(c_t^l\right).o_t^l \tag{8}$$

3.2 *Bi-directional LSTM*

The output of a bidirectional LSTM at time t depends not only on the frames that have come before it but also on the frames that will follow [20]. Two recurrent units are piled on top of one another in bidirectional LSTMs, which are relatively straightforward. There are two recurring units: one that travels forward, and the other that travels backward. The hidden state of both recurrent units is then used to compute the combined output. The input data is sent into the bidirectional LSTM, and the output layers combine the hidden forward pass and backward pass states. Following back-propagation adjustments to the output layer, weights, and biases, the validation and cost are determined.

3.3 Gated Recurrent Unit

Researchers first introduced the LSTM and GRU, two prominent RNN variations [21]. The input, forget, output, and memory cells are among the GRU found in the LSTM, whereas the reset gate, update gate, and activation unit are found in the GRU. The LSTM has a more intricate structure with gates and memory cells, requiring more computation to process a sequence. In contrast, GRU only uses two gates, which enables it to analyze real-time sequential data. But numerous studies have found that LSTM is more efficient than GRU [22].

4 Emotion Recognition Techniques Using Sequential Learning

Researchers have presented a variety of hand-crafted and deep nets-based techniques for emotion recognition over the past 10 years. In previous work, a performer would express some emotion in an audio file with a plain background using hand-crafted elements for non-realistic emotion. In order to recognize emotions, these systems capture low-level features from the audio data and input them to a classifier like SVM, decision tree, or KNN. However, the current techniques for behavior and emotion recognition need actual acoustic data. For instance, Mustaqeem et al. [23] proposed an emotion identification framework using CNN features and deep Bi-directional LSTM (DB-LSTM) for sequence learning. From every audio segment of the input sound, features were retrieved. The DB-LSTM network receives the sequential information that has been retrieved from these segments. In the DB-LSTM, multiple layers are stacked together in both the forward and backward passes to boost the network depth and enable long-term sequence learning. Figure 4 depicts the DB-structural GRUs layout.

Another new technique for real-time emotion identification was put out by Mustaqeem et al. [24]. Their methodology is depicted in Fig. 5, where they initially used CNNs to propose human emotion traits to select one significant segment from each cluster after grouping the continuous voice signals into clusters. Next, a Resnet101 model's convolutional layers are used to extract the temporal aspects of emotion in the sequence of segments. The last step is the presentation of a multi-layer Bi-directional long short-term memory (Bi-LSTM) for learning long-term sequences in the CNN features for emotion recognition.

Figure 5 illustrates how the method suggested by Kwon et al. [22] chooses discriminative components in the spatial dimension. They use sliding windows to create local tensors in the spatial direction after splitting the audios into several tensors with fixed-length segments in the temporal direction. They create a part selection process to choose useful features, and employ a multi-learning trick approach to create an emotion representation. The SoftMax layer and fully

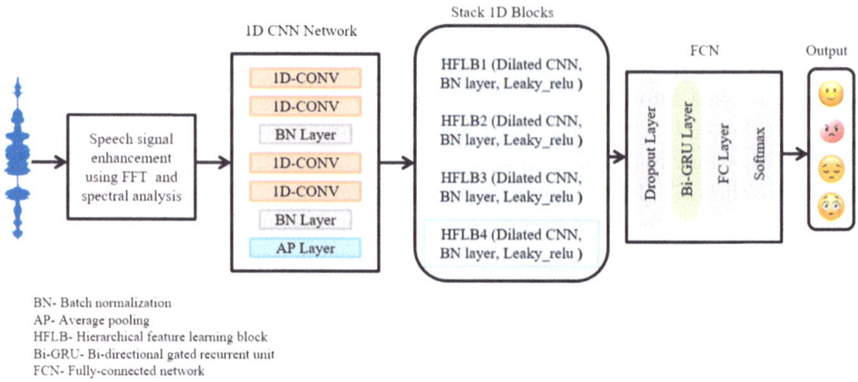

Fig. 4 A comprehensive description of the SER's one-dimensional planned architecture. The local characteristics from the speech signals are taken and fed into the 1D-CNN networks after passing from the stacked blocks. The blocks have links to the previous fully connected network, where the global weight was changed and temporal or sequential signals were retrieved using GRU and fully connected (FC) layers with Softmax to determine the likelihood of each class [23]

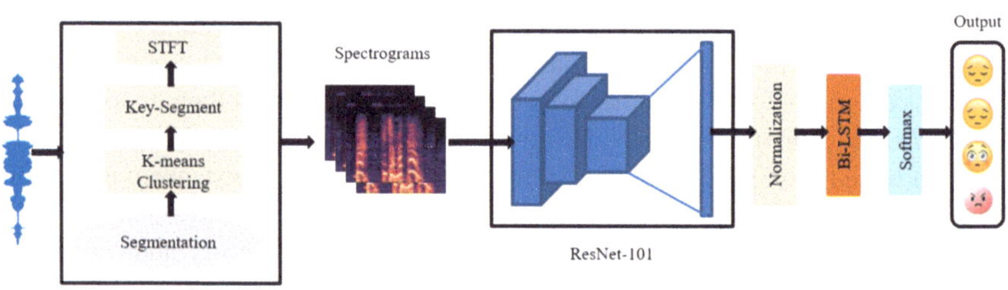

Fig. 5 The suggested architecture for voice emotion identification combines deep bidirectional LSTM and normalized CNN features with dynamic clustering-based key segment selection [24]

connected layers are then given the representation of emotion, which produces the prediction by using the feature vector produced by the spatial and sequence network.

The linguistic system and voice signals were thoroughly researched by the authors. The suggested system ignores the time-varying characteristics of the voice data because it doesn't have a sequential module. So, using a basic network method, they created a new SER system and added the GRUs network between the FC layer and the remaining blocks. Figure 6 displays the system's visual organization. The structure of this model was the same as that of the planar model, but the GRUs layers were also added in order to take advantage of the time-varying characteristics of the speech signal and to recognize the sequential or temporal information. Utilizing the suggested method, the constructed network learned numerous temporal cues and spatial emotion components from the speech data. Therefore, the authors developed a new SER system employing a multi-learning technique (MLT), which is depicted in Fig. 7, and these systems are used as a baseline. The first models that were offered had some drawbacks, which the proposed architecture addresses and gets

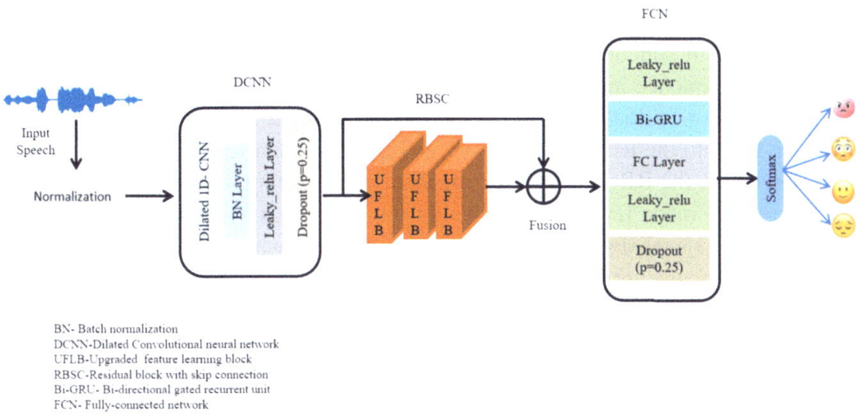

Fig. 6 Developed dilated CNN framework for SER without Multi-Learning Trap (MLT) utilizing residual blocks with a skip connection using GRUs network [22]

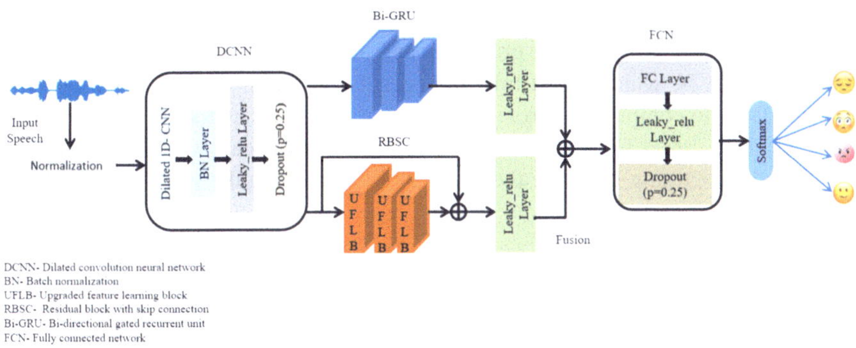

Fig. 7 Overview of the suggested system, which includes a sequence module to identify the temporal cues in the speech, one-dimensional residual blocks with a skip connection to improve the local learnt features, and a dilated CNN to learn the initial local information from the raw audio speech. A completely linked network passes spatial and temporal information to the Softmax layer, which makes the ultimate decision, using the fusion layer to integrate data [22]

over by using a new feature learning technique. Experiments demonstrate that the suggested system learns many incidental emotions and traits to accurately identify the emotional state of the speakers.

The primary feature extraction network for the SER is the dilated convolution neural network (DCNN), which uses a 1D structure to extract data directly from the unprocessed speech inputs. The SER, which concentrates on the local features in the input, is explained first before the suggested DCNN system. To improve the learnt features and understand emotional cues, these local features are also fed into feature learning blocks (UFLBs). The sequential model with GRUs, on the other hand, learns the global representation that uses the local input features through the complicated nonlinear changes between the time steps. As a result, the CNN and

GRU working together is essential for the SER system to recognize local and global information with concurrent contextual dependencies. The suggested system uses modules for sequence learning (Seq-L) and residual blocks with a skip connection (RBSC) to learn multi-type features. A speech segment's locally learned feature maps are fed into the RBSC, which acts as a convolutional features extractor. For a given utterance, the RBSC network is denoted as extracting features from the input learned feature maps for each speech segment, which comprises [s (1), s (2) ... s (T)]. The letters s (t) stands for the original audio file, while the numerals T show how many segments are in each speech.

The second method of learning is called sequence learning, and it entails relating each segment to the time step that corresponds to the initial input. In this work, they used the GRUs structure, which is the most effective for addressing long-term dependencies, to extract and recognize the temporal cues in extended sequences. The network gathers knowledge from the CNN characteristics and generates outputs that are sequential, such as [r (1), r (2) ... r (T)]. They implemented a common layer leaky Relu for richer statistical cues of the GRUs networks after gathering information from both modules. In order to create a dimension for a fully connected network, they finally fused these properties and passed them from the flattened layer (FCN). The FCN, which has a fully connected layer with a leaky Relu and one Softmax layer to compute the probability and forecast the category of emotion, was then fed to the fused feature vectors. Additionally, they dealt with the model and minimized over fitting by using the dropout and kernel regularization (L1 and L2) methods [25].

5 Result and Discussion

In this part, we evaluate the outcomes of several emotion identification techniques using benchmark datasets. In Table 1, the in-depth comparisons of cutting-edge methods are provided. The sequence-based approaches are mentioned in chronological order, with the most current method being shown first. In the most recent method [26], which combined CNNs and LSTM with a dilated mechanism, recurrent attention CNN was adopted. They achieved good accuracy with this method for IEMOCAP [27], RAVDESS [28], and EMO-DB [29], respectively. On the IEMOCAP, RAVDESS, and EMO-DB datasets, the CNN based temporal features followed by multilayer LSTM [30] achieved accuracy scores of 94.45%, 55%, and 72.1% respectively. By using a lightweight CNN model for feature extraction, this technique enhanced the emotion identification system's computational performance. Using relational LSTM, a different method [31] captures both long and short-range relationships between the frequencies in the audio data. On the IEMOCAP and EMO-DB, they attained accuracy of 94.8% and 71.9% respectively. In Table 2, a list of common emotion datasets is provided, along with examples of some of the categories. These datasets include human emotion gathered from YouTube and other sources, giving categories of actual emotional experiences.

Table 1 Comparison of different LSTM based emotion recognition methods on different challenging benchmark datasets

References	EMO-DB	SAVEE	IEMOCAP	RAVDESS
Zheng et al. [32]	–	–	40.02%	–
Han et al. [33]	–	–	57.01%	–
Luo et al. [37]	–	–	61.60%	–
Chen et al. [35]	82.82%	–	66.50%	–
Sugan et al. [46]	–	78.60%	–	–
Guo et al. [2]	84.49%	–	57.01%	–
Meng et al. [34]	84.99%	–	69.32%	–
Zhao et al. [36]	–	–	63.98%	–
Badshah et al. [39]	84.53%	–	–	–
Zeng et al. [41]	–	–	–	64.48%
Jalal et al. [42]	–	–	–	69.40%
Bhavan et al. [43]	–	–	–	75.69%
Avots et al. [50]	–	77.40%	–	–
Özseven et al. [48]	–	72.39%	–	–
Hnoohom et al. [51]	–	65.83%	–	–
Assunção et al. [53]	–	70.40%	–	–
Mustaqeem et al. [23, 30]	91.14%	–	72.75%	78.1%
Sajjad et al. [24]	85.57%	–	72.25%	77.02%
Kwon et al. [22]	95%	85%	–	82%

Table 2 List of popular benchmark emotion recognition datasets

Dataset	Language	Subject	Sample	Type	Length (hour)	Class
EMODB	German	10	535	Acted	0.22	7
IEMOCAP	English	10	10,039	Acted & improvised	12.0	10
SAVEE	British	4	480	Acted	0.42	7
RAVDESS	English	24	1440	Acted	3.5	8

Further a thorough account of emotion recognition using deep learning and ML, the implementation and security constraints can be seen from [2, 32–54].

For the IEMOCAP, EMO-DB, and RAVDESS speech emotion datasets, Kwon et al. [23] employed a 1D CNN model for segment-level feature extraction followed by DB-GRUs and achieved accuracy rates of 91.21%, 87.64%, and 92.84% respectively. The EMO-DB dataset's accuracy was improved by using this strategy. In order to assess the impact of Spatial-temporal corresponds to prominent motion patterns for emotion recognition, the DRNN [55] adopted differential gating technique employing LSTM network. This method's accuracy was tested on the very simple and old dataset SAVEE [56], yielding a result of 92.12%. Sixty-five percent accuracy on IEMOCAP was reported by an end-to-end long-term RNN [57]. After employing this method, [58] improved IEMOCAP by 20% by using sequential and dynamical features to address the spatial and temporal domain problem. On the IEMOCAP, EMO-DB, and RAVDESS tests, the accuracy was

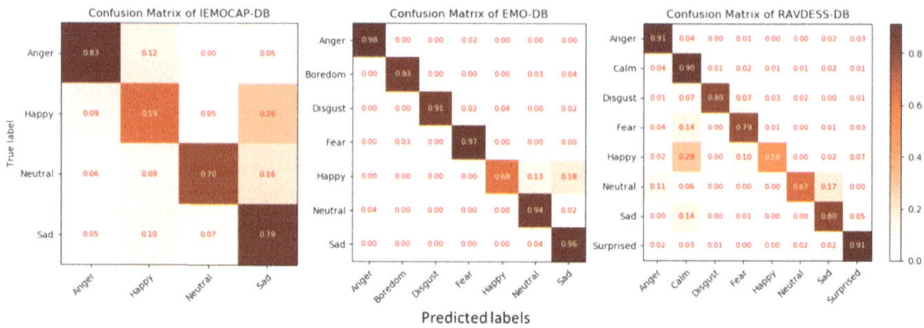

Fig. 8 Confusion matrices of IEMOCAP, EMO-DB and RAVDESS datasets obtained by stacked features leaning blocks using GRU for emotion recognition [23]

85%, 95%, and 58.2% respectively. In addition to these approaches, others shown in Table 2 offered additional strategies using LSTMs and RNNs to accurately identify the emotions. Most of these techniques enhanced accuracy by 3–5% while testing the performance over the IEMOCAP and RAVDESS datasets. The highest accuracy on the IEMOCAP dataset is 75%, attained by 78% on the RAVDESS dataset with DB-LSTM [59].

Figure 8 shows the confusion matrices and classification report of each emotion using the IEMOCAP, EMO-DB, and RAVDESS corpora respectively. The confusion matrix shows the recognition rate and confusion rate of each class in corresponding rows. The x-axes represent the expected labels and y-axes represent real labels of data and the diagonal values illustrate the average recall value of the model. The prediction accuracy of each dataset represented illustrate the average recall of each emotion. The overall prediction of the proposed model is improved, which secure the recognition rate of 78.01%, 80.00%, and 93.00% for IEMOCAP, RAVDESS, and EMO-DB datasets respectively. The obtained experimental results show the significance, robustness, and effectiveness of the proposed SER model, which is much improved than the baseline state-of-the-art approaches. In another method Kwon et al. [24] conducted speaker dependent and independent experimentation to illustrate the model significance and robustness. The confusion matrices of dependent and independent emotion/classes are show in Figs. 9 and 10 respectively.

In the proposed framework [24], the efficient sequence selection using K-mean clustering based on RBF similarity and normalized discriminative features with deep Bi-LSTM are major contributions for SER utilizing speech signals. They performed speaker-dependent and speaker independent experiments over three benchmarked datasets for recognizing the emotional state of his/her speech as well as developed a new technique for SER, and to select a sequence from utterance using RBF based K-mean clustering technique. The authors selected one segment from each cluster which is near to centroid as a key segment that represents the corresponding clusters and converted all key segments into spectrograms applying STFT for 2-D representations. Furthermore, they extracted the high level discriminative (CNN) features from spectrograms utilizing the "FC-1000" layer of

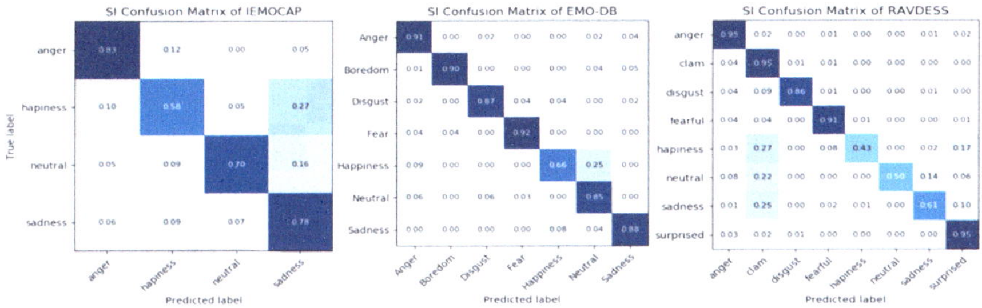

Fig. 9 Confusion between actual and expected emotions are shown in the confusion matrix of speaker-independent emotions prediction on IEMOCAP, EMO-DB, and RAVDESS, corpora [24]

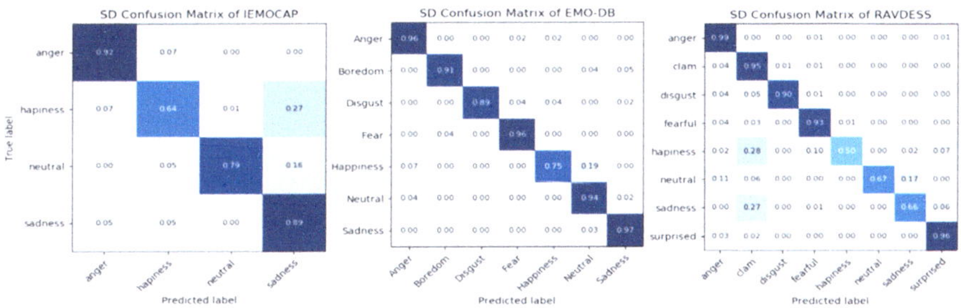

Fig. 10 Confusion between actual and expected emotions are shown in the confusion matrix of speaker-dependent emotions prediction on IEMOCAP, EMO-DB, and RAVDESS, corpora [24]

the Resnet101 model. They normalized the extracted features using average mean and standard deviation algorithm and passed from deep Bi-LSTM. They used this novel approach for emotion to improve the classification accuracy and reduce the processing time as compared to other traditional CNN_ELM [60] and DNN_KLM networks [61]. They obtained better results on three benchmarks: IEMOCAP, EMO-DB and RAVDESS datasets using this novel approach for speaker-dependent and speaker-independent experiments. The proposed system boosts the overall accuracy up to (6.14%), (2.14%) and (7.01%) in speaker-dependent and (3.07%), (1.57%) and (2.41%) in speaker-independent experiments on IEMOCAP, EMO-DB, and RAVDESS datasets to recognize the emotional state of speaker respectively (see Fig. 11).

In Fig. 11, the confusion matrices for both datasets are displayed by the authors. The EMO-DB result is depicted by Fig. 11b, which displays the confusion ratio between the actual and predicted labels, whereas Fig. 11a indicates the IEMOCAP result. The model's recall values for the IEMOCAP dataset, which is diagonally displayed in Fig. 11a, are 83%, 57%, 69%, and 78%. The model outperforms the baseline [62] models for all emotion classes, while the happy emotion performs substantially worse than the others. Figure 11b, which shows the overall effectiveness of the proposed model, illustrates the EMO-DB mismatch between the actual and expected labels. The recall of the suggested SER system

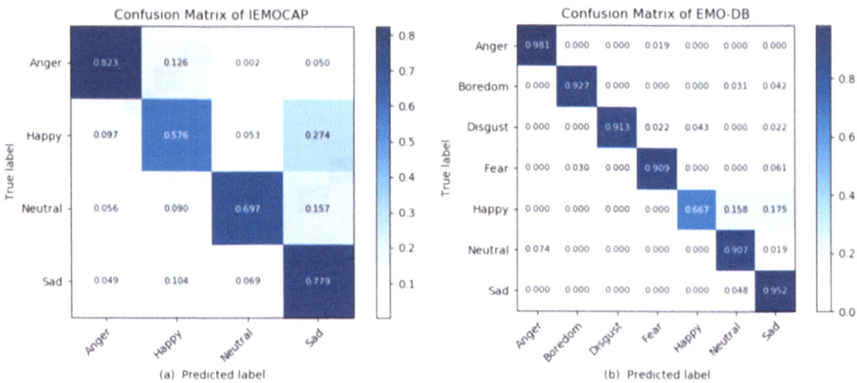

Fig. 11 The suggested model's confusion matrix illustrates how well it performs predictions on the IEMOCAP and EMO-DB corpora [22]

was represented by the diagonal numbers 98%, 92%, 91%, 91%, 67%, 90%, and 95%. The suggested model generates strong results for both datasets, demonstrating the excellent generalizability of the emotion recognition system. For the IEMOCAP dataset, the model indicated a 73% average recall, and for the EMO-DB dataset, a 90% average recall. For extracting characteristics from the speech signals, the suggested model utilized a novel sort of deep learning architecture.

Besides, the proposed model [22] can deal with the arbitrary length of the speech and can handle a speech with multiple emotion classes. Using a GeForce-GTX 1070 GPU and a Core i7 CPU, the suggested system's experiments for salient feature extraction, training, and testing of the proposed model for emotion recognition are carried out. In this configuration, the model uses RBSC to extract prominent characteristics from a 1-second audio sample in 0.19 seconds using GPU and 1.4 seconds using basic CPU (residual blocks with skip connection). Second, the sequence learning (Seq L) module holds emotion patterns to a plan for sequential features in 0.0.74 seconds using GPU and 2.8 seconds using CPU. The classifier divides the given collection of characteristics into emotion categories in only 0.12 and 0.4 seconds respectively. The system takes 1.05 seconds for processing 1 second audio signal stream, showing nearly real-time processing using high computational device (GPU) based on the statistics shown in Fig. 12. The visual analysis demonstrates that the GPU processes information more than four times faster than the CPU. The proposed system must be implemented on powerful computing hardware to be used in practical applications. Because of its quick computation and great accuracy, the recommended system is a better option that may be used in any kind of real-time speech processing.

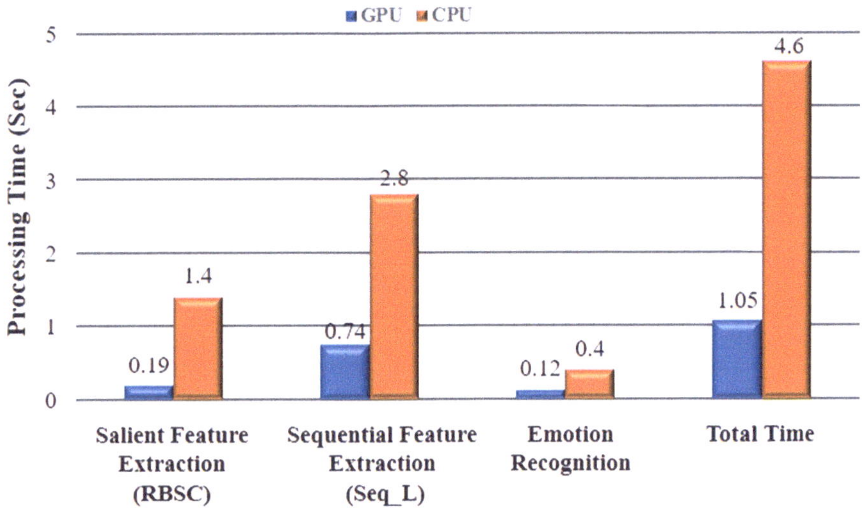

Fig. 12 The suggested SER system's CPU and GPU time complexity for real-time emotion identification on a 1 second audio file

6 Future Research Directions

It has been noted that numerous studies have approached the issue of emotion detection by utilizing various sequential learning methodologies in accordance with the requirements in distinct domains. According to the most recent research, numerous researchers have employed various characteristics for sequence representations, followed by RNN or LSTM. For instance, some researchers have also used deep features for sequence representation, although those techniques have a very high time complexity. Therefore, cutting-edge methods that can balance both time complexity and accuracy are required. The following are suggestions and future directions for emotion recognition:

- Modern results for spectrogram classification using CNN have surpassed human error rates in more recent deep models [62, 63]. It is possible to express frames in sequential audio data using these current spectrogram-based models. Mobile Net [64], Squeeze Net [65], and Shuffle Net [66] are just a few of the lightweight CNN models that can aid in real-time processing for emotion recognition. In numerous other tasks, including object detection and localization [67], image and video retrieval [68], video summarization [69], etc., these CNN models have already produced state-of-the-art results. The key feature of these models is that they learn discriminative hidden local and global representations from visual data by being trained on large-scale visual datasets.
- Even though there are many techniques for recognizing emotions, it is observed that these techniques work well for processing information about a single person's emotions. IEMOCAP, EMO-DB, RAVDESS, and SAVEE datasets, where only one person is reading a script in sample audio, are used to train the

existing techniques. From the available literature it can be stated that there are no solutions to the issue of multiple emotion recognition in audio data. Therefore, in the future, this work needs to be accomplished by first locating the individual in the audio data, then determining their emotions using further processing. This is a difficult task that can be handled using innovative research techniques.
- Since the datasets currently used by academics are so limited, they cannot be used to create deep learning models from start. Nevertheless, there are several sizable audio-based emotion identification datasets that can be well recognized. Some of the well-liked datasets in the research on emotion recognition are shown in Table 2. Numerous researchers have previously used the first three datasets, and as already stated, they produced positive findings. However, using data from YouTube, this kind of information can be gathered and a single recognition algorithm can be used to identify a greater variety of emotions.

7 Conclusion

The challenge of evaluating and forecasting a human's behavior from a series of audio clips is human emotion recognition. The idea of sequence learning for emotion recognition utilizing LSTM, GRUs, and its variations such as multi-layer or deep LSTM and bi-directional LSTM networks have been discussed in this chapter. The achievements and shortcomings of modern LSTM/GRUs-based systems for emotion recognition have also been evaluated and debated. We also discussed the shortcomings of conventional RNNs and why LSTM is superior to RNN. The chapter ends with a discussion and suggestions for further research that might aid in resolving the current difficulties in speech-based emotion recognition utilizing sequential models.

Acknowledgments This research was made possible thanks to a grant from the National Research Foundation of Korea, which was sponsored by the Korean government through the Ministry of Science and ICT under Grant NRF-2020R1F1A1060659.

References

1. Akçay MB, Oğuz KJSC (2020) Speech emotion recognition: emotional models, databases, features, preprocessing methods, supporting modalities, and classifiers. Speech Commun 116:56–76
2. Guo L, Wang L, Dang J, Liu Z, Guan HJIA (2019) Exploration of complementary features for speech emotion recognition based on kernel extreme learning machine. IEEE Access 7:75798–75809
3. Al Dujaili MJ, Ebrahimi-Moghadam A, Fatlawi AJIJOE, C. Engineering (2021) Speech emotion recognition based on SVM and KNN classifications fusion. Int J Electr Comput Eng 11(2):1259
4. Ali MS, Islam MS, Hossain MAJIJOCS, Engineering, and I. Technology (2012) Gender recognition system using speech signal. Int J Comput Sci Eng Inf Technol 2(1):1–9

5. Amodei D et al (2016) Deep speech 2: end-to-end speech recognition in English and Mandarin. In: International conference on machine learning. PMLR, pp 173–182
6. Roh M-C, Shin H-K, Lee S-W (2010) View-independent human action recognition with volume motion template on single stereo camera. Pattern Recogn Lett 31(7):639–647
7. Xin M, Zhang H, Wang H, Sun M, Yuan D (2016) Arch: adaptive recurrent-convolutional hybrid networks for long-term action recognition. Neurocomputing 178:87–102
8. Weinland D, Ronfard R, Boyer E (2006) Free viewpoint action recognition using motion history volumes. Comput Vis Image Underst 104(2–3):249–257
9. Baccouche M, Mamalet F, Wolf C, Garcia C, Baskurt A (2010) Action classification in soccer videos with long short-term memory recurrent neural networks. In: International conference on artificial neural networks. Springer, Berlin, Heidelberg, pp 154–159
10. Kovashka A, Grauman K (2010) Learning a hierarchy of discriminative space-time neighborhood features for human action recognition. In: 2010 IEEE conference on computer vision and pattern recognition (CVPR). IEEE, San Francisco, pp 2046–2053
11. Sekma M, Mejdoub M, Amar CB (2015) Human action recognition based on multi-layer fisher vector encoding method. Pattern Recogn Lett 65:37–43
12. Hou J, Wu X, Sun Y, Jia Y (2018) Content-attention representation by factorized action-scene network for action recognition. IEEE Trans Multimed 20(6):1537–1547
13. Ullah FUM, Ullah A, Muhammad K, Haq IU, Baik SW (2019) Violence detection using spatiotemporal features with 3D convolutional neural network. Sensors 19(11):2472
14. Sherstinsky AJPDNP (2020) Fundamentals of recurrent neural network (RNN) and long short-term memory (LSTM) network. Phys D: Nonlinear Phenom 404:132306
15. Yu Y, Si X, Hu C, Zhang JJNC (2019) A review of recurrent neural networks: LSTM cells and network architectures. Neural Comput 31(7):1235–1270
16. Huang Z, Xu W, Kai Y (2015) Bidirectional LSTM-CRF models for sequence tagging. arXiv preprint arXiv:1508.01991
17. Greff K, Srivastava RK, Koutník J, Steunebrink BR, Schmidhuber JJITONN, l. Systems (2016) LSTM: a search space odyssey. IEEE Trans Neural Netw 28(10):2222–2232
18. Gers FA, Schraudolph NN, Schmidhuber JJJOMLR (2002) Learning precise timing with LSTM recurrent networks. J Mach Learn Res 3(1):115–143
19. Zhang S, Liu X, Xiao J (2017) On geometric features for skeleton-based action recognition using multilayer LSTM networks. In: 2017 IEEE Winter conference on applications of computer vision (WACV). IEEE, Santa Rosa, pp 148–157
20. Graves A, Fernández S, Schmidhuber J (2005) Bidirectional LSTM networks for improved phoneme classification and recognition. In: International conference on artificial neural networks. Springer, Berlin, Heidelberg, pp 799–804
21. Dey R, Salem FM (2017) Gate-variants of gated recurrent unit (GRU) neural networks. In: 2017 IEEE 60th international midwest symposium on circuits and systems (MWSCAS). IEEE, Piscataway, pp 1597–1600
22. Kwon SJESWA (2021) MLT-DNet: speech emotion recognition using 1D dilated based on multi-learning trick approach. Expert Syst Appl 167:114177
23. Mustaqeem SJC-C, Kwon M (2021) 1D-CNN: speech emotion recognition system using a stacked network with dilated CNN features. Comput Mater Contin 67(3):4039–4059
24. Sajjad M, Kwon SJIA (2020) Clustering-based speech emotion recognition by incorporating learned features and deep BiLSTM. IEEE Access 8:79861–79875
25. Zhang Y, Li R, Tsai C-LJJOTASA (2010) Regularization parameter selections via generalized information criterion. J Am Stat Assoc 105(489):312–323
26. Muhammad K et al (2021) Human action recognition using attention based LSTM network with dilated CNN features. Future Gener Comput Syst 125:820–830
27. Busso C et al (2008) IEMOCAP: interactive emotional dyadic motion capture database. Lang Resour Eval 42(4):335–359
28. Livingstone SR, Russo FAJPO (2018) The Ryerson audio-visual database of emotional speech and song (RAVDESS): a dynamic, multimodal set of facial and vocal expressions in North American English. PLoS One 13(5):e0196391

29. Burkhardt F, Paeschke A, Rolfes M, Sendlmeier WF, Weiss B (2005) A database of German emotional speech. In: Ninth european conference on speech communication and technology, Lisbon
30. Mustaqeem M, Kwon S (2021) Speech emotion recognition based on deep networks: a review. In: Proceedings of the Korea information processing society conference. Korea Information Processing Society, pp 331–334
31. Ishaq M, Son G, Kwon S (2021) Utterance-level speech emotion recognition using parallel convolutional neural network with self-attention module. In: 7th international conference on next generation computing 2021. Korean Institute of Next Generation Computing
32. Zheng W, Yu J, Zou Y (2015) An experimental study of speech emotion recognition based on deep convolutional neural networks. In: 2015 international conference on affective computing and intelligent interaction (ACII). IEEE, Piscataway, pp 827–831
33. Han K, Yu D, Tashev I (2014) Speech emotion recognition using deep neural network and extreme learning machine. In: Fifteenth annual conference of the international speech communication association
34. Meng H, Yan T, Yuan F, Wei H (2019) Speech emotion recognition from 3D log-mel spectrograms with deep learning network. IEEE Access 7:125868–125881
35. Chen M, He X, Yang J, Zhang H (2018) 3-D convolutional recurrent neural networks with attention model for speech emotion recognition. IEEE Signal Process Lett 25(10):1440–1444
36. Zhao Z et al (2019) Exploring deep spectrum representations via attention-based recurrent and convolutional neural networks for speech emotion recognition. IEEE Access 7:97515–97525
37. Luo D, Zou Y, Huang D (2018) Investigation on joint representation learning for robust feature extraction in speech emotion recognition. Interspeech 2018:152–156
38. Bhat GM, Mustafa M, Parah SA, Ahmad J (2010) Field programmable gate array (FPGA) implementation of novel complex PN-code-generator-based data scrambler and descrambler. Maejo Int J Sci Technol 4(1):125–135
39. Badshah AM et al (2019) Deep features-based speech emotion recognition for smart affective services. Multimed Tools Appl 78(5):5571–5589
40. Sarosh P, Parah SA, Bhat GM, Heidari AA, Muhammad K (2021) Secret sharing-based personal health records management for the Internet of Health Things. Sustain Cities Soc 74:103129
41. Zeng Y, Mao H, Peng D, Yi Z (2019) Spectrogram based multi-task audio classification. Multimed Tools Appl 78(3):3705–3722
42. Jalal MA, Loweimi E, Moore RK, Hain T (2019) Learning temporal clusters using capsule routing for speech emotion recognition. Proc Interspeech 2019:1701–1705
43. Bhavan A, Chauhan P, Shah RR (2019) Bagged support vector machines for emotion recognition from speech. Knowl-Based Syst 184:104886
44. Parah SA, Rashid M, Varadarajan V (2022) Artificial intelligence for innovative healthcare informatics. Springer, Cham
45. Hafiz AM, Parah SA, Bhat RU (2021) Attention mechanisms and deep learning for machine vision: a survey of the state of the art. arXiv preprint arXiv:2106.07550
46. Sugan N, Srinivas NS, Kar N, Kumar L, Nath MK, Kanhe A (2018) Performance comparison of different cepstral features for speech emotion recognition. In: 2018 International CET conference on control, communication, and computing (IC4). IEEE, Piscataway, pp 266–271
47. Sarosh P, Parah SA, Mansur RF, Bhat GM (2020) Artificial intelligence for COVID-19 detection–a state-of-the-art review. arXiv preprint arXiv:2012.06310
48. Özseven T (2019) A novel feature selection method for speech emotion recognition. Appl Acoust 146:320–326
49. Parah SA, Sheikh JA, Ahad F, Bhat GM (2018) High capacity and secure electronic patient record (EPR) embedding in color images for IoT driven healthcare systems. In: Internet of things and big data analytics toward next-generation intelligence. Springer, Cham, pp 409–437
50. Avots E, Sapiński T, Bachmann M, Kamińska D (2019) Audiovisual emotion recognition in wild. Mach Vis Appl 30(5):975–985

51. Mekruksavanich S, Jitpattanakul A, Hnoohom N (2020) Negative emotion recognition using deep learning for thai language. In: 2020 Joint international conference on digital arts, media and technology with ECTI northern section conference on electrical, electronics, computer and telecommunications engineering (ECTI DAMT & NCON). IEEE, Piscataway, pp 71–74
52. Parsa S, Parah SA, Bhat GM, Khan M (2021) A security management framework for big data in smart healthcare. Big Data Res 25:100225
53. Assunção G, Menezes P, Perdigão F (2020) Speaker awareness for speech emotion recognition. Int J Online Biomed Eng 16(04):15–22
54. Kwon SJIJOIS (2021) Optimal feature selection based speech emotion recognition using two-stream deep convolutional neural network. Int J Intell Syst 36:5116
55. Zhai Y-J, Yu D-W, Guo H-Y, Yu D-LJEAOAI (2010) Robust air/fuel ratio control with adaptive DRNN model and AD tuning. Eng Appl Artif Intell 23(2):283–289
56. Jackson P, Haq SJUOSG (2014) Surrey audio-visual expressed emotion (savee) database. University of Surrey, Guildford
57. Anvarjon T, Kwon SJS (2020) Deep-net: a lightweight CNN-based speech emotion recognition system using deep frequency features. Sensors 20(18):5212
58. Kwon SJS (2020) A CNN-assisted enhanced audio signal processing for speech emotion recognition. Sensors 20(1):183
59. Kwon SJM (2020) CLSTM: deep feature-based speech emotion recognition using the hierarchical ConvLSTM network. Mathematics 8(12):2133
60. Duan M, Li K, Yang C, Li KJN (2018) A hybrid deep learning CNN–ELM for age and gender classification. Neurocomputing 275:448–461
61. van Kleef D (2012) Redesign of the control model of the catering distribution network of KLM. Delft University of Technology, Delft
62. Paz-Ares L et al (2019) Durvalumab plus platinum–etoposide versus platinum–etoposide in first-line treatment of extensive-stage small-cell lung cancer (CASPIAN): a randomised, controlled, open-label, phase 3 trial. Lancet 394(10212):1929–1939
63. Hestness J, Ardalani N, Diamos G (2019) Beyond human-level accuracy: computational challenges in deep learning. In: Proceedings of the 24th symposium on principles and practice of parallel programming. Association for Computing Machinery, New York, pp 1–14
64. Howard AG et al (2017) Mobilenets: efficient convolutional neural networks for mobile vision applications. arXiv
65. Iandola FN, Han S, Moskewicz MW, Ashraf K, Dally WJ, Keutzer KJAPA (2016) SqueezeNet: AlexNet-level accuracy with 50x fewer parameters and < 0.5 MB model size. arXiv
66. Zhang X, Zhou X, Lin M, Sun J (2018) Shufflenet: an extremely efficient convolutional neural network for mobile devices. In: Proceedings of the IEEE conference on computer vision and pattern recognition. IEEE, pp 6848–6856
67. Zou Z, Shi Z, Guo Y, Ye JJAPA (2019) Object detection in 20 years: a survey. arXiv
68. Sebe N, Lew MS, Zhou X, Huang TS, Bakker EM (2003) The state of the art in image and video retrieval. In: International conference on image and video retrieval. Springer, Berlin, Heidelberg, pp 1–8
69. Apostolidis E, Adamantidou E, Metsai AI, Mezaris V, Patras IJPOTI (2021) Video summarization using deep neural networks: a survey. Proc IEEE 109(11):1838–1863

Comparison of Deep Learning Model Performance for Handwritten Character Recognition of Schoolchildren

Siti Azura Ramlan, Iza Sazanita Isa, Nurul Nur Aqirah Jiad, Ahmad Puad Ismail, Muhammad Khusairi Osman, and Zainal Hisham Che Soh

1 Introduction

Despite the availability of modern writing tools, many schoolchildren prefer to take notes in the old-fashioned way, which involves a pen and paper. The traditional way of writing appears more at ease and accustomed to this. The practicality of paper and pen over keyboards in various day-to-day circumstances is one of the reasons handwriting has survived in the digital era [1]. Handwritten text, on the other hand, has limitations. It might be challenging to store, retrieve, search through, and exchange physical documents with others. As a result, unintended consequences occur, such as the loss of many essential information that were not evaluated because physical papers could not be transferred or converted into digital format. Handwritten recognition on scanned documents or shot images remains the most demanding and challenging problem in computer science, and it is a well-known unresolved issue in various domains. Due to the reason that historical records have emerged forcefully in recent years, handwriting classification and identification are most significant and valuable.

Nowadays, handwriting recognition is proliferating because of its ability to translate human writing into digital form. This technology can be used to recognize handwritten characters and may be used on a variety of different devices, such as tablet PCs. The benefit of handwriting recognition is its ability to use electronic storage, which requires fewer people to sort and organize documents and retrieve

S. A. Ramlan · I. S. Isa (✉) · N. N. A. Jiad · A. P. Ismail · Z. H. C. Soh
Electrical Engineering Studies, College of Engineering, Universiti Teknologi MARA, Cawangan Pulau Pinang, Permatang Pauh, Penang, Malaysia
e-mail: izasazanita@uitm.edu.my

M. K. Osman
Centre for Electrical Engineering Studies, Universiti Teknologi MARA, Cawangan Pulau Pinang, Permatang Pauh, Penang, Malaysia

S. A. Parah et al. (eds.), *Intelligent Multimedia Signal Processing for Smart Ecosystems*, https://doi.org/10.1007/978-3-031-34873-0_14

electronic data [2]. In addition, handwritten recognition can be more helpful in preserving old family records and personal diaries that may be damaged by accidents, natural disasters such as floods or earthquakes, or other factors. Input data on handwriting styles can come from various places, including images and scanned documents, and this data can then be analyzed by a system known as handwriting recognition. A handwriting recognition system is a promising area for research because it could dramatically enhance automated processes and facilitate more natural communication between humans and machines. It is a process that converts machine-printed, hand-printed, or handwritten documents into digital form documents.

Image processing that combines mathematical operations with signal processing is known as image recognition. The input in this situation could be a scanned document, a video, an image, or something else. A set of parameters for images or an image itself may result from the process. There are various components to image processing, such as document image processing. It has been used to produce handwritten documents for analysis, identification, and preparation of secondary information for further use in document image processing. Handwriting recognition systems are classified into two types. The first one is offline, in which the handwritten input is in the form of an image. The second one is online, in which the input contains a recorded list of pen tip movements representing the handwritten character. Offline handwriting recognition involves converting a picture into text, which is less complex than many computer vision issues. Offline handwritten recognition is an exchange of the image into text which is text graphics are more superficial than many computer vision problems. However, developing a word classifier becomes prohibitively expensive beyond the limited vocabulary size due to many classes and a lack of data. Therefore, the character type of handwriting recognition is commonly used. Convolutional Neural Network (CNN), which has been extensively studied for the handwriting recognition process, offers great promise for identifying and classifying input images of handwriting.

The CNN has recently shown remarkable accuracy in image classification [3]. CNN is employed efficiently and successfully in many patterns and image recognition applications, including gesture detection, face recognition and object classification. Deep learning extracts information from image datasets using several layered designs that include convolutional layers, pooling layers, and finally, a fully connected layer to provide the output. Many researchers are currently studying CNN and applying it to a wide range of recognition problems. CNNs have been used in various studies related to handwriting recognition. They have achieved detection rates of 99.77% using the MNIST database of handwritten digits [4] proposed back in 1989, which contains 60,000 and 10,000 images for training and testing. Recently, many researchers have built various CNN models that can be used to classify or recognize the image dataset. For instance, the AlexNet model is a well-known trained CNN implementation that employs ImageNet [5]. The model maintains a low error rate when trained on millions of images in the ImageNet dataset.

To date, much popular research has merged into handwritten character recognition in documents. Many studies have proposed various CNN networks to recognize and detect handwriting images, and the performance results are widely varied [6–8].

The development of handwritten character recognition is simple, and it involves a digitized image as input and desired character as output [9]. The use of the system will help improve people's efficiency in recognizing handwriting. Deep learning architectures have recently been applied [10] to improve performance and produce significant output in recognizing handwriting images among schoolchildren. However, there is still minimal research on recognizing schoolchildren's handwriting. As a result, it is imperative to measure the performance of various models that may be used as indicators in identifying children's handwriting images using deep learning, especially CNN models.

Therefore, in order to validate the comparisons of the CNN models, this chapter uses images of schoolchildren's handwriting to compare several deep learning networks, specifically CNN models. The performance of the CNN models is then examined in terms of accuracy and loss, revealing the contribution of the CNN models.

The content of this chapter is divided into five main sections: related studies, methodology, results, discussion, and conclusion. The description of related studies Using CNN architecture and deep learning models to recognize handwriting in multiple ways based on prior research is also provided. All techniques and algorithms used for investigating the performance of several CNN architecture implementations on offline handwriting images have been described in the Sect. 3. Results and discussion show the experimental findings and accomplishments, while the whole chapter is summarised in the conclusion.

2 Related Studies

CNN is one of the machine learning methods most frequently used for image classification. CNN is adept at handling datasets as massive nodes and parameters [11] and has shown remarkable accuracy for image classification [3]. Many studies have been conducted to improve accuracy and reduce loss. It is a feedforward network because information flows only one way, from its input to the output. CNN is like the Artificial Neural Network (ANN), which is biologically inspired. Their architecture is inspired by the visual cortex of the brain, which is made up of alternating layers of simple and complicated cells [12]. CNN architectures come in a variety of variations; however, in general, they are made up of several layers that can learn high-level features from labeled training data efficiently. Following these modules are one or more fully connected layers, as in a typical feedforward neural network. Modules are frequently piled on top of one another to create a deep model. In addition, its central concept is to use convolutional and pooling layers to gradually extract more abstract patterns that can be performed locally [13].

In earlier studies, there were concerns with image classification that most frequently relied on the CNN network. From the studies by Neha Sharma et al. [14], the architectural performance of AlexNet, GoogleNet, and ResNet-50 was used in the study by training and testing on an extensive data set by using the commonly found benchmark dataset for evaluating CNN performance which are ImageNet, CIFAR10, CIFAR100, and MNIST data sets in order to identify and classify objects in real-time video feeds. A comparison has been made, and it was discovered that GoogleNet and ResNet50 could recognize the object with better accuracy than AlexNet. ResNet-50 was found to have the highest accuracy in the CIFAR10 data set at 78.10%, while Alexnet yielded the lowest accuracy at 36.12%.

There is also research by Feiyang Chen et al. [15] that compares four different CNN models, namely CNN, ResNet, DenseNet, and CapsNet model, in which improvement has been made to the CNN baseline. This project uses the MNIST dataset to observe which models have better performance. They have demonstrated that, while previous models do a pretty good job in this regard, their retrofitting can also be used to improve the performance of the CNN model. The results obtained by CapsNet have the highest accuracy compared to other models, even using only a small amount of data. For future work, the author has decided to use CapsNet's capabilities to generalize other image recognition fields.

Another study compares CNN models on dyslexia handwriting classification and handwritten digit recognition. In [10], the studies used different deep learning models, specifically the CNN models, to evaluate the performance of each model in classifying the dyslexia handwriting of schoolchildren. There is a critical need in deep learning, specifically for CNN, to validate the performances of different models so that the best performing CNN can be used as a benchmark in detecting dyslexia symptoms among schoolchildren. The result is that CNN with two convolutional layer models has the highest accuracy and the least validation loss. In contrast, research by [16] analyzed the impact of different layers of CNN on overall network performance using the MNIST dataset. The system is trained using stochastic gradient and backpropagation algorithms and tested using the feedforward technique. The result obtained showed that the simplest model yields the best performance. As a result, both papers discovered that the model with a minimum number of hidden layers performs well.

Next, the studies of [17] employed four various CNN models with the MNIST dataset to identify the most effective algorithm for reading handwritten numbers in real time. They also compared the train and assessment times, memory consumption, and other important parameters for each model. 50,000 inputs were used to train the models, except for the CapsNet model, with 40,000 inputs. However, the results show that CapsNet has the highest accuracy with a smaller number of inputs. It also works quickly in real-time applications and detects handwritten numerals much faster, within 1–2 s, compared to other models. It is the best model for real-time applications because the other three models lagged during evaluation and did not achieve the same level of precision as the CapsNet model.

Besides, research by [8, 9] has compared the different deep learning models for other handwriting character recognition by evaluating eight deep learning models using Urdu handwritten characters because Urdu is one of the world's major languages and has received little attention in the existing research area. This study employed a combination of data augmentation and transfer learning approaches to recognize digits and characters with an accuracy of 98.94% for the Xception model and 99.08% for MobileNet. In addition, they also discovered that advanced models might not outperform simpler models for this problem. Next, with the same study aims, a comparison network of GoogleNet and the AlexNet model was investigated by [18], which used the TU-Berlin sketch data set, a public resource collection of 20,000 common subject doodles arranged into 250 classes. The classification domains are image samples, including airplanes, birds, cars, cats, dogs, frogs, and others. Then, the study provides a better understanding of CNN's usability in different but connected domains with training sets. This study has shown that neither AlexNet nor GoogLeNet can classify the sketches displayed consistently. Drawing lines are frequently confused with objects that have simple and thin lines, such as safety pins. Nonetheless, both networks performed random guesses in several categories and performed random guesses in others. Lastly, different concerns of study by [3] that utilized CNN for the ImageNet classification addressed the problem of image classification in which the input image may belong to 1000 different classes, and the output would be a vector of 1000 numbers.

Since then, various methodologies and strategies have been proposed to improve CNN performance to yield better results [19, 20]. Many studies were suggested to improve image classification accuracy results using various approaches. Therefore, the suggested approach is proposed for various applications such as image recognition, segmentation, and other tasks. Table 1 tabulates the related papers using different deep learning models for various types of image classification.

3 Methodology

This section discusses the overall process of the methodology in detail. Data acquisition, image preprocessing and data augmentation process, CNN model development, and performance evaluation are the four stages of character recognition in general.

3.1 Data Acquisition

Databases are used on the system for data acquisition. In this process, the data can be in an image format such as Joint Photographic Experts Group (JPEG), Portable Network Graphics (PNG) or scanned image, or other suitable digital devices, or it can be generated on a user interface canvas. Schoolchildren's handwriting is the type of characters involved in the project. Handwritten data consisting of image forms have been collected among primary schoolchildren aged 7–9 years. The images were saved as PNG files in grayscale, with a resolution of 3456×4608 pixels. Then, the images are split into two groups, with 70% of the images for the training set and 30% for the validation set, as shown in Table 2. The total number of databases in data collection that were obtained is 520 image characters which were entered as input into the generated system.

Table 1 Summary of related paper using different deep learning models for other types of image classification

References	Categories	Dataset	Architecture	Results
Neha Sharma et al., 2018 [14]	Object classification	ImageNet, CIFAR10, CIFAR100, MNIST	AlexNet, GoogleNet, ResNet-50	Accuracy ResNet-50 is the highest (78.1%) using CIFAR10
Feiyang Chen et al., 2018 [15]	Digit recognition	MNIST	ResNet-50, DenseNet, CapsNet, CNN	Accuracy CapsNet is the highest (99.75%)
Iza Sazanita Isa et al., 2021 [10]	Dyslexia classification	NIST Special Database 19, A-Z handwritten alphabets, Dyslexia kids	CNN-1, CNN-2, CNN-3, LeNet-5	The accuracy of CNN-2 is the highest (87%)
Akmaljon Palvanov et al., 2018 [17]	Digit recognition	MNIST	ResNet-50, CapsNet, CNN	The accuracy of CapsNet is the highest (98.1%)
Weiwei Jiang, 2020 [9]	Character recognition	Urdu character	12 different CNN models	Accuracy of 98.94% and 99.08%
Pedro Ballester et al., 2016 [18]	Object recognition	ImageNet	GoogleNet, AlexNet	Both models are mainly unable to classify the shown sketches consistently

Table 2 Dataset Details

Dataset	Each Class
Training Data	364
Validation Data	156
Overall Data	520

3.2 *Image Preprocessing and Data Augmentation*

The preprocessing process reduces the system's complexity and increases its accuracy using the algorithm it employs. Developing a decent algorithm for every situation when an image is captured is tough. Therefore, after obtaining the images, they must be translated into a format allowing a generic algorithm to be better understood and solved. The overall work in the dataset image preprocessing method is represented in Fig. 1. Pre-processing is a process of converting an image dataset to a more common form before entering it into classifiers as the original datasets [10]. These datasets often come in various sizes, resolutions, and forms.

First, the dataset image preprocessing process will be done by cropping the image to the writing part. This phase removes unnecessary portions of the image from the top, bottom, right, and left, allowing the image to focus just on the letters

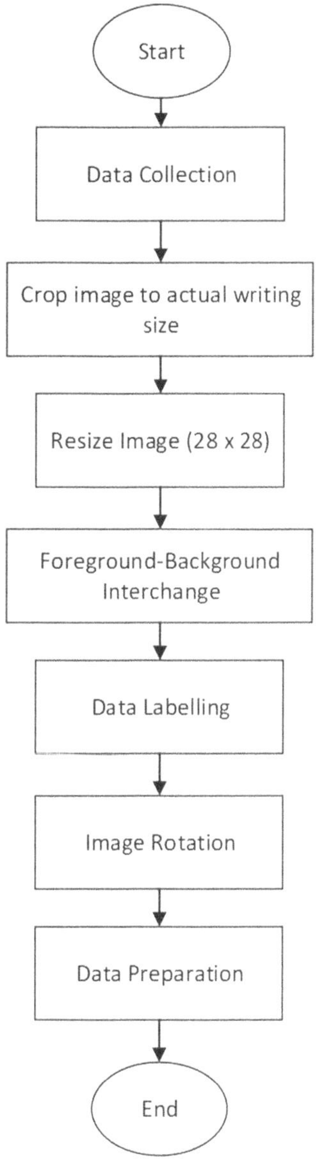

Fig. 1 The flowchart method used for image preprocessing and data augmentation

in the middle. Then, all the images will be resized to 28 × 28 pixels to get a consistent dataset size for the CNN model. The process is done by choosing an interchange background to reduce the computational overhead due to the image's more significant proportion of white than black points. As a result, more power and memory would be used to train the image [10]. The background will turn black when the handwriting is changed to white during this process. The final image after

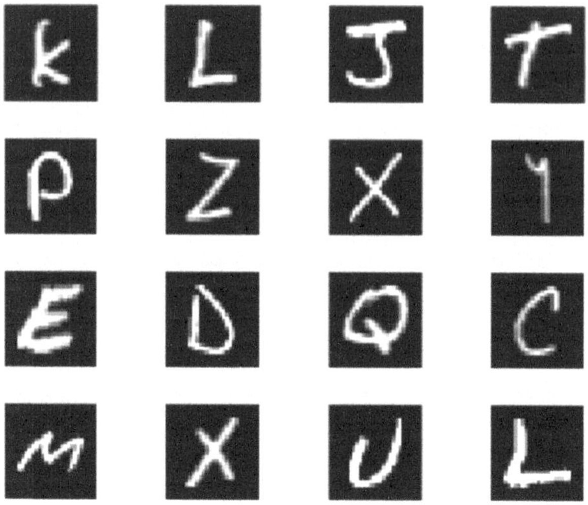

Fig. 2 Example of randomly displayed images

preprocessing is shown in Fig. 2. The image of this character is then distributed evenly into 26 different classes and labeled from A to Z, then each class consists of 20 images.

Next, a large dataset must be trained using CNN models to obtain satisfactory performance. In practice, gathering a high-quality image takes much time. With some image transformation techniques, data augmentation has expanded the training dataset. Training, validation, and test data will be transformed into an augmented image datastore with optional preprocessing like resizing, rotation, and reflection. This technique could help to increase the validation performance since it can produce a better image for data training. Image rotation was used as augmented sets for the dataset in this process. The image is randomly rotated between −30 and +30 degrees, with the center fixed. Figure 3 shows an example of data augmentation where the images were rotated. Data augmentation is also used to reduce the model's overfitting [15] and memorize the exact details of the training images, especially when using small quantities of a dataset.

3.3 Development of CNN Models

Pre-trained models will be used to compare different deep learning networks, particularly CNN models, in terms of identifying and classifying the handwritten characters of schoolchildren with a high accuracy rate and a low loss rate. CNN architectures consist of convolutional, pooling, and fully connected (FC) layers. In this study, four models have been used: AlexNet, GoogleNet, CNN-2, and ResNet-

Fig. 3 Example data augmentation images

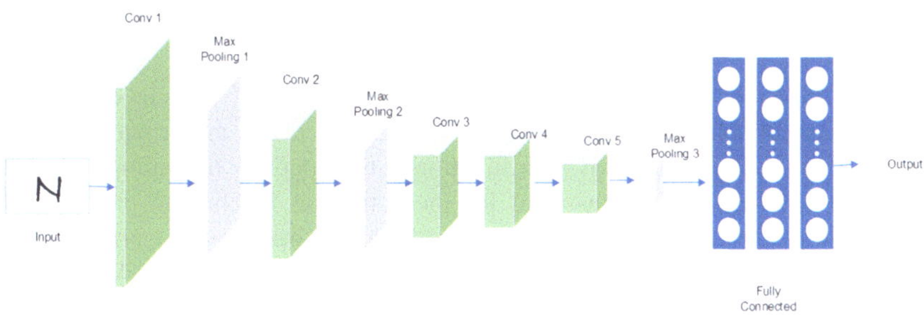

Fig. 4 AlexNet architecture

50. Utilizing pre-trained CNN models aims to make the dataset training process quicker and simpler while improving accuracy and loss rates.

There are many pre-trained models available, but the AlexNet model has been selected as the one to focus on because it was the first leading model to include the convolutional layers that define a CNN, which will help provide a better understanding of how CNN models are constructed and function. Eight layers make up the AlexNet architecture, with three fully connected layers and five convolutional layers [3] as shown in Fig. 4. The innovations made were that datasets were trained on multiple Graphical Processing Units (GPUs), and additional versions such as inverted, scaled, and noise of images was used for training. Moreover, the model uses the Rectified Linear Unit (ReLU) activation function instead of tanh (hyperbolic tangent), which was standard at the time, and helped to reduce the time training for the model. The overlapping pooling was also introduced in this model, and it was found slightly more challenging to overfit when it was used. Since the pre-trained model was used to train the dataset, the last three layers were replaced with a fully connected layer, a softmax layer, and a classification output layer. The final fully connected layer is set in order to obtain the same size as the number of classes in the training dataset. Lastly, the fully connected layer's learning rate factors are increased so the network can train faster.

The GoogleNet model has a total of 9 inception modules used in this study, and it has 27 layers overall, including pooling and 22 deep layers. This model has inception modules that carry out convolutions of various sizes and incorporate filters for the following layer [17]. The input layer of this model takes in an image with dimensions of 224 × 224 pixels. The first one is an average pooling layer with a 5 × 5 filter size and stride 3, which is the precise structure for GoogleNet with the auxiliary classifier [21]. Next, a 1 × 1 convolution with 128 filters was employed for dimension reduction and rectified linear activation. 1024 linear activation units with rectified connections and fully connected layers were also applied. This model also reduced 70% of the output ratio for the dropout layer and uses a linear layer with softmax loss. A new fully connected layer with the same number of outputs as the new data set's number of classes was substituted for the last layer of the fully connected layer in order to use GoogleNet as the pre-trained model for this study. Subsequently, a new convolutional layer with an equal number of filters as classes in place of the convolutional layer was added.

In this study, the CNN-2 model constructed consisted of four layers which are two convolutional layers, one fully connected layer, and an output layer, as shown in Fig. 5. After each activation function, a max-pooling operation of 2 × 2 was used. The ReLU layer was used for the activation function to prevent the computation required to run the neural network from growing exponentially [22]. Dropout operations are also used during training datasets, so the network does not experience overfitting. The mini-batch size used for this model is 10.

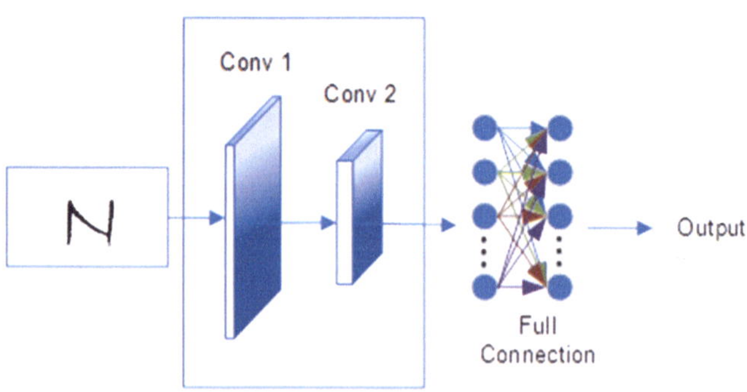

Fig. 5 CNN-2 architecture

Lastly, Residual Network (ResNet) was also used for this study. This model has been used by numerous researchers in other visual recognition tasks, and they have benefited from it. The need to delve deeper into these architectures arises from the need to solve more challenging problems. Five stages comprise the ResNet-50 model, each with its convolution and identity block. Each convolution block has three convolution layers, and each identity block has three convolution layers [23]. The ResNet-50 has about 23 million trainable parameters. Using the ResNet-50 as a pre-trained network involves replacing the fully connected layer with a new layer, followed by the last layer of the model with a new layer with the same number of outputs as the number of classes in the dataset. This step is similar to those for AlexNet and GoogleNet models.

3.4 Performance Evaluation

The performance of each CNN model will be evaluated from the accuracy percentage rate and losses as the performance indicator for this study. Since classification is the main focus of this study, accuracy—the proportion of test samples that are correctly classified—is the main evaluation metric for this research. The accuracy measures the proportion of actual classifications among all the datasets. Formally, accuracy can be defined as follows:

$$Accuracy = \frac{TP + TN}{TP + FP + TN + FN} \quad (1)$$

TP is true positive, representing the total number of schoolchildren's handwriting images that the model correctly classifies. Meanwhile, TN is a true negative, which refers to the number of images incorrectly classified by the model.

Next, the losses for the performance of each CNN model can be defined as follows:

$$\mathrm{Loss} = \frac{1}{\mathrm{N}} \sum_{\mathrm{i}=1}^{\mathrm{N}} (\mathrm{y_i} - \hat{\mathrm{y}}_\mathrm{i})^2 \quad (2)$$

Equation 2 expresses the total sum of the square of the difference between the actual value, y_i, and the predicted value, $\hat{y}_i$, while N is the total number of data sets used for the study.

Besides, a confusion matrix, a summary of prediction results on a classification problem, was used to evaluate the performance of each CNN model used in this study. A confusion matrix is a standard measure used while solving classification problems. This evaluation approach can be used for binary and multiclass classification problems by predicting any object in the classifier to visualize the model classification accuracy [24]. The number of correct and incorrect predictions is

		Predicted			
	Classess	A	B	C	D
Actual	A	TN	FP	TN	TN
	B	FN	TP	FN	FN
	C	TN	FP	TN	TN
	D	TN	FP	TN	TN

Fig. 6 Example of Confusion Matrix for Four Classes

Table 3 CNN configuration

CNN Parameter	Default Value
Learning Rate	0.0001
MaxEpochs	20
Validation Frequency	3 iterations
MiniBatch Size	10
Verbose	False

summarized with count values and grouped by class, as shown in Fig. 6. All of the evaluation metrics for a multiclass classification model are understandable in the context of a binary classification model in which the classes are simply positive and negative. These metrics are derived from the four categories listed below:

True Positives (TP): Characters whose true label is positive and whose class is correctly predicted to be positive.
False Positives (FP): Characters whose true label is negative and whose class is incorrectly predicted to be positive.
True Negatives (TN): Characters whose true label is negative and whose class is correctly predicted to be negative.
False Negatives (FN): Characters whose true label is positive and whose class is incorrectly predicted to be negative.

3.5 Configuration

Table 3 shows the default parameters for specifying the training option for CNN in MATLAB. Firstly, the learning rate is used because it is the main parameter that will control the speed of training of the model. A low learning rate can produce a more accurate outcome, but it may take longer to train the network. The next training option used in this study is Max epochs, where an epoch represents a complete pass of the training algorithm across the whole training set. The longer the network training takes, the more epochs are utilized.

Nonetheless, the precision may increase with each epoch. In addition, a Mini-batch size is employed in the training option, which simultaneously processes a fraction of the training data set. The bigger the Mini-Batch size, the faster the training. However, the maximum size is limited by the memory of the graphics processing unit (GPU). If a memory error occurs during training, it is advised to decrease the mini-batch size. Lastly, stochastic gradient descent with momentum (SGDM) optimizer is chosen by configuring 0.0001 as the learning rate. SGDM is a method for iteratively improving an objective function with sufficient smoothness attributes. The optimization algorithm is used to train a machine learning model to reduce its error rate.

4 Results and Discussion

This section comprises two main findings: image preprocessing and overall performance of classification.

4.1 Image Pre-processing

The pre-processing image procedure is a technique of preparing a database so that it can refer to relevant aspects that match and improve performance. The image is cropped, resized, foreground-background interchanged, and labeled by each class in this section before the data sets, were used for the training process [28–30]. Each original image had a resolution of 960 × 1280 pixels and was in Red, Green, and Blue (RGB) format. As the image size was too large to be used for training, the image size was reduced and adjusted to 28 × 28 pixels; the same as the actual handwriting. This process is also essential to obtain a consistent data set size for the CNN model. If the images used for training are too large, the training process will be slowed and take longer to complete. Figure 7 shows the results obtained after the images have been cropped and resized to the size of the actual handwriting, which is 28 × 28 pixels. Then, the process is continued by interchanging the foreground background of the image in order to reduce the computational overhead. In this process, the handwriting part in the images changed from black to white, while the black background color was changed to white. The result of the interchange background is illustrated in Fig. 8 This process may consume more power and memory when an image has more black than white color during training.

Next, CNN models must be trained on a large dataset to achieve better recognition performance. However, collecting high-quality images in practice takes much time. Therefore, data augmentation is a technique that has been used to enhance the training dataset by utilizing image processing techniques. It can help increase the training dataset size; however, it can still perform a random transformation, such as rotation on each training image, and present that to the network instead of the original training images. It also rarely sees an exact copy of the actual training

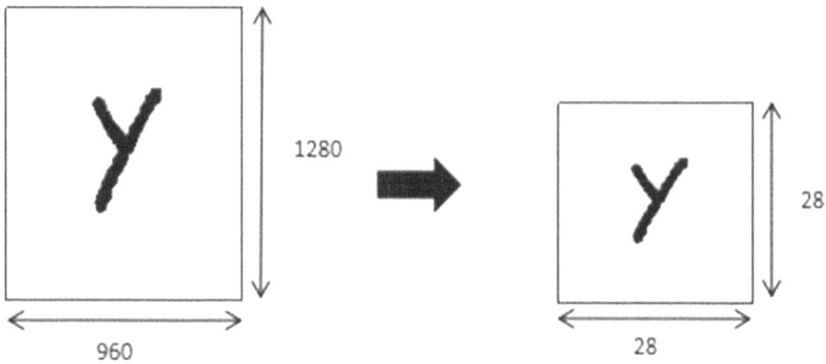

Fig. 7 The result after the image was cropped and resized

Fig. 8 The result after foreground-background interchange

images. Then, the random generator must rotate the image at a volume close enough to 0 degrees so that the image is rotated well when discretization is considered. It differs from what is expected because, for instance, it is anticipated that data augmentation will allow for the generation of a more crucial training data set that can be split up into mini-batches and presented to the network repeatedly every epoch, resulting in more iterations per epoch.

4.2 Performance Evaluation

CNN model has been applied to the handwritten data set of schoolchildren to validate the comparison between CNN models, which are AlexNet, GoogleNet, CNN-2, and ResNet-50 models, by analyzing the performance of each model in terms of accuracy and loss. Training and validation accuracy for 20 epochs were observed for the performance of different types of deep learning methods to be used in recognizing the handwritten images of schoolchildren by taking the mini-batch size of 10 for all CNN models. A different CNN model was used when predicting the handwriting of schoolchildren. The performance of the models for the handwritten image of schoolchildren recognition was evaluated using overall recognition accuracy, learning curves, and confusion matrix.

Table 4 shows the results of validation accuracies, losses, and training time of each model after the simulation of recognizing schoolchildren's handwritten characters. The dataset's image was divided randomly into 70% for training and 30% for validation. It is observed that the AlexNet has the highest accuracy

Table 4 Performance comparison for each model

CNN model	Accuracy (%)	Rate Loss (%)	Training Time
AlexNet	92.95	3.79	178 min 7 s
GoogleNet	80.77	5.72	540 min 25 s
CNN-2	83.98	9.46	5 min 40 s
ResNet-50	87.82	4.29	1181 min 6 s

compared to other models, followed by ResNet-50, CNN-2, and lastly, GoogleNet, with an accuracy of 92.95%, 87.82%, 83.98%, and 80.77%, respectively. The validation accuracy is the percentage of correctly recognized schoolchildren's handwritten characters. In addition, it is observed that the least rate loss performance was obtained by AlexNet, followed by ResNet-50, GoogleNet, and ResNet-50, with rate loss of 3.79%, 4.29%, 5.72%, and 9.46%, respectively. The rate loss is the percentage of incorrectly recognized schoolchildren's handwritten characters. The outcome indicates that the method's performance is linked to the network's complexity. Adding a layer or a more complicated architecture will not ensure a higher prediction of accuracy percentage.

The training time of each CNN model obtained by AlexNet, GoogleNet, CNN-2, and ResNet-50 is also shown in Table 4, with 178 min 7 s, 540 min 25 s, 5 min 40 s, and 1181 min 6 s, respectively. CNN-2 model showed the least training time to complete the learning methods because of the lowest network dept. However, it did not yield the best performance. CNN-2 performed better than GoogleNet on all divided datasets. In a previous study, the CNN model provided the most substantial baseline. However, it could not yield the best result on a dataset [19]. On the other hand, ResNet-50 used the most prolonged time among the other three models to train the data because it is a deep architecture and uses a large amount of memory to store the model and make predictions. Each CNN model has four architectural layers: ResNet-50 has 50 layers, GoogleNet has 22, and AlexNet has eight layers.

Figures 9 and 10 illustrate the training and validation for accuracy and losses for each CNN model. In Fig. 9, it is shown that the AlexNet model has the most excellent and stable accuracy. For the AlexNet model, the training and validation accuracy for 20 epochs were 100% and 92.95%, respectively. Moreover, it is also apparent that the loss of AlexNet is the lowest among the three models, which is 0.3% for 20 epochs for training losses and 3.79% for validation losses, as shown in Fig. 10. Next, as shown in Fig. 9, it can be seen that the training accuracy for GoogleNet is 90%. The overall validation accuracy is 80.77%, where the GoogleNet model has the lowest performance among the other models. The graph also shows that the graph plots of CNN-2 and ResNet-50 are remarkably similar. In addition, AlexNet displays the best results; despite the CNN-2 model having the quickest training time to finish the learning methods, AlexNet was still able to perform better than CNN-2. The AlexNet model's accuracy is low at the beginning of the iteration, but it rises after eight epochs and then stabilizes with an average accuracy of about

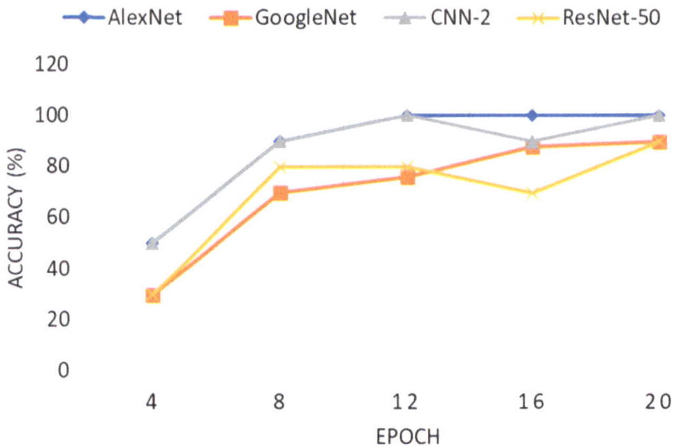

(a)

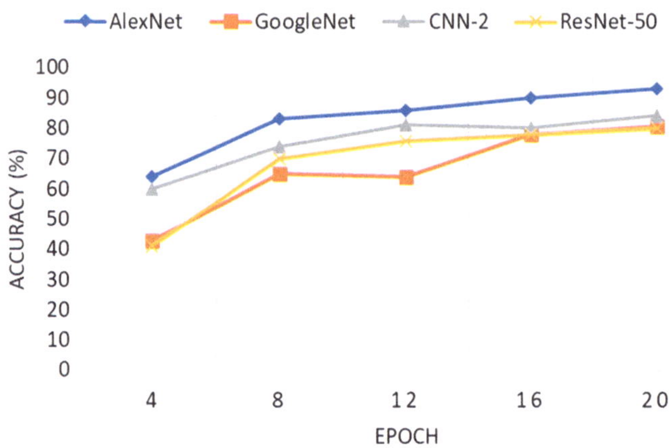

(b)

Fig. 9 (**a**, **b**) Training and validation losses against the number of epochs

92.95%. The other model, though, does not exhibit a high degree of accuracy compared to the AlexNet model.

Next, Figs. 11, 12, 13, and 14 illustrate the multiclass classification model with the confusion matrix in order to understand which characters are mistaken for the others. On the x-axis are the target class labels, while on the y-axis are the output class predictions. The correct categories are displayed along the first diagonal portion, while all other entries exhibit incorrect classifications. The confusion matrix, the bottom at the right-side cell, shows the overall classification accuracy. Figure 11 shows that letter H has three letters that are mistakenly predicted as letter K and letter M because it belongs to the positive class by the model.

Furthermore, it can be noticed that two letters are incorrectly projected as letter R for letter P. Next, in Fig. 12, letters E are misclassified as letters B, letter D, and letter

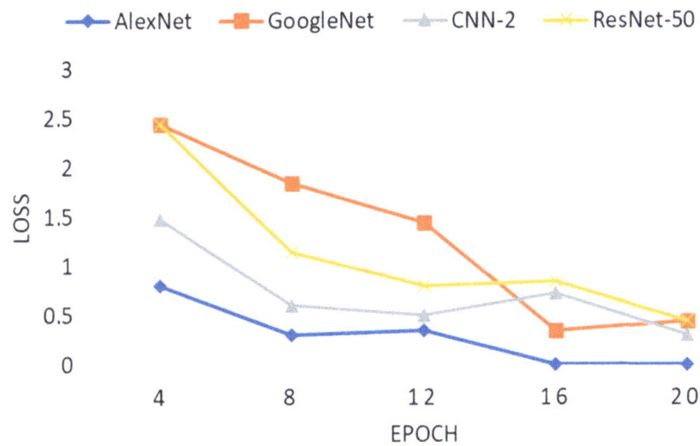

(a)

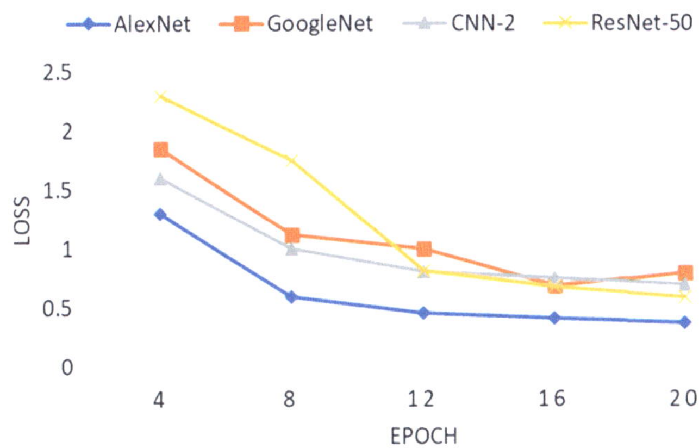

(b)

Fig. 10 (**a**, **b**) Training and validation losses against the number of epochs

G. There are also three letters of the letter W that the model incorrectly predicts as belonging to the positive class. Figure 13 indicates that the model correctly classified just two letters as letter N. In comparison, the other four letters, two for letter A, one for letter D, and one for letter V were mistakenly identified as belonging to the negative class by the model. Finally, in Fig. 14, it can be observed that none of the letters M is correctly recognized by the model, where the letters are mistakenly classified as the letter N, and the letter belongs to a negative class.

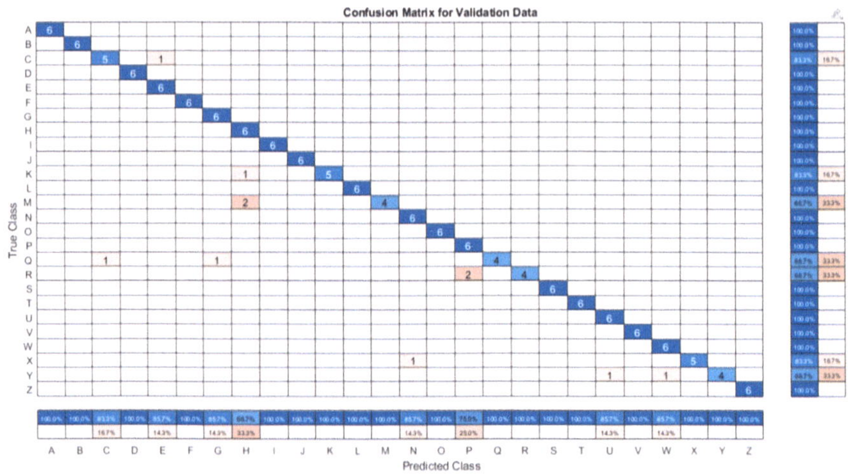

Fig. 11 Confusion matrix for all classes and all attributes for the AlexNet model

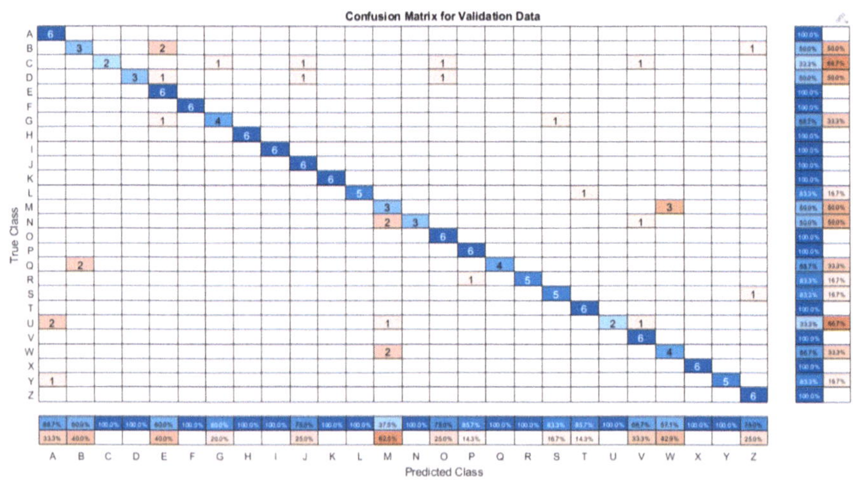

Fig. 12 Confusion matrix for all classes and all attributes for the GoogleNet model

4.3 Performance Comparison

Finally, experiments are conducted to compare the results to the previous findings, as presented in Table 5. This table summarises the performance results with various CNN model configurations and different sources of dataset domains. Zin T. et al. [25] have revealed the performance of the traditional network model for a combined self-collect dataset with 91.63% accuracy. In contrast, N. Saqib et al. [26] have investigated the performance of the MNIST dataset for different values of learning

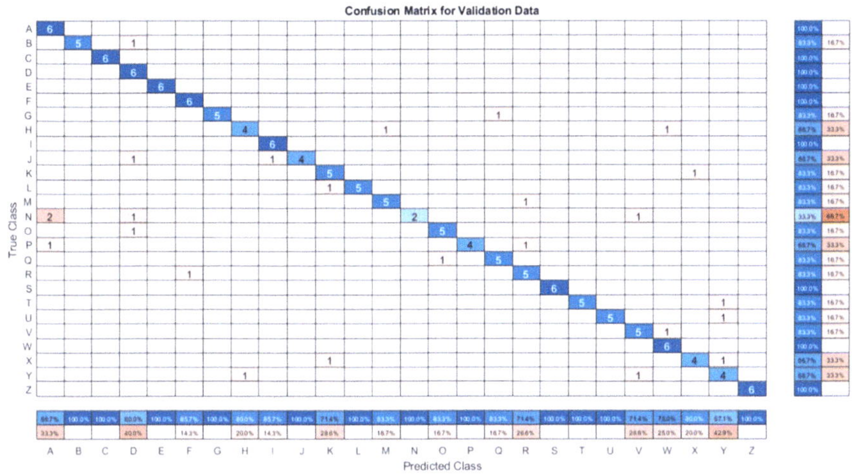

Fig. 13 Confusion matrix for all classes and all attributes for CNN-2 model

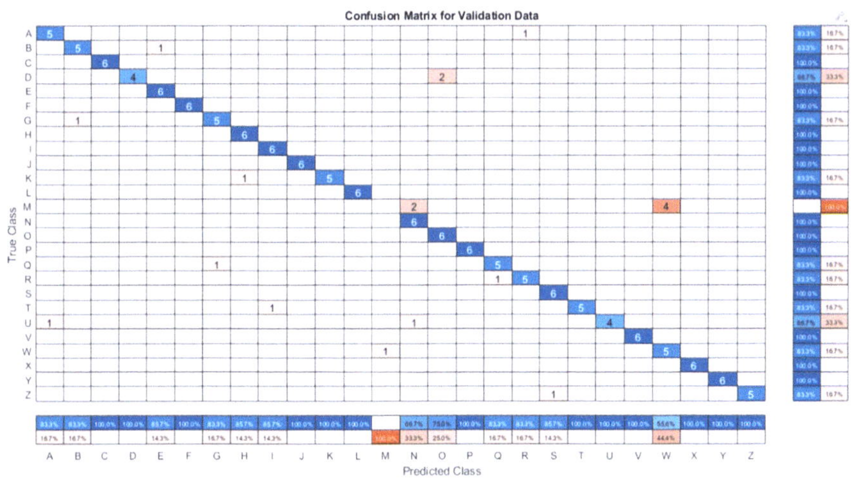

Fig. 14 Confusion matrix for all classes and all attributes for the ResNet-50 model

rates and two types of optimizers; ADAM and RMSprop. Meanwhile, different types of pretrained networks have been explored by S. Corbille et al. [27] and obtained VGG-11 achieved the highest accuracy compared to LeNet-5 and ResNet-18 with 95% accuracy. In this study, the pretrained networks model and plain sequence CNN-2 have been implemented, and an SGDM optimizer is relatively effective in handwriting.

Figure 15 illustrates the scattered accuracy performance for other research in handwriting recognition and clearly shows that the CNN-based model is significant

Table 5 Performance comparison of current research in handwriting recognition

Study	CNN Model	Dataset Domain	Classification Category	Performance
Zin, T. et al. (2021) [25]	Three convolutional layers, three max-pooling layers, and one fully connected layer.	Handwritten image: 72 characters (digit, alphabet and special characters).	6 categories: Digit and Special Characters, Uppercase, Lowercase, Uppercase and Lowercase, Digit and Lowercase, All combine	ADAM optimizer Digit and Special Characters 99.25%, Uppercase 98.8%, Lowercase 98.67%, Uppercase and Lowercase 91.97%, Digit and Lowercase 95.46%, All combined 91.63%
N. Saqib et al. (2022) [26]	Four convolution layers, three max-pooling layers, two fully connected layers	Kaggle English handwritten dataset was sourced from the (NIST), NMIST, and other google images.	26 categories of character	ADAM optimizer (99.516%, 99.511%, and 99.563% for LR 0.001, LR 0.0001, and LR 0.00001). RMSprop optimizer (99.292%, 99.108%, and 99.191% by LR 0.001, LR 0.0001, and LR 0.00001).
S. Corbille et al. (2020) [27]	LeNet5, VGG and ResNet-18	Handwritten Latin Alphabet	26 categories of character	ADAM optimizer LeNet-5: 92.66% VGG 11 = 95.00% and ResNet-18 = 94.64%
This study	AlexNet, GoogleNet, CNN-2, ResNet-50	Handwritten English Alphabet	26 categories of character	SGDM optimizer AlexNet = 92.95% GoogleNet = 80.77% CNN-2 = 83.98% ResNet-50 = 87.82%

LR learning rate

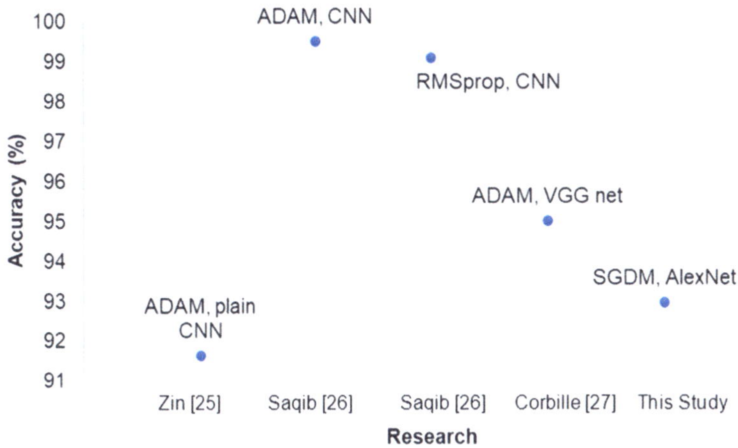

Fig. 15 Accuracy comparison of current research in handwritten recognition

in various domains of handwriting. This study specifically used schoolchildren's handwriting images as input and showed relatively impressive performance accuracy using SGDM optimizer at 0.0001 learning rates. Overall, all findings have depicted the significance of handwriting recognition, especially the exploration of schoolchildren's handwriting products.

5 Conclusion

In conclusion, the study was implemented to compare the deep learning model specifically for CNN for handwriting recognition of schoolchildren. This study aims to evaluate the different CNN models for the handwritten character recognition system among schoolchildren that have been completed. Fundamentally, the study is needed to collect the images of schoolchildren's handwritten characters as much as possible from primary schoolchildren aged 7–9. Then, the images go through an image preprocessing process and data augmentation technique before the images are used for training and validating using different CNN models. There are several CNN models, but generally, CNN comprises layers of convolution and pooling organized into modules. CNN is a deep neural network often used in Artificial Intelligence (AI) for computer vision problems. CNN is a deep learning technique that can train and classify data inputs such as handwritten images by assigning learnable weights and biases to numerous features.

In this study, four different models have been used: AlexNet, GoogleNet, CNN-2, and ResNet-50 model.

The result shows that the AlexNet model has the best performance, with the most incredible and stable accuracy. It can successfully reach the highest accuracy within a short period. Furthermore, AlexNet has the lowest losses among the other three models. Moreover, the performance of different deep learning CNN models in recognizing schoolchildren's handwritten images in terms of the percentage of accuracy and losses are also successfully compared. In this research, the AlexNet, GoogleNet, CNN-2, and ResNet-50 models were tested for their accuracy and error rates in classifying the handwritten characters of schoolchildren. The AlexNet model showed the highest validation accuracy compared to other models that achieved an accuracy of 100% for training and 92.95% for validation. The results show that the method's performance is related to the complexity of the network, and the addition of a layer or a more complicated architecture will not lead to a high percentage of prediction accuracy. This study is extensive for each architecture, including overall recognition accuracy, learning curve, and confusion matrix. An automatic character recognition system will help improve the efficiency of recognizing handwriting among schoolchildren, such as their notes, exercises, and more.

It is advised that the CNN model is trained with a large dataset of handwritten images of schoolchildren for subsequent studies in order to obtain more accurate results from the model. Additionally, training with more data can help to improve the model's accuracy and prevent the overfitting issue.

Acknowledgments This research work is financially supported by the Ministry of Higher Education Grant Scheme (FRGS) - (Ref: FRGS/1/2021/ICT02/UITM/02/4). The authors would like to express their gratitude to members of the Research Intervention for Dyslexia: Learning & Technology (RIDyLT), and Centre for Electrical Engineering Studies, Universiti Teknologi MARA, Cawangan Pulau Pinang, for their assistance and guidance during the fieldwork. Finally, the authors are thankful to Universiti Teknologi MARA, Cawangan Pulau Pinang, for their immense administrative and financial support.

References

1. Chakraborty R et al (2020) Online handwritten Bangla and Devanagari character recognition by using CNN: a deep learning concept. In: 2020 IEEE Int. Conf. Converg. Eng. ICCE 2020 – Proc., pp 68–71. https://doi.org/10.1109/ICCE50343.2020.9290566
2. Nikitha A, Geetha J, Jayalakshmi DS (2020) Handwritten text recognition using deep learning. In: Proc. – 5th IEEE Int. Conf. Recent Trends Electron. Inf. Commun. Technol. RTEICT 2020, pp 388–392. https://doi.org/10.1109/RTEICT49044.2020.9315679
3. Alex Krizhevsky GEH, Sutskever I (2012) ImageNet classification with deep convolutional neural networks. Commun ACM 60(6):84–90
4. Cireşan D, Meier U, Schmidhuber J (2012) Multi-column deep neural networks for image classification. In: Proc. IEEE Comput. Soc. Conf. Comput. Vis. Pattern Recognit. https://doi.org/10.1109/CVPR.2012.6248110
5. Alzubaidi L et al (2021) Review of deep learning: concepts, CNN architectures, challenges, applications, future directions, vol 8, no 1. Springer International Publishing, J Big Data. https://doi.org/10.1186/s40537-021-00444-8
6. Cires DC, Meier U, Masci J, Gambardella LM (2003) Flexible, high performance convolutional neural networks for image classification. In: Proceedings of the twenty-second international joint conference on artificial intelligence flexible, pp 1237–1242. https://doi.org/10.5591/978-1-57735-516-8/IJCAI11-210

7. Koyuncu B, Koyuncu H (2019) Handwritten character recognition by using convolutional deep neural network; review. Int J Eng 5(1):1–5. https://doi.org/10.19072/ijet.528775
8. Yuan A, Bai G, Yang P, Guo Y, Zhao X (2012) Handwritten English word recognition based on convolutional neural networks. In: Proceedings – international workshop on frontiers in handwriting recognition, IWFHR, pp 207–212. https://doi.org/10.1109/ICFHR.2012.210
9. Jiang W (2020) Evaluation of deep learning models for Urdu handwritten characters recognition. J Phys Conf Ser 1544(1). https://doi.org/10.1088/1742-6596/1544/1/012016
10. Isa IS, Zahir MA, Ramlan SA, Li-chih W, Sulaiman SN (2021) CNN comparisons models on dyslexia handwriting classification. ESTEEM Acad J 17(March):12–25
11. Mahmud M, Kaiser MS, Hussain A, Vassanelli S (2018) Applications of deep learning and reinforcement learning to biological data. IEEE Trans Neural Netw Learn Syst 29(6):2063–2079. https://doi.org/10.1109/TNNLS.2018.2790388
12. Waseem Rawat ZW (2018) Deep convolutional neural networks for image classification: a comprehensive review. 2733:2709–2733. https://doi.org/10.1162/NECO
13. Eltay M, Zidouri A, Ahmad I (2020) Exploring deep learning approaches to recognize handwritten Arabic texts. IEEE Access 8:89882–89898. https://doi.org/10.1109/ACCESS.2020.2994248
14. Neha Sharma AM, Jain V (2018) An analysis of convolutional neural networks for image classification. Procedia Comput Sci 132(Iccids):377–384. https://doi.org/10.1016/j.procs.2018.05.198
15. Chen F, Chen N, Mao H, Hu H (2018) Assessing four neural networks on handwritten digit recognition dataset (MNIST). Chuangxinban J Comput June:1–4
16. Arif RB, Siddique MAB, Khan MMR, Oishe MR (2018) Study and observation of the variations of accuracies for handwritten digits recognition with various hidden layers and epochs using convolutional neural network. In: 4th International conference on electrical engineering and information and communication technology, iCEEiCT 2018, pp 112–117. https://doi.org/10.1109/CEEICT.2018.8628078
17. Palvanov A, Cho YI (2018) Comparisons of deep learning algorithms for MNIST in real-time environment. Int J Fuzzy Log Intell Syst 18(2):126–134. https://doi.org/10.5391/IJFIS.2018.18.2.126
18. Ballester P, Araujo RM (2016) On the performance of googlenet and alexnet applied to sketches. In: 30th AAAI Conf. Artif. Intell. AAAI 2016, pp 1124–1128
19. Ranzato M, Huang FJ, Boureau YL, LeCun Y (2007) Unsupervised learning of invariant feature hierarchies with applications to object recognition. Proc IEEE Comput Soc Conf Comput Vis Pattern Recognit:1–8. https://doi.org/10.1109/CVPR.2007.383157
20. Hinton NSG, Sutskever AKI, Salakhutdinov R (1993) Dropout: a simple way to prevent neural networks from overfitting. Phys Lett B 299(3–4):345–350. https://doi.org/10.1016/0370-2693(93)90272-J
21. Zeng G, He Y, Yu Z, Yang X, Yang R, Zhang L (2016) Going deeper with convolutions. J Chem Technol Biotechnol 91(8):2322–2330. https://doi.org/10.1002/jctb.4820
22. Kwak NS, Müller KR, Lee SW (2017) A convolutional neural network for steady state visual evoked potential classification under ambulatory environment. PLoS One 12(2):1–20. https://doi.org/10.1371/journal.pone.0172578
23. Kaiming (2006) Deep residual learning for image recognition. Indian J Chem – Sect B Org Med Chem 45(8):1951–1954. https://doi.org/10.1002/chin.200650130
24. Grandini M, Bagli E, Visani G (2020) Metrics for multiclass classification: an overview. pp 1–17. [Online]. Available: http://arxiv.org/abs/2008.05756
25. Zin TT, Thant S, Pwint MZ, Ogino T (2021) Handwritten character recognition on android for basic education using convolutional neural network. Electronics 10(8). https://doi.org/10.3390/electronics10080904
26. Saqib N, Haque KF, Yanambaka VP, Abdelgawad A (2022) Convolutional-neural-network-based handwritten character recognition: an approach with massive multisource data. Algorithms 15(4). https://doi.org/10.3390/a15040129

27. Corbille S, Fromont E, Anquetil E, Nerdeux P (2020) Integrating writing dynamics in CNN for online children handwriting recognition. In: Proc. Int. Conf. Front. Handwrit. Recognition, ICFHR, vol 2020, pp 270–275. https://doi.org/10.1109/ICFHR2020.2020.00057
28. Parah SA, Seikh JA et al (2018) Utilizing neighborhood coefficient correlation: a new image watermarking technique robust to singular and hybrid attacks. Multidim Syst Sign Process 29:1095–1117
29. Parah SA, Sheikh J, Dey N, Bhat GM (2017) Realization of a new robust and secure watermarking technique using DC coefficient modification in pixel domain and chaotic encryption. J Glob Inf Manag (JGIM) 25(4):80–102
30. Parah SA, Rashid M, Vijaykumar V (2022) Artificial intelligence for innovative healthcare informatics. Springer. ISBN: 978-3-030-96568-6

Handwritten Urdu Recognition Using BERT with Vision Transformers

Aejaz Farooq Ganai and Farida Khurshid

1 Introduction

Urdu Language is a very famous language and widely spoken and written in most of the countries like Bangladesh, Afghanistan, UAE, Guyana, Pakistan, and in the Indian subcontinents [1, 2]. Recognition of Urdu has gained much importance due to enormous applications of optical character recognition (OCR) tools. Urdu OCR has been exhaustively explored from last few years after the emergence of various machine learning and deep learning techniques, in an effort to develop OCR tool for Urdu text. The exploration of printed Urdu OCR has achieved some remarkable recognition rates with these deep learning algorithms [3], but handwritten Urdu has been least explored because of several challenges encountered in handwritten Urdu text recognition such as unavailability of a standard handwritten Urdu dataset, huge variation among writing styles of different Urdu writers, the cursive nature of script, the similarity in shape of few Urdu handwritten characters and many more difficulties (Fig. 1) and complexities [4] in Urdu writing styles. Only a few researchers have contributed significant achievements towards the development of handwritten Urdu OCR. Furthermore, for printed Urdu text, there are commercially available Urdu OCR systems [3], but there is no OCR system available for the recognition of handwritten Urdu till date.

A. F. Ganai (✉) · F. Khurshid
Department Electronics & Communication Engineering, NIT Srinagar, Srinagar, India
e-mail: fklone@nitsri.net

S. A. Parah et al. (eds.), *Intelligent Multimedia Signal Processing for Smart Ecosystems*, https://doi.org/10.1007/978-3-031-34873-0_15

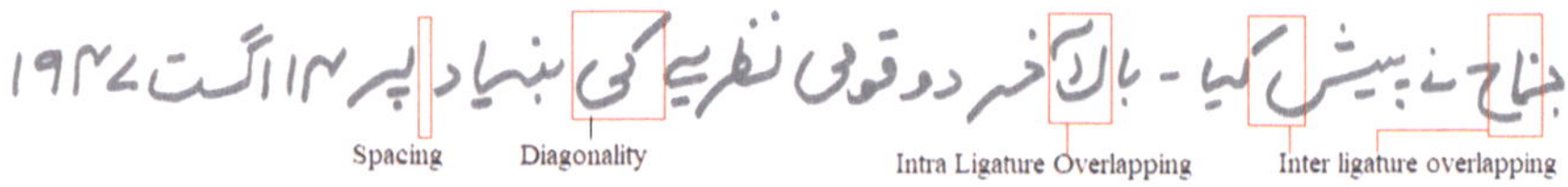

Fig. 1 Challenges in handwritten Urdu recognition

1.1 Challenges in Handwriting Recognition

The generalized challenges encountered in the development of Handwritten Urdu OCR include:

1. Extreme stroke variation and ambiguity from author to author.
2. Each writer's handwriting style is inconsistent and changes from time to time.
3. Deterioration in the original document's or image's quality over time.
4. The printed text is written in a straight line whereas handwritten text usually fluctuates while writing on a plain paper.
5. Diacritic separation and identification are difficult in cursive handwriting.
6. In addition to above challenges, the Urdu text recognition poses some basic challenges also that include ligature overlapping, diagonality and uneven spacing between ligatures (Fig. 1).
7. The collection and formulation of a labeled handwritten Urdu dataset for learning is also a challenge.

Because of these challenges, handwritten Urdu has been least explored, and hence this research mainly emphasizes on recognition of handwritten Urdu. These challenges have been addressed in the proposed work and a novel Transformer based methodology using Bidirectional Encoder Representations from Transformers (BERT) architecture [5] has been proposed in this research.

BERT [5] is a recently developed popular Natural Language Processing (NLP) model developed by Google. It is a pre-training methodology used in various NLP tasks such as sentiment analysis, text classification, masked language modelling and multi-label classification. We have employed BERT model for recognizing handwritten Urdu text because attention localization using BERT outperforms multidirectional LSTM models [6]. The main contributions of this study encompass:

1. A novel contextual attention-based Encoder-Decoder model is presented for handwritten Urdu text recognition.
2. The research covers most of the Urdu ligature corpus including six and seven character words/ligatures by evaluating all the available handwritten Urdu datasets Urdu Nastaliq Handwritten dataset (UNHD) [7], Urdu Handwritten Ligature dataset (UHLD) [our own dataset] and NUST-UHWR [8], for experimental evaluation in the proposed handwritten Urdu recognition system.
3. The research proposes a new approach of employing Multilayer perceptron (MLP) [9] for extraction of features from handwritten Urdu dataset images

(UNHD, UHLD, NUST-UHRW), and these features train BERT model for attention localization with context.
4. The accuracy at ligature/word level instead of character level have been reported in the proposed work.

The broad sections of this paper are categorized as follows: Section 2 describes notable research in the related field. In Sect. 2, the methodology employed for the recognition of handwritten Urdu text has been discussed in detail. Section 3 demonstrates the subjective/objective results obtained using the proposed approach and comparison with various state-of-the-art techniques to prove the superiority of the approach. Section 4 presents the conclusion of the paper in brief.

2 Related Work

The exploration of handwritten Urdu started in 2014 by Saad et al. [10]. The authors proposed a new dataset for the Urdu language written in Nastaliq writing style called UCOM Dataset. The UCOM dataset comprises 6400 text lines written by 100 writers. The authors extend their database up to 300 writers and labelled it with a unique identity of users. Later in 2018 [7], this dataset was extended to 500 writers and renamed as UNHD dataset. The authors performed training and classification using a single dimensional BLSTM classifier and realized an approximate error of 6.04–7.93%. In early 2019, Mujtaba et al. [11] employed Convolutional Neural Network (CNN) for the recognition and classification of handwritten Urdu characters. These authors developed a new dataset of handwritten Urdu characters and numerals and claimed an accuracy of 96% on character classification and 98% on numeral classification. Since their work remains confined to the character and numeral recognition only, hence it can't be applied to real-world handwritten Urdu text recognition. In mid-2019, Hassan et al. [12] developed the Urdu handwritten database of 6000 text lines [not publicly available] and employed 7 CNN layers for feature extraction and 2 BLSTM layers for classification and recognition of Urdu characters. Although the authors explored their work on handwritten character recognition from several thousand text lines but the recognition rate obtained was only 83%. In December 2019, Saad et al. [13] proposed another approach of exploiting transfer learning experience of similar patterns on handwritten Urdu text examination. The authors employed an MNIST pre-trained network by transferring its learning experience on UNHD samples. The features were extracted using CNN and experimental evaluation was performed using deep multi-dimensional long short-term memory (MDLSTM) networks. Although the authors reported a good character accuracy of approximately 93%, yet their work remained confined to the UNHD database that doesn't cover the significant variation with other datasets and their work doesn't explore recognition of handwritten Urdu Ligatures. Zia et al. [8] proposed their database called NUST-UHWR that includes around 18,000 unique ligatures and presented a hybrid CNN-RNN architecture using N-gram

Table 1 Summarized literature review of handwritten Urdu OCR

S.No.	Study	Database proposed/used	Framework employed	Unit of recognition	Recognition rate
1	Ahmed S.B. et al. [10]	UCOM	–	–	–
2	Ahmed S.B. et al. [7]	UNHD	BLSTM classifier	Character	92%
3	Husnain M. et al. [11]	Proposed a database of Urdu characters.	Convolutional neural network	Individual characters & numerals	96% characters 98% numerals
4	Hassan S et al., 2019 [12]	Claimed their own database, not publicly available	7 CNN layers for feature extraction and 2 BLSTM layers for classification and recognition.	Character	83.6%
5	Saad et al. [13]	UNHD	The features are extracted using CNN, experimental evaluation implemented using deep MDLSTM memory networks	Character	92%
6	Zia et al. [8]	NUST-UHWR	Presented a hybrid CNN-RNN architecture using N-gram language models.	Ligature	92.75%

Language models, but N-gram [14] models poorly capture longer-distance context and increase computational complexity. Moreover, the dataset doesn't include six and seven character ligatures.

The summarised literature review in the development of handwritten Urdu OCR is shown in Table 1.

After an exhaustive literature survey, it has been revealed that word recognition for handwritten Urdu text has been least explored to date, and efforts are made on segmentation based handwritten Urdu recognition approaches with maximum character recognition rate of 92% only. Hence in this research, an attempt has been made for complete word recognition of handwritten Urdu without implicit segmentation. Several accuracy parameters such as precision, recall, F1-Score, and

Receiver Operating Characteristic (ROC) curve have been evaluated for several ligature/word clusters. The elaborated methodology to achieve all these objectives has been discussed in the next section.

3 Methodology

The proposed methodology is based on BERT model for recognizing handwritten Urdu text. The publicly available handwritten Urdu datasets UNHD [7] and NUST-UHWR [8] have been employed for feature extraction using MLP [9]. The features are extracted from ligature patches in transformer encoder and are used to train the BERT model. The validation is performed on our own unconstrained handwritten Urdu dataset UHLD [our own dataset]. The trained model recognizes any unconstrained handwritten Urdu text. The various phases of the proposed model are discussed in following details.

3.1 Ligature Patch Extraction

The Urdu text line images from different handwritten Urdu datasets (UNHD [7], UHLD and NUST-UHWR [8]) are obtained. Since it is very difficult to separate words in a text line due to uneven spacing between words, hence recognizing a word is challenging [4]. However, a ligature being the longest connected component in a word, can be separated from another ligature [15, 16]. Therefore, localization of a ligature in a text line is possible. It must be noted that localizing an Urdu character as a part of a ligature is also a challenging task as characters vary their shapes with respect to position in a word due to the context-sensitive nature of Urdu script [4]. Hence, the research aims at localizing ligature images (with associated diacritics) in an Urdu text line. It is also worth to mention here that the process of extracting ligature images from handwritten Urdu text lines to undergo feature extraction in the transformer, has been performed using the connected component labelling [17] algorithm.

3.2 Patch and Position Embedding

The ligature image patches are unrolled into a sequence of images associated with positional embedding numbers like 1,2,3,4 and so on as shown in Fig. 2. This positional embedding assist the transformer to understand the sequence of ligatures for reconstruction of exact Urdu sentence. The positional embedding numbers are learnable vectors stored in the look-up table for each number representing position of the patch. Hence, for first patch, the first vector from look-up table is grabbed

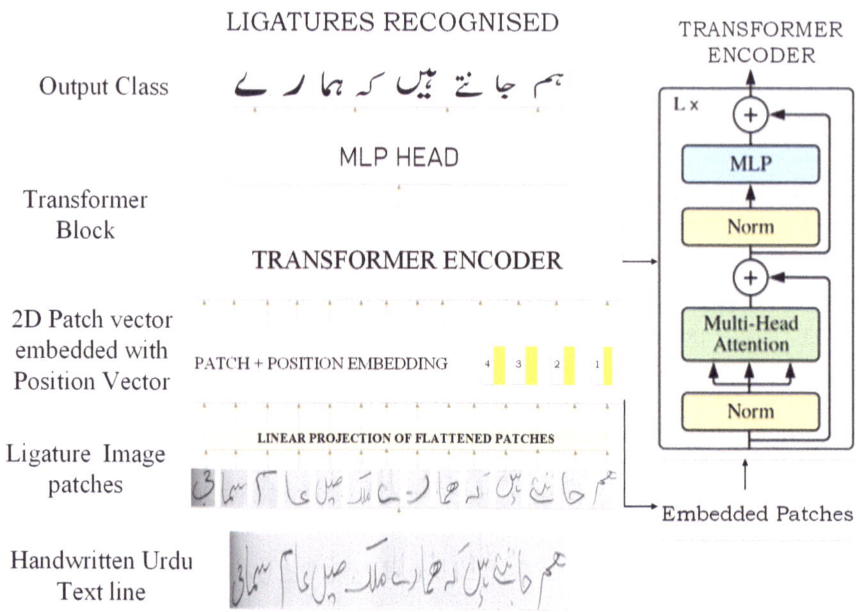

Fig. 2 Architecture of proposed Handwritten Urdu recognition model

and attached with the patch into the transformer, and so on. The image patch is fed to transformer by converting it into 256-dimension vector (16 × 16 pixels). Since the linear projection of flattened images are used, the image patch is unrolled into a linear vector. This linear vector is then multiplied with embedding matrix E. The resultant matrix is fed to transformer along with positional embedding as shown in Fig. 2.

3.3 Feature Extraction in Transformer Encoder

We employ MLP [9] for extracting features from embedded patches. Since the image is transformed into a vector, there might be very less loss of spatial information by using MLP for feature extraction. A Dense Net MLP architecture is used for extracting high level features. Filters can be applied to all of the prior layers in a block rather than just the most recent one by utilizing dense architecture. Because dense blocks minimize information loss during forward and backward propagation, they offer more features.

The Transformer encoder reads the entire string of words at once, in contrast to directional models that read the text input sequentially (from right to left or left to right). As a result, it is referred to as bidirectional, even though it would be more accurate to refer to it as non-directional. Because of this feature, the model

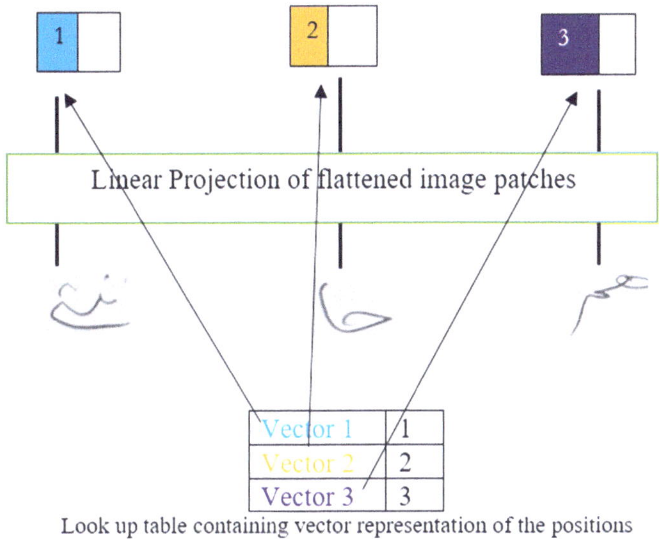

Fig. 3 Embedding vector matching

can understand a word's context by considering all of its surroundings (left and right of the word). The input image $X \in R^{L \times B \times C}$ is restructured into a series of compacted two dimensional patches $X^T \in R^{(N \times T2)\,C}$. The dimensions of image obtained is L × B by C channels while as dimensions of patch used is T × T. The patch sequence obtained has length N. In all of its layers, the transformer encoder uses features and an embedding with a constant width D. To match this embedding of size D, each compacted patch is transformed into an embedding of size D using a linear projection, which is depicted as a yellow box in Fig. 2. The patch sequence has a learnable class embedding matrix of the equal dimension D embedded in it. An exclusive position encoding of identical dimension D, is appended to every embedding (numbered in white patch in Fig. 2) to get a resulting vector which is given as input to Transformer encoder. The Fig. 3 shows the layers inside the transformer block.

Initially, each input is normalized because the styles of writers differ with respect to skew, slant, height and width of ligatures. The extracted input ligatures are not clear and contain noise and gaps within strokes caused due to loss of data. The Normalisation Layer (NORM) translates input images into constant and fixed size and makes feature extraction feasible. We have employed Multi-Head Self-Attention layer (MSA) in the transformer encoder that makes it possible for our proposed model to address the data simultaneously from diverse representations at diverse positions [18]. The relationships between feature vectors are determined by the MSA. We have used 'H' heads in our model. The information collected from all 'H' heads is again normalized before input to feature extraction phase. The MLP module extracts features from normalized input embedded patches. The MLP is

Table 2 The various transformer configurations

Transformer version	Patch size T	Depth L	Embedding size / Hidden size D	No. of heads H	Sequence length S
Tiny	16	12	192	3	27
Small	16	12	384	6	27
Large	16	12	768	12	27

composed of two sub layers with GELU activations [9]. An outstanding connection is located among output of NORM and MSA layers.

Mathematically, we can show the input to encoder (Z_i) as:

$$Z_i = \left[X_{class}; X_t^1 E; X_t^2 E \ldots; X_t^N\right] + E_{pE} \tag{1}$$

Where $E \in R^{T^2 C \times D}$, $E_{pE} \in R^{(N+1) \times D}$

The output obtained from MSA block can be written as:

$$Z_l' = \text{MSA}\left(\text{NORM}\left(Z_{l-1}\right)\right) + Z_{l-1} \tag{2}$$

For l = 1,2, 3...L where L is number of encoder blocks that the transformer is stacked of.

The output obtained from MLP block can be written as:

$$Z_l = \text{MLP}\left(\text{LN}\left(Z_l'\right)\right) + Z_l' \tag{3}$$

For l = 1,2,3 . . . L.

At the end, the MLP Head is composed of a series of linear projections that forms the basis of ligature prediction.

$$Z_i = \text{MLP}\left(\text{LN}\left(Z_l\right)\right) + Z_l \tag{4}$$

for i = 1...G. where G is the highest ligature length added by Start [S] and last [G] symbols. The summarized transformer configurations used in the proposed work are displayed in Table 2.

4 Experimental Results and Discussions

In order to determine the performance of various Transformer configurations, we have evaluated different handwritten Urdu datasets. Also we are importing ‘Simple

Table 3 The statistics of various handwritten Urdu datasets evaluated in the proposed work

Dataset	Text-lines	Words	Unique words	Unique ligatures
UNHD [7]	10,000	64,000	11,000	1100
NUST-UHWR [8]	10,608	110,940	24,327	8237
UHLD	6000	136,000	21,000	2100
Train dataset in proposed work	20,608	174,940	35,327	9337
Test dataset in proposed work	6000	136,000	21,000	2100

Table 4 The samples of handwritten Urdu datasets evaluated in the proposed work

Name of dataset	Text sample
UNHD [7]	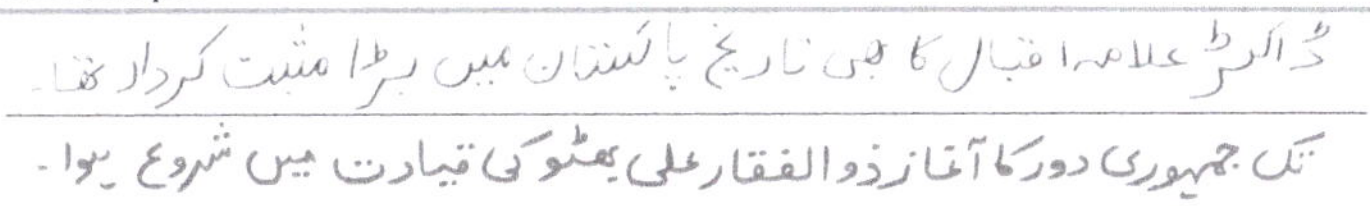
NUST-UHWR [8]	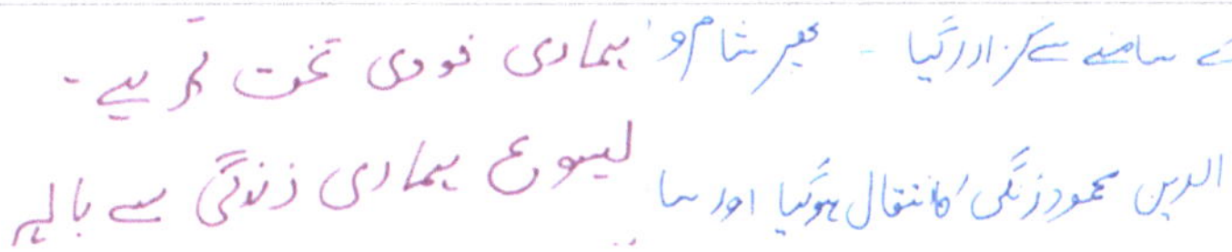
UHLD [our own dataset, not published yet]	

Transformers' library as this library offers a lot of additional features in addition to BERT. The experimental results obtained on different datasets are explained in the following sections.

4.1 Dataset

We have employed three different datasets for training our model. The statistics of these datasets along with their word length is given in Table 3. Few of the text lines of these datasets are given in Table 4.

In the proposed handwritten Urdu text recognition framework, each dataset (UNHD and NUST-UHWR) contributes 50% of total training dataset and the dataset used for testing purpose is our own dataset UHLD.

Table 5 Network parameters employed for training

Train dataset: 50% UNHD [7] + 50% NUST-UHWR [8]	Batch size: 192
Epochs: 30	No. of parameters: 110 M
Optimizer: Jais et al. [19]	Learning rate α: 0.001
beta1 = 0.9	*beta2 = 0.999, epsilon = 0.1*
Loss: logistic loss	Clip norm: 4
Image dimensions: 16 × 16	Greyscale channels: 1

4.2 Training Conditions

The network is trained using the $BERT_{BASE}$ with network parameters shown in Table 5. The features extracted from around 20,000 handwritten Urdu text lines, train the proposed network. The network is initially trained on 10 epochs, followed an increase in number of epochs to determine the impact on performance of network.

4.3 Recognition Rates

The Ligature Error Rate (LER) can be calculated by comparing output and target text and can be determined using the following formula.

$$\text{LER} = [(N_{ins} + N_{subs} + N_{del}) \times 100] \div n \tag{5}$$

where

N_{ins} = no. of insertions required to transform the target into output.
N_{subs} = no. of substitutions required to transform the target into output.
N_{del} = no. of deletions required to transform the target into output.
And n = no. of ligatures in the target text.

The different variations of LER obtained at different epoch levels while training and validating handwritten Urdu datasets are plotted in Fig. 4. It must be noted here that training and validating similar datasets produces less validation LER while as in dissimilar datasets, the validation LER is comparatively larger. The obvious reason behind this variation is some ligatures that are present in one dataset are not available in another dataset and vice versa is also true. Hence training one dataset and validating another dataset yields less accuracy of recognition. It must also be noted here that training NUST-UHWR and validating NUST-UHWR yields lowest LER of 3% only. It is also worth to mention here that the LER decreases with an increase in number of epochs and found to be lowest in range the of 30–40 epochs.

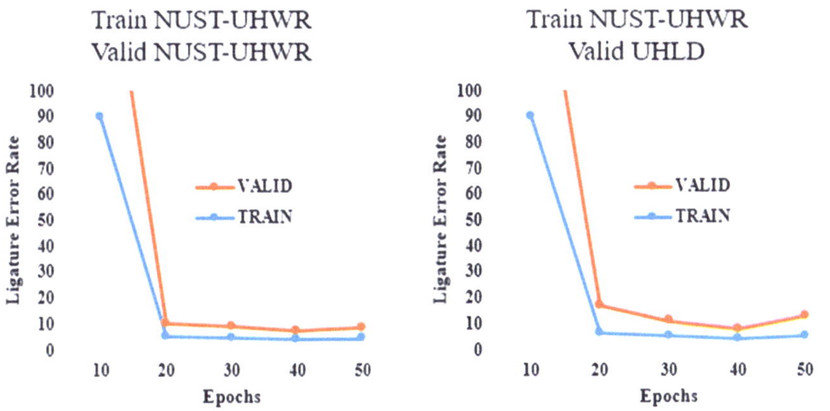

Fig. 4 Ligature Error Rate obtained on different datasets at different epochs

Table 6 Performance parameters (%age) of the proposed system

Training dataset	Testing dataset	Performance parameter				
		Sensitivity	Specificity	Precision	F1-Score	Accuracy
UNHD	UNHD	96	98	97	97	**96**
UNHD	NUST-UHWR	88	91	92	90	**92**
UNHD	UHLD	89	93	91	92	**93**
NUST-UHWR	UNHD	87	90	90	91	**91**
NUST-UHWR	NUST-UHWR	96	98	97	97	**97**
NUST-UHWR	UHLD	94	96	96	95	**94**

4.4 *Performance Parameters*

The various accuracy parameters such as Sensitivity, Specificity, Precision, F1-Score and Accuracy are determined by comparing the no. of ligatures recognized in target text. The higher values of these parameters determines low LER as shown in Table 6.

4.5 *Analysis*

The Table 7 displays the visualisation of the multi-head self-attention on a particular ligature in a handwritten text. The right column of the Table 7 indicates attention focussed on a particular part of text (connected component in a word), the central column indicate the text component read and the left column indicate the text component recognised.

Table 7 Visualization of multi-head self-attention as it reads out the test images of handwritten Urdu text

Text recognized	Text component read	Attention on a particular text component

The Fig. 5 below shows the ligatures correctly recognized and misrecognized. The main cause of wrong prediction is confusion between similar ligatures and uneven position of dots. As is clear from Fig. 5 that the attention on a particular ligature recognizes that ligature with very less LER. In Fig. 5a, the LER is determined to be 0% as while reading the given Urdu text line, all the ligatures are correctly recognized. In Fig. 5b, the LER is determined to be only 5% as while reading the given Urdu text line, all except one ligatures are correctly recognized. We have tested around 6000 handwritten Urdu text lines and found an approximate LER of 0–10%. Hence the accuracy of the proposed system outperforms the state-of-the-art [8] in handwritten Urdu recognition.

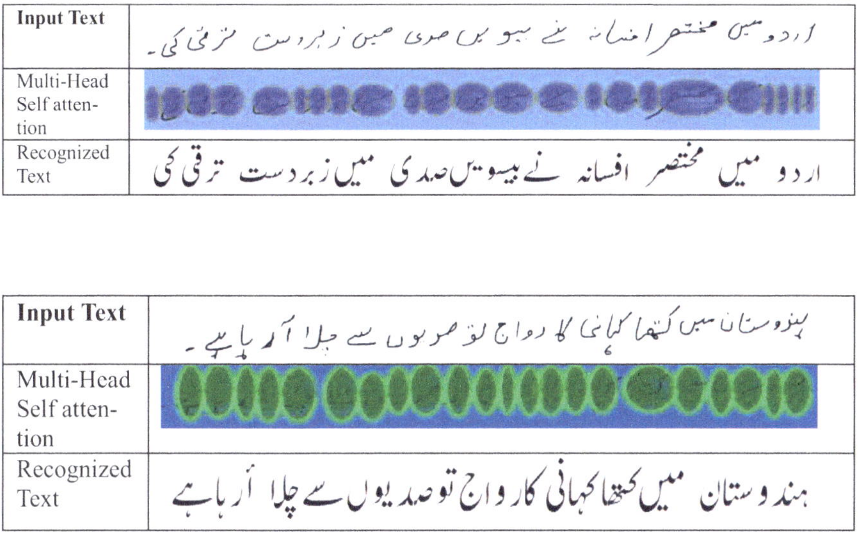

Fig. 5 (**a**, **b**) Handwritten Urdu recognition results obtained. In each panel, top row is input image, second row visualises imposition of multi-head attention on input image, third row is the recognised text string. (**a**) Multi-head self attention, LER = 0%. (**b**) Multi-head self attention, LER = 5%

4.6 Comparison of Proposed Work with the State-of-the-Art

A comparison of latest notable researches in handwritten Urdu and that of proposed work is summarized in Table 8. It is clear from table that the proposed technique outperforms the state-of-the-art techniques in relation to lowest LER for handwritten Urdu recognition and registers the new state-of-the-art in unconstrained handwritten Urdu recognition.

5 Conclusion

We have proposed a transformer based self-attention model for recognizing unconstrained handwritten Urdu text. In the suggested model, the MLP extracts feature from handwritten Urdu datasets and these features train BERT model for training and recognition. The test images' findings demonstrate that the suggested model learns to read and identify Urdu text by focusing on just one connected component at a time. Various accuracy parameters such as sensitivity, specificity, precision and f1-score are evaluated on test images using the proposed model. The results of these parameters determine that the LER is reduced by 20% than that of the state-of-the-art approaches. The proposed model achieves better accuracy on two publicly available handwritten Urdu datasets UNHD and NUST-UHWR and enrols

Table 8 Comparison of proposed work with notable researches in Handwritten Urdu

Study	Database	Training set	Test set	Framework used	Unit of recognition	Error rate	True ligature recognition
Ahmad S.B. et al. [10]	UCOM	50 text lines	20 text lines	Only dataset is proposed			NO
Ahmad S.B. et al. [7]	UNHD	6400 text lines	1840 text lines	BLSTM classifier	Character	6–7%	NO
Husnain M. et al. [11]	Proposed a database of Urdu characters.	Information unavailable	Information unavailable	CNN classifier	Individual characters & numerals	4% on characters 2% on numerals	NO
Hassan S et al., 2019 [12]	Claimed their own database, not publicly available	4000 handwritten lines	1000 text lines	7 CNN layers and 2 BLSTM layers	Character	17%	NO
Ahmad S.B. et al. [13]	UNHD, MNist pre-trained network	10,000 text lines	2000 text lines	CNN and MDLSTM	Character	7%	NO
Zia et al. [8]	NUST-UHWR	8000 ligatures	1300	Presented a hybrid CNN-RNN architecture using N-gram language models,	Ligature	7.25%	NO
Proposed work	**UNHD, NUST-UHWR, UHLD**	**20,608 text lines**	**6000 text lines**	**BERT using MLP**	**Ligature**	**0–10%**	**YES**

the new state-of-the-art in handwritten Urdu OCR. We will test the proposed approach with additional text recognition tasks, like 'patwari Urdu text recognition' and handwritten Urdu scene recognition, in further studies. We also intend to add data augmentations and language models to the suggested model to enhance its functionality.

Acknowledgement We thank Dr. Saad Bin Ahmed for providing UNHD Database. We also thank Dr. Zia for providing NUST-UHWR database.

References

1. Naz S, Umar AI, Shirazi SH, Khan SA, Ahmed I, Khan AA (2014) Challenges of Urdu named entity recognition: a scarce resourced language. Res J Appl Sci Eng Technol 8(10):1272–1278
2. Daud A, Khan W, Che D (2017) Urdu language processing: a survey. Artif Intell Rev 47(3):279–311
3. Khan NH, Adnan A (2018) Urdu optical character recognition systems: present contributions and future directions. IEEE Access 6:46019–46046
4. Satti DA, Saleem K (2012) Complexities and implementation challenges in offline Urdu Nastaliq OCR. In: Proceedings of the conference on language & technology, pp 85–91
5. Devlin J, Chang MW, Lee K, Toutanova K (2018) Bert: pre-training of deep bidirectional transformers for language understanding. arXiv preprint arXiv:1810.04805. 11 Oct 2018
6. Ezen-Can A (2020) A comparison of LSTM and BERT for small corpus. arXiv preprint arXiv:2009.05451. 11 Sept 2020
7. Ahmed SB, Naz S, Swati S, Razzak MI (2019) Handwritten Urdu character recognition using one-dimensional BLSTM classifier. Neural Comput & Applic 31(4):1143–1151
8. ul Sehr Zia N, Naeem MF, Raza SMK et al (2022) A convolutional recursive deep architecture for unconstrained Urdu handwriting recognition. Neural Comput & Applic 34:1635–1648. https://doi.org/10.1007/s00521-021-06498-2
9. Ramchoun H, Ghanou Y, Ettaouil M, Janati Idrissi MA (2016) Multilayer perceptron: architecture optimization and training
10. Ahmed SB, Naz S, Swati S, Razzak I, Umar AI, Khan AA (2017) UCOM offline dataset-an Urdu handwritten dataset generation. Int Arab J Inf Technol (IAJIT) 14(2)
11. Husnain M, Saad Missen MM, Mumtaz S, Jhanidr MZ, Coustaty M, Muzzamil Luqman M, Ogier JM, Sang Choi G (2019) Recognition of Urdu handwritten characters using convolutional neural network. Appl Sci 9(13):2758
12. Hassan S, Irfan A, Mirza A, Siddiqi I (2019) Cursive handwritten text recognition using bi-directional LSTMs: a case study on Urdu handwriting. In: 2019 international conference on deep learning and machine learning in emerging applications (deep-ML). IEEE, pp 67–72
13. Ahmed SB, Hameed IA, Naz S, Razzak MI, Yusof R (2019) Evaluation of handwritten Urdu text by integration of MNIST dataset learning experience. IEEE Access 7:153566–153578
14. Pauls A, Klein D (2011) Faster and smaller n-gram language models. In: Proceedings of the 49th annual meeting of the Association for Computational Linguistics: human language technologies 2011 June, pp 258–267
15. Ganai AF, Koul A (2016) Projection profile based ligature segmentation of Nastaleeq Urdu OCR. In: 2016 4th international symposium on computational and business intelligence (ISCBI). IEEE, pp 170–175
16. Lehal GS (2013) Ligature segmentation for Urdu OCR. In: 2013 12th international conference on document analysis and recognition. Available at: https://doi.org/10.1109/icdar.2013.229

17. Gazit H (1991) An optimal randomized parallel algorithm for finding connected components in a graph. SIAM Journal on Computing. 20(6):1046–67
18. Vaswani A, Shazeer N, Parmar N, Uszkoreit J, Jones L, Gomez AN, Kaiser Ł, Polosukhin I (2017) Attention is all you need. Adv Neural Inf Proces Syst 30, pp 1–11
19. Jais IK, Ismail AR, Nisa SQ (2019) Adam optimization algorithm for wide and deep neural network. Knowledge Engineering and Data Science. 2(1): 41–6

Correction to: Analysis of Ensemble Methods for Phishing Detection

Deepak Gupta, Ekta Gandotra, Yogesh Mohan, and Sukhvir Singh

Correction to:
Chapter 4 in: S. A. Parah et al. (eds.), *Intelligent Multimedia Signal Processing for Smart Ecosystems*, https://doi.org/10.1007/978-3-031-34873-0_4

The original version of Chapter 4 was published with incorrect author names

Deeapk Gupta
Sukhvir Thakur

The names have now been corrected as

Deepak Gupta
Sukhvir Singh

The updated original version for this chapter can be found at
https://doi.org/10.1007/978-3-031-34873-0_4

S. A. Parah et al. (eds.), *Intelligent Multimedia Signal Processing for Smart Ecosystems*, https://doi.org/10.1007/978-3-031-34873-0_16

The manufacturer's authorised representative in the EU is Springer Nature Customer Service Centre GmbH, Europaplatz 3, 69115 Heidelberg, Germany. If you have any concerns regarding our products, please contact ProductSafety@springernature.com

Printed and bound by CPI Group (UK) Ltd, Croydon, CR0 4YY
01/10/2026
02225851-0012